Understanding and Servicing CD Players

Understanding and Servicing CD Players

Ken Clements

NEWNES

Newnes
An imprint of Butterworth-Heinemann Ltd
Linacre House, Jordan Hill, Oxford OX2 8DP

A member of the Reed Elsevier plc group

OXFORD LONDON BOSTON
MUNICH NEW DELHI SINGAPORE SYDNEY
TOKYO TORONTO WELLINGTON

First published 1994

© Ken Clements 1994

All rights reserved. No part of this publication may be reproduced in any material form (including photocopying or storing in any medium by electronic means and whether or not transiently or incidentally to some other use of this publication) without the written permission of the copyright holder except in accordance with the provisions of the Copyright, Designs and Patents Act 1988 or under the terms of a licence issued by the Copyright Licensing Agency Ltd, 90 Tottenham Court Road, London, England W1P 9HE. Applications for the copyright holder's written permission to reproduce any part of this publication should be addressed to the publishers

British Library Cataloguing in Publication Data
Clements, Ken.
Understanding and Servicing CD Players
I. Title
621.38932

ISBN 0 7506 0934 6

Library of Congress Cataloguing in Publication Data
Clements, Ken.
Understanding and servicing CD players/Ken Clements.
p. cm.
Includes index.
ISBN 0 7506 0934 6
1. Compact disc players – Maintenance and repair. I. Title.
TK7881.85.C56
621.389'32–dc20
93–47069
CIP

Typeset by Graphicraft Typesetters Ltd., Hong Kong
Printed and bound in Great Britain by Bath Press, Avon

CONTENTS

	Preface	vii
1	Introduction	1
2	The compact disc player	18
3	Optical assemblies	36
4	Servo systems	55
5	Circuit descriptions	73
6	Test mode and adjustment procedures	109
7	System control	135
8	In-car compact disc players	141
9	Fault diagnosis	150
	Appendix 1: Analogue and digital processing	179
	Appendix 2: Health and safety	193
	Index	197

PREFACE

Over the years that the compact disc has been around, the actual servicing of the relevant players has at times been the cause of much frustration insomuch that when a CD player is not working it frequently just sits there doing not very much, which in fact can be the same condition when the player may be working quite satisfactorily whilst awaiting the disc to be inserted. As a result the thought that has frequently arisen is 'Where do we go from here?'; the answer is to read on McDuff.

Many types of players have a numerous range of adjustments, whilst others, and especially the later models, have hardly any adjustments at all, but unfortunately the CD player in the past has become a 'tweakers paradise', on the basis of 'if it has a preset give it a tweak and see what happens'.

Of all the adjustments that engineers regard with suspicion, probably the most dreaded is diffraction grating, which in my experience has been the cause of much individual frustration as it is often an adjustment that requires the touch of a watchmaker, and not that of a mechanic adjusting the tappets of a car engine. Fortunately this adjustment is now disappearing in the relevant players, with the new types of optical block currently making their appearance.

The CD player is not necessarily the beast that it occasionally appears to be, but requires a certain approach and an element of confidence, and then there is not really any problem at all.

The aim of this book is to provide a readable understanding of CD technology, and also to supplement the relevant service manual for the CD player that may be undergoing some form of service.

Over the years that I have been personally involved in CD technology, usually by way of the many technical training courses that I have been privileged to provide for engineers servicing CD players, I have made many friends throughout the service industry, and I wish to extend my personal thanks to those many friends who have made many contributions to my understanding of the subject with their own 'coal face' knowledge from the bench, especially as I leave Pioneer, due to early retirement, literally as this book is completed.

Also, this book could not have been completed without the support of engineers who have kindly loaned or donated a wide range of service manuals from which I have been able to provide most of the information contained within the book. It would be most unfair to mention engineers by name – I am liable to forget someone and that would be unforgiveable – but they know who they are, and I wish them to accept my appreciation.

I have endeavoured to extend the information over a wide range of manufacturers and would acknowledge my appreciation to those manufacturers that have been mentioned in the text.

Finally, I must mention my wife, Olive, who has tolerated my long and frequent absences from home whilst providing training courses, and who has also maintained the required pressure, not really nagging, to ensure that this book reaches completion.

My best wishes to those readers who gain some form of benefit from the following pages.

Ken Clements
West Drayton, Middlesex
Formerly Technical Training Manager with Pioneer High Fidelity (GB) Limited

Throughout this book the following abbreviations for units are used:
ms^{-1} (= metres per second)
$V\ cm^{-1}$ (= volts per centimetre).

1
INTRODUCTION

The compact disc (CD) first appeared on the market in 1982, and since that time a large number of CD players have made their appearance. Whilst the CD player is a world-wide standard that utilises a laser beam to extract the information from the disc, the methods of mechanical and electronic circuitry that eventually enable the information to be effectively extracted and reproduced are various. Some CD players can prove to be quite reasonable to service, whilst others are not, usually because of accessibility, or rather the lack of it, numerous adjustments, or the lack of them, and especially some mechanical adjustments that mystify or frustrate when the required results fail to materialise.

The aim of this book is to help engineers to achieve a reasonable understanding of CD technology, especially those who have recently decided 'to give CD a go' and engineers newly into the profession. A down-to-earth approach has been applied wherever possible throughout the book. The book is not intended to replace the service manual (no complete circuit diagrams have been included), but to complement the service manual so that the engineer can, hopefully, achieve an appreciation of some of the circuit intricacies.

Before delving into the depths of CD players themselves, it is worthwhile to consider what is actually on the CD itself, as this will provide a reasonable appreciation of some of the relevant circuits as they are approached in terms of terminology and as to why certain things 'happen', either on the disc or in the player.

As various generations of CD players have appeared, whether for the domestic hi-fi or the in-car markets, numerous improvements, innovations and developments have continually taken place, but whatever those various developments may have been, the CD itself still remains in the same original format.

The compact disc: some basic facts

There are two sizes of CD, i.e. 8 cm and 12 cm, with the starting speed of either size of disc being in the region of 486–568 rpm, and the scanning of the disc commencing from the centre. As the CD continues to be played towards the outer edge of the disc, the actual rotation speed slows down so that the disc maintains a constant surface speed as it passes a laser beam which retrieves the data information from the disc.

The constant surface speed of the CD passing the laser beam is in the region of $1.2–1.4$ m s^{-1}, and this is referred to as constant linear velocity (CLV), whereas the fixed speed of $33\frac{1}{3}$ rpm of a vinyl record was referred to as constant angular velocity (CAV). The important difference between the two types of drive is related to the fact that more data or information can be effectively 'packed' onto a CLV disc compared to the CAV system. Again comparison with a vinyl record reveals that the surface speed of the record at the outer edge was much faster than at the centre, indicating that the analogue audio information was effectively more 'stretched out' at the outer edge compared to that nearer the centre.

The normally accepted maximum playing time for the 12 cm disc is in the region of 74 min, whilst the maximum playing time for the 8 cm disc is around 20 min, but by pushing the limits of the manufacturing tolerances increased maximum playing times can be made possible.

The finishing speeds of the two sizes of disc are approximately 300–350 rpm for the 8 cm disc, and 196–228 rpm for the 12 cm disc, with the actual finishing speed being related to whichever surface speed was chosen, within the range of $1.2–1.4$ m s^{-1}, at the time of recording the disc.

Introduction

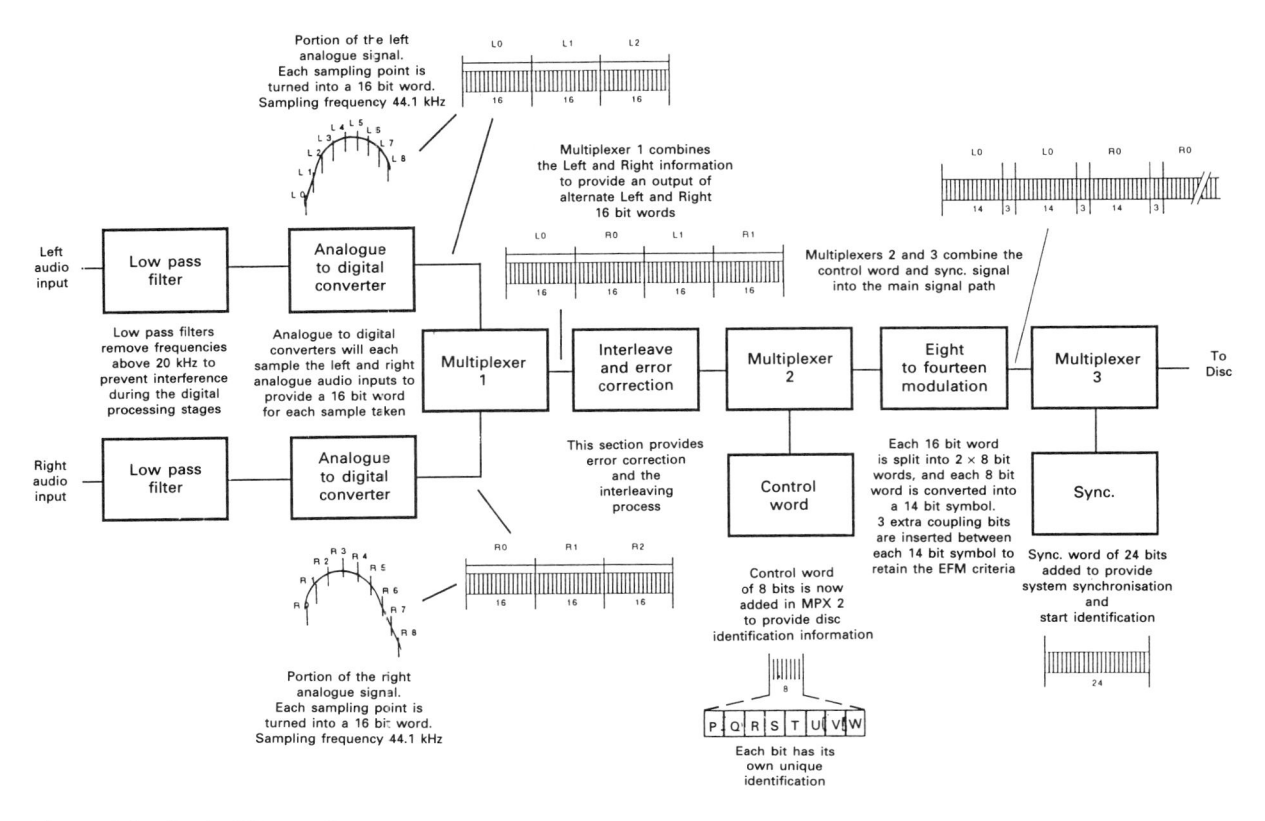

Figure 1.1 *Basic CD recording process*

Basic compact disc technology

The information to be placed on a CD is in the form of digital data, i.e. 0's and 1's to be contained in a series of 'pits' or 'bumps' placed along a fine helical or spiral track originating from the centre of the disc. These 'pits' or 'bumps' will vary in length, relating to when the 1's appear in the digital data stream, with the start or finish of a 'pit' or 'bump' being the moment when a digital 1 occurs. The 0's in the data stream are in effect not recorded onto the CD, but are retrieved within a decoder section of the CD player.

The track dimensions on the CD can appear somewhat mind boggling when trying to find a suitable physical comparison with a track width of 0.5 μm and a track spacing or pitch of 1.6 μm. The thickness of a human hair could contain approximately 30 tracks of a CD, whilst the actual track length on a maximum play 12 cm disc could extend to more than three miles. Another interesting comparison is that if one of the 'pits' that represents a small element of digital information were to be enlarged to the size of a grain of rice, then the equivalent diameter of a CD would be in the region of half a mile.

The compact disc: what information it contains

The audio analogue information to be recorded onto the disc is regularly sampled at a specific rate in an analogue to digital converter (A to D converter) with each sample being turned into digital data in the form of 16 bit binary data words (Fig. 1.1). There are 65 536 different combinations of the 16 bit binary code from all the 0's through to all the 1's, and the sampling frequency used for the CD system is 44.1 kHz. With a stereo system of recording there will be two A to D converters, with one for the left channel and one for the right.

Introduction

```
Quantisation    1001 = 1.7V
  levels        1100 = 1.6V
(4 bit data     1011 = 1.5V
  words)        1010 = 1.4V
                1001 = 1.3V
                1000 = 1.2V
                    Sampling points or frequency
```

The sampling points a and d occur at precise quantisation levels, and therefore the digital value will be an accurate representation of those specific samples, whilst those at b, c, e and f fall between particular quantisation levels.

Whichever quantisation level is chosen to represent these particular samples, it will not enable the original levels to be reproduced when the digital values are eventually converted back into analogue.

This illustrates quantisation error, which in effect is limited to one-half LSB, but will improve by increasing the number of bits of the data word for each quantisation level.

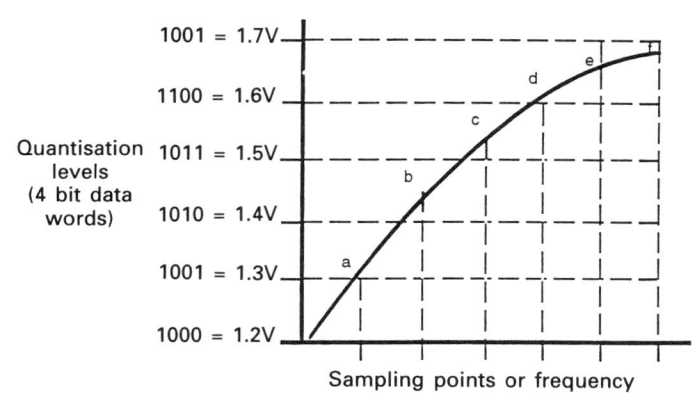

Figure 1.2 *Quantisation errors*

The process of sampling the analogue information is referred to as quantisation, and the overall operation can produce certain effects such as quantisation errors and aliasing noise.

Quantisation errors

Whenever an analogue signal is sampled, the sampling frequency will represent the actual times of measuring the signal, whereas quantisation will represent the level of the signal at the sample time (Fig. 1.2). Unfortunately, whatever the length of the data word that portrays or depicts a specific sample of the analogue signal, quantisation errors will occur when the level of the sampled signal at the moment of sampling lies between two quantisation levels.

As an example, consider a portion of an analogue signal, which for convenience of description is converted into a 4 bit data signal. The analogue values shown in Fig. 1.2 each relate to a specific 4 bit data word. The sampling points indicate the moment when the waveform is sampled in order to determine the relevant 4 bit data word.

The example indicates that at the moment of sampling, the instantaneous value of the waveform lies between two specific data words, and thus whichever data word is selected will comprise a quantisation error, which in turn will reflect an analogue error when that digital information is converted back into analogue form at a later stage. This problem will be related to the specific frequency of the analogue signal, and also the length of the data words.

The lower pass filter and digital to analogue converter

There is no intention of delving into the intricacies of converting the original analogue information into digital data, as there are many excellent reference books available that pursue this particular topic. During the process of recording analogue information onto the CD, the audio frequencies are first of all passed through a low pass filter before being applied to the A to D converter as shown in Fig. 1.3 for the left audio channel. The purpose of this filter is to remove all audio frequencies above 20 kHz.

When the audio signal is converted from analogue to digital, the signal is sampled at regular intervals, and the value that is obtained at each sample interval is in turn converted into a 16 bit digital signal.

The sampling frequency for CDs is 44.1 kHz, which is the standard determined to provide the lowest acceptable sampling frequency that will enable the audio spectrum of 0–20 kHz to be reproduced as faithfully as possible, following a criterion stated by Nyquist, an American engineer, in 1928.

Nyquist criterion

Basically, if an analogue signal is digitally sampled, and providing you sample that signal at a rate of at least twice the highest frequency in the bandwidth that is being used, it is possible to reproduce faithfully all the frequencies within that

Introduction

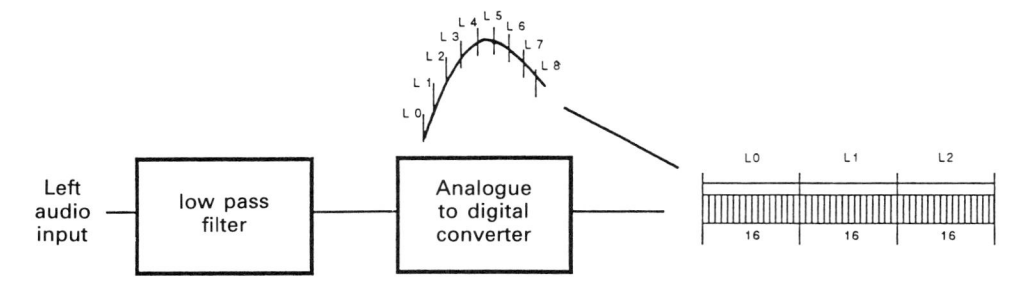

Figure 1.3 *Low pass filter and analogue to digital conversion*

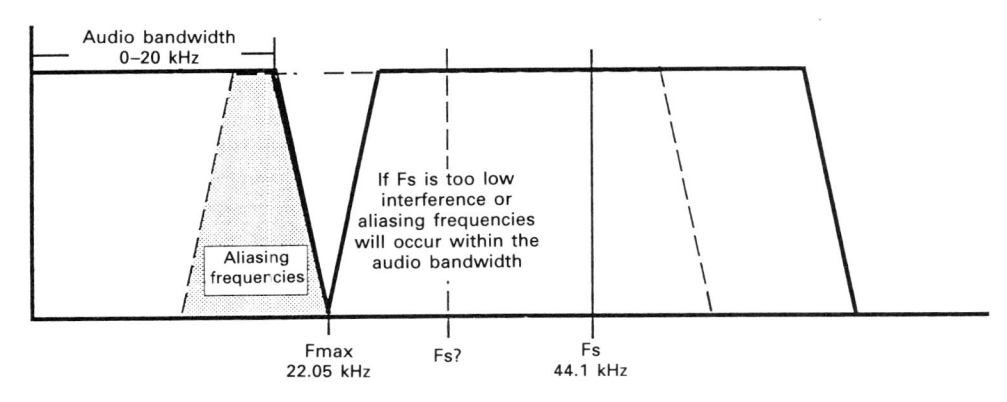

Figure 1.4 *Frequency spectrum showing the effects of aliasing if the sampling frequency is too low*

bandwidth. The reason for sampling at a rate of twice the highest frequency is to ensure that at least one sample takes place during the positive half cycle, with the same applying for the negative half cycle.

Thus with CDs, the audio bandwidth being from 0–20 kHz, the logical sampling frequency would be 40 kHz, but to ensure that effects such as aliasing are minimised, the standard was set at 44.1 kHz (Fig. 1.4). As previously mentioned, the audio signal is first filtered via a low pass filter to remove frequencies in excess of 20 kHz, the reason for this process being to minimise the effect of aliasing, which can be caused when frequencies above 20 kHz are passed through to the A to D converter.

Aliasing frequencies

The A to D converter can, in effect, sample frequencies above the audio range to produce resultant frequencies that can occur within the audio range.

Consider a sampling frequency in the A to D converter of 44 kHz. If a frequency of say 33 kHz, as a result of harmonics outside of the normal listening range from a particular musical instrument, were to be present and therefore sampled, a resultant aliasing frequency of 11 kHz would become apparent. In fact as these harmonics approached the sampling frequency, a range of descending frequencies could be present.

As the analogue information is processed into digital data, in the form of 16 bit data words, there will in effect be a 'natural' clock frequency regarding this data information, which is related to the sampling frequency and the number of bits in a data word. For the A to D converter processing either channel of the stereo audio information, this clock frequency will become 44.1 kHz × 16 = 705.6 kHz.

During the process of preparing the analogue information for recording onto the CD, and after each of the left and right analogue signals have

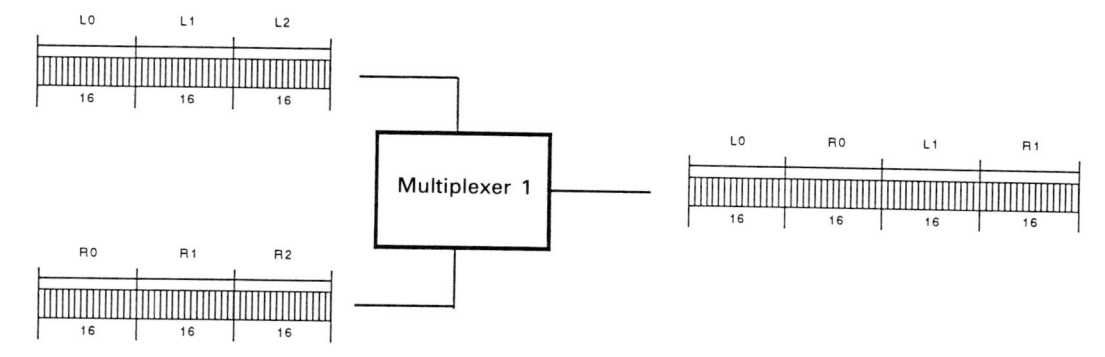

Figure 1.5 *Purpose of Multiplexer 1*

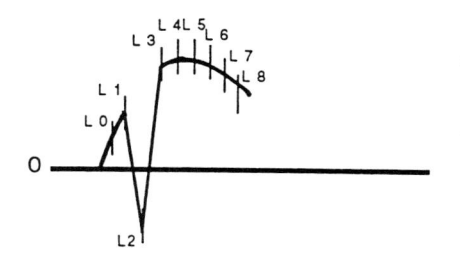

Small section of the left signal, with an erroneous L2 data word causing an undesirable signal or 'glitch' at that moment in time as a result of incorrect data due to scratches or dirt on the surface of the CD

Figure 1.6 *The effect of erroneous data (words)*

been sampled and converted into a series of 16 bit data words, there are now effectively two separate channels of data information which now have to be placed onto what will become a single track on the CD. This is achieved by passing the separate left and right data words into a multiplexer, the output of which will be a series of alternate left and right data words, with the left data being the first data word from the output, and thus the first audio data word to appear from the CD (Fig. 1.5).

It is a requirement of the CD specification that the first audio information data word to be laid on the CD track is to be a left data word. This is to enable the decoder within the CD player to 'know' that the first audio data word output from the decoder will be a left data word, so that this data word will be 'directed' towards the left audio channel for subsequent processing. Accordingly the next data word will be a right data word, which will in turn be 'directed' towards the right audio channel.

As a result of this alternating procedure, the natural clock frequency of the data stream must be increased in order that the audio information can be maintained within the correct time-scale. The bit or clock frequency will now increase by two times to 705.6 kHz × 2 = 1.411200 MHz.

Error correction

Whilst there are benefits to be gained from the process of digital audio recording as used with the CD, such as the improvements in the dynamic range, signal to noise ratio and also stereo separation, when compared to the good old-fashioned vinyl analogue recordings, problems do arise when some of the data information become corrupted for some reason or other.

Corruption of the digital data can occur during the manufacturing process of a CD, or as a result of improper care of the disc causing scratches and dirt in various forms to be present on the surface of the disc. These effects can cause incorrect data words which in turn provide an unsuitable output.

Unfortunately the CD is not as indestructible as one was first led to believe; in fact, the treatment of a CD should be on a par with the treatment

Introduction

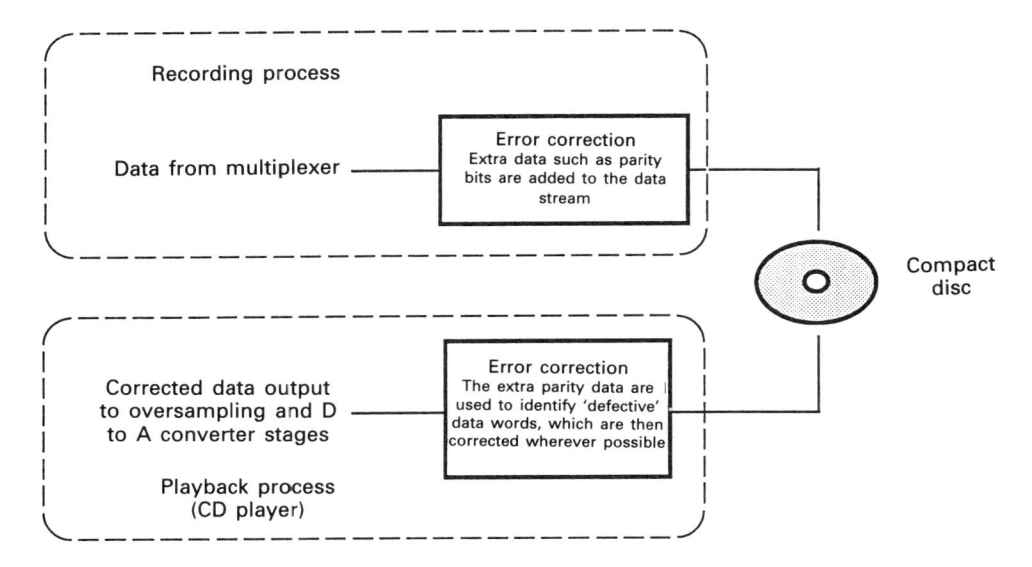

Figure 1.7 *The error correction process*

of a vinyl long-playing record. Those original 'raspberry jam' demonstrations when CDs were first unveiled have a lot to answer for.

The main physical advantage of the CD is that it does not wear in use as there is no mechanical 'stylus' touching the surface of the disc, with the data being read from the disc via a reflected beam of laser light.

So how does error correction take place? Basically, extra information, such as parity bits, is added to the data stream and eventually recorded onto the CD (Fig. 1.7). When the disc is being played in the CD player, the extra information will assist in identifying any defective words as the data stream is being processed. The error correction stage within the CD player will apply certain types of correction depending upon how many errors are present. This is an extremely complex process, and there are excellent references available to provide the enquiring mind with further information. Here it is sufficient to say that by the addition of this extra information, it is possible for the CD player to identify and endeavour to 'correct' the errors if they are within certain criteria; or if the errors exceed these criteria, the audio output can be muted. If the errors prove to be too difficult to handle, the CD player will usually 'throw in the towel' and shut down as far as the operation of that particular disc is concerned.

How are the errors corrected?

It is useful at this stage to consider, in a very basic form, how the errors are effectively 'corrected' within the CD player.

In Fig. 1.8 we consider the sampled section of an analogue signal, compared to the same sample when one of the data words has gone on 'walkabout', which could provide an incorrect output. The overall processing of the digital information within CD technology is very much a complex mathematical operation. But one of the the aims of this book is to minimise the mathematical content, and endeavour to consider as much of the technology in a more 'low key' approach, and fortunately this approach can be applied to the principles of error correction.

Within the CD player there are three main methods of overcoming errors as they occur, depending upon which error occurs at a specific moment in time.

Muting

It is quite a logical concept that when an error occurs, the result of that error is, in real terms, a different level data word, which in turn produces an undesirable sound or 'glitch' as the output.

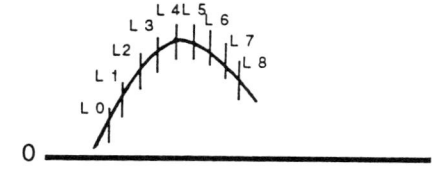

Introduction

Small sampled section of the left analogue signal, showing the stream of data words to be correct

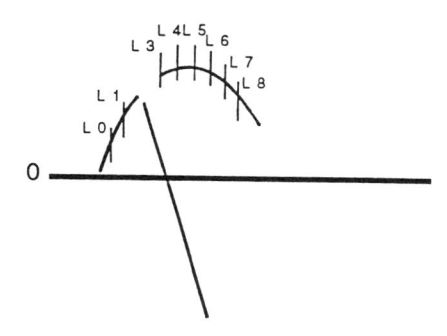

The same sampled section of the left analogue signal, showing that one of the data words, L2, is missing, or on 'walkabout', due to an error in the stream of data words

Figure 1.8 *Comparison between correct and erroneous data stream*

Again it would be logical to mute the sound output when an error has been identified, so that the recipient does not hear undesirable sounds. Unfortunately life rarely proves easy in the most logical of concepts.

Consider the sampled signal and the application of muting when the error occurs in Fig. 1.9. If the portion of the analogue waveform shown is the motion of the loudspeaker as it responds to the audio signal, then when the muting is applied, a bigger and better 'glitch' could occur, which would probably be even more undesirable than the original error if that had been allowed to be reproduced. In fact, history reveals that in the early design stages of CDs, when this form of error correction was applied at high output levels, the final effect proved detrimental to loudspeakers whereby the speech coil and the cone parted company quite effectively. However, muting does play its part in the error correction process, and is normally applied at the times when an extensive series of errors occurs and other methods prove unacceptable.

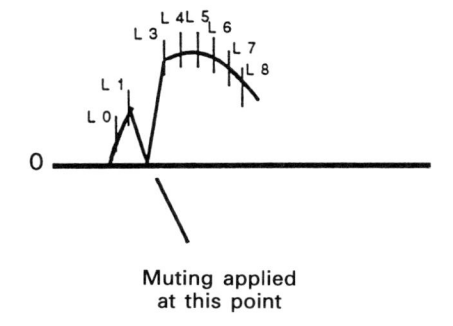

Previous word hold

Due to the large number of operations that take place when manipulating the digital information, whether when putting that information on to the disc, or during digital processing within the CD player, a substantial number of memory circuits are used, especially within the CD player itself. In fact, as the audio data are being processed within the player, there is a period of time when all the digital audio information is sequentially held in memory for a brief period of time.

It could therefore be logical to replace a doubtful data word with the nearest one that is similar in value. When the original analogue signal is sampled, and each sample turned into a 16 bit word, there are in fact 65 536 different combinations of the 16 bit code from all the 0's to all the 1's. Thus it is reasonable to accept that the data word preceding the error word could be approximately 1/65 000th different in level to the original correct data word. Again, as it is quite logical to assume that there is no person in existence who is capable of determining such differences in sound levels, then the previous word hold is an acceptable substitute (Fig. 1.10).

Muting applied at this point

Figure 1.9 *The effect of applying muting*

7

Introduction

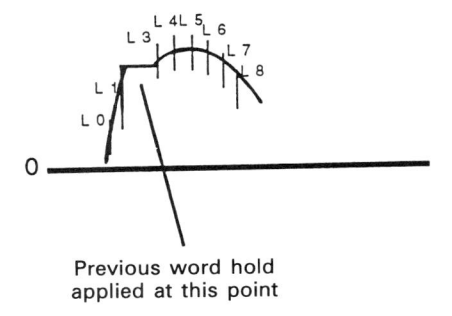

Figure 1.10 *Previous word hold*

Linear interpolation

The previous word hold method of error correction can be further improved with linear interpolation, where it is possible to compare the data word before the error word and the data word after the error word and, by taking the average of the sum of these two words and using this as the substitute for the error word, it is possible to achieve a more accurate assumption of the missing word (Fig. 1.11).

Interleaving

Interleaving is an extremely complex process and is complementary to the error correction process, where the basic operation is to 'jumble up' the audio data words on the CD in a defined arrangement, and to re-arrange those data words back into their original order within the CD player before the process of restoring the digital data back into the original analogue information.

The interleaving process is achieved by inserting the data words into a series of delays. These delays are in multiples of one sampling period or data word in length, and are in the form of random access memory, from where they are initially stored and then extracted in a strictly defined sequence. This actually causes data words to be delayed by varying amounts, which in turn causes the data words to be effectively 'jumbled up' on the disc. Providing these data words are 'unjumbled' within the CD player in the reverse manner, the data words will re-appear in the correct sequence.

Fig. 1.12 illustrates the basic concept of the interleaving process. During the recording process the serial data words are turned into parallel data words by the first serial to parallel converter, and then passed into each delay with the result that the output will be the data words appearing at differing times. The parallel data words are then converted back into serial data by the subsequent parallel to serial converter before eventually arriving on the CD.

Within the CD player this process is reversed through a series of delays which are set in the reverse order, and it is this operation that will effectively 'restructure' the data sequence back into the original order.

Control word

After the error correction and interleaving processes, it is now necessary to insert an extra 8 bit data word, which is referred to as the control word, or sub-code, with another multiplexer performing this task at regular and specific intervals (Fig. 1.13).

The purpose of the control word is to provide disc identification information, and each of the eight bits that comprise the control word has its own unique identification. As the data information is 'laid' along the track on the CD, the control word is inserted immediately before a group or block of audio data information, and immediately after a sync. word, the purpose of which is described at the end of this chapter.

As the disc rotates within the CD player, the control word is identified within the processing circuits of the player. Each bit, with its own identification, is inserted into a specific memory location, and the relevant memory location is analysed at regular intervals to determine the data information that has 'built up' over a period of time, which is in effect after 98 control words have been stored.

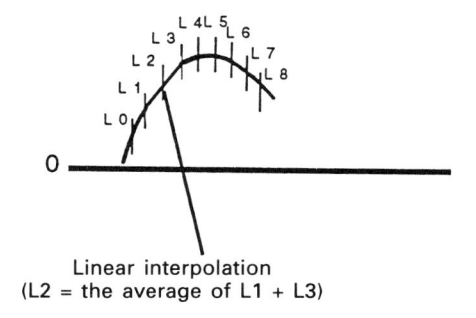

Figure 1.11 *Linear interpolation*

Introduction

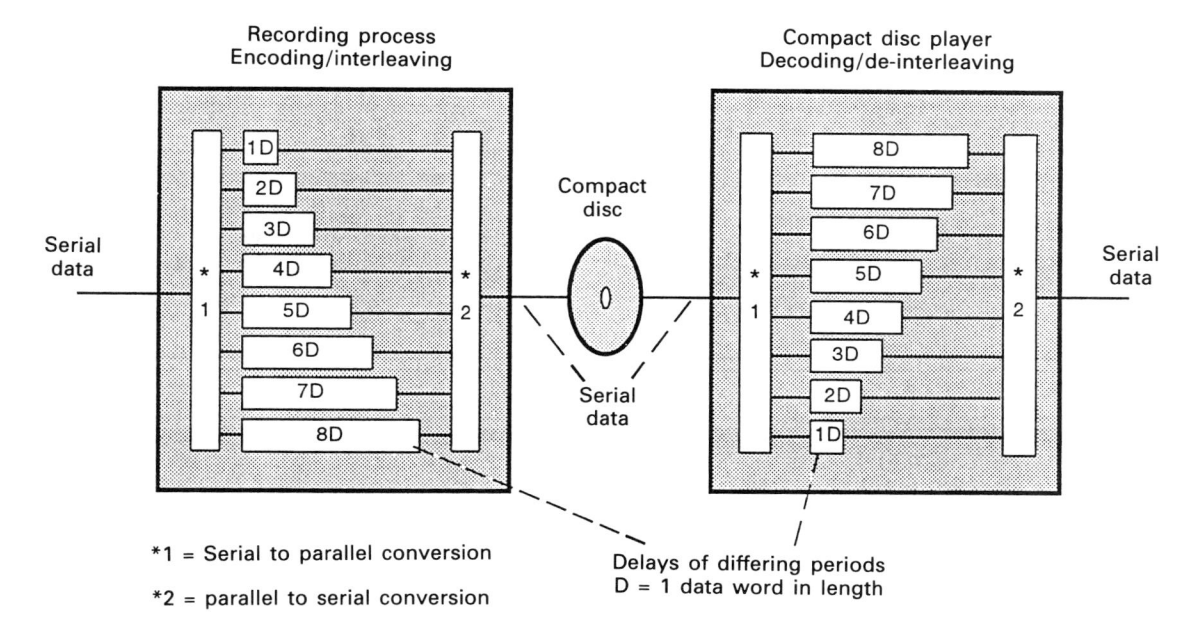

*1 = Serial to parallel conversion
*2 = parallel to serial conversion

Delays of differing periods
D = 1 data word in length

The basic delay concept, where the serial data are turned into parallel data, in terms of blocks of data words. Each data word is delayed by differing amounts during the recording process, and delayed again during the playback process. However, providing the decoding is in effect the reverse of the encoding, the original data word arrangement will be restored.

Figure 1.12 *Basic interleaving/de-interleaving process*

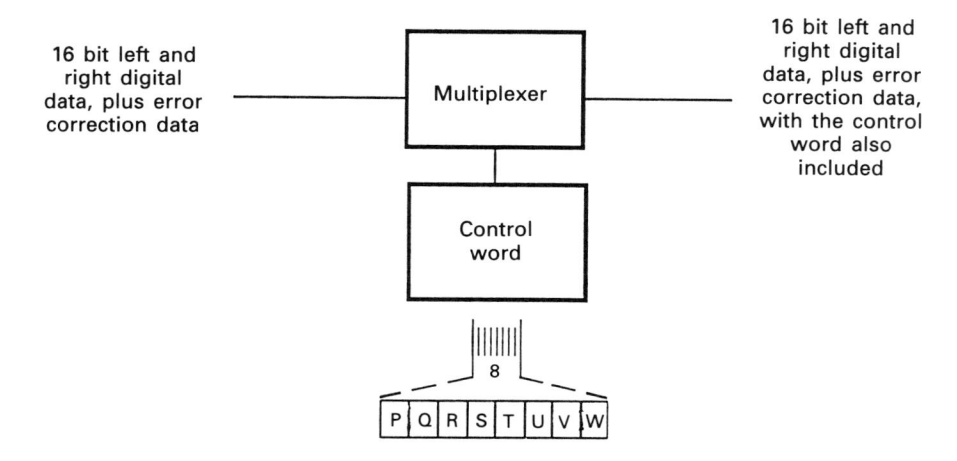

Figure 1.13 *The control word*

Introduction

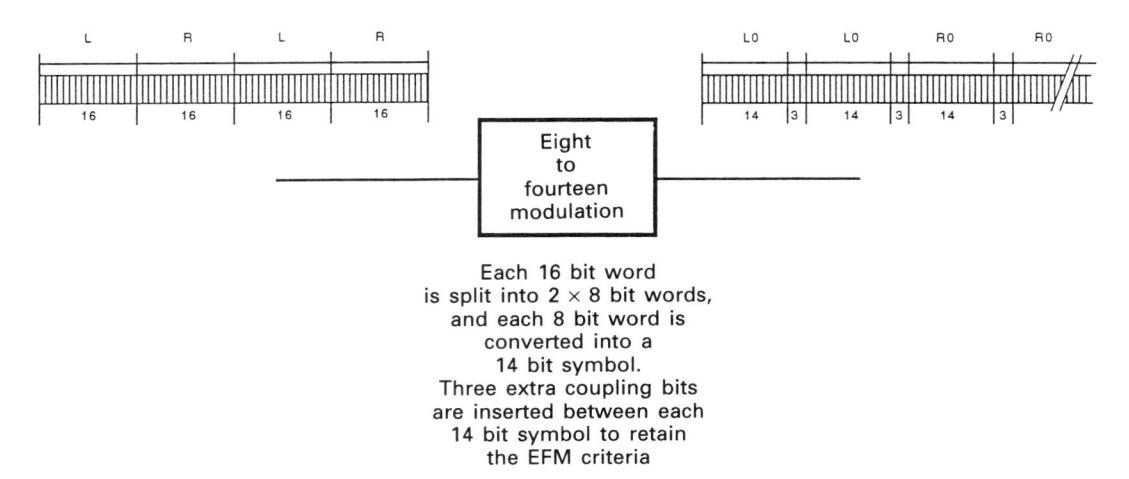

Figure 1.14 *Eight to fourteen modulation*

Only the P and Q bits are in general use with CD players. The R through to W bits are intended for use with computer displays and the relevant software to display graphics, such as the words of the song being sung on the disc, a new concept of 'Singalong a Max' (though not to be confused with Karaoke).

The P bit is used to 'inform' the CD player that the music from the disc is about to commence, and enables the internal muting circuits on some of the early CD players to be switched off so that the analogue information can be applied to the audio circuits. The later and more sophisticated players tend to release the mute circuits in relation to the availability of the data ready for audio processing.

The Q bit contains an extensive range of information including:

1 total playing time,
2 total number of music tracks,
3 individual music track identification,
4 elapsed playing time for each music track,
5 end of playing area information to enable the player to be stopped, and
6 de-emphasis information – a requirement by the recording authorities to be able to apply additional emphasis on any specific track and therefore enable the relative de-emphasis circuits to be switched in or out accordingly.

Other information can be available such as copyright, catalogue number, as well as recording date and serial number, none of which is used within most domestic players currently available.

The total playing time and the total number of music tracks comprise the table of contents (TOC) information which all CD players are required to 'know' before commencing to play a CD.

The control word is processed within the decoder section of the CD player.

Eight to fourteen modulation

After the process of interleaving the data, and after adding the control word, the next major operation is the technique of eight to fourteen modulation (EFM) (Fig. 1.14). Within this stage of operation during the recording processes onto CD, the 16 bit data words are changed into two 8 bit data words, and each 8 bit data word is then changed into a 14 bit symbol. The term 'symbol' is used to differentiate between a data word of 8 or 16 bits, and its transformation into a 14 bit word or symbol.

To appreciate the reasons for the technique of EFM, it is useful to determine the problems that can occur when the data are actually put onto the CD, and therefore the basic process of disc manufacture is now worth considering.

Compact disc construction

When a CD is manufactured, a disc of glass, of extremely high precision, is coated with a photo-resist material which is sensitive to laser light (Fig.

Introduction

Glass disc with photo-resist coating

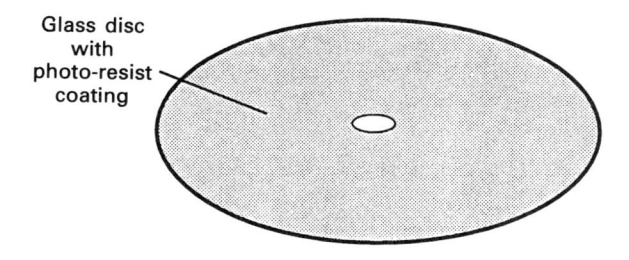

Figure 1.15 *CD construction*

1.15). This glass disc is placed into the recording machinery at the time of manufacture to expose a fine helical track onto the photo-resist material, the width of the track being 0.5 μm (Fig. 1.16). The laser will expose the photo-resist surface with a 0.5 μm width spiral track as the disc rotates, commencing from the centre and working towards the outer edge. The laser beam will be operated in relation to the data information, whereby the laser is switched on and off as the 1's information appears, creating a track of exposed photo-resist material that will comprise a series of dashes, the length of which and the intervals between which will be related to when the 1's occur. No 0's are recorded onto the disc; they are re-generated within the CD players.

Whenever a 1 occurs in the digital data, the laser beam, recording the information onto the photo-resist surface, will alternately switch ON and OFF as each 1 occurs, subsequently exposing the photo-resist surface of the disc and creating a series of 'dashes' 0.5 μm wide, the length being dependent on the period or distance between the 1's.

On completion of the recording process, the unexposed photo-resist material is chemically removed, leaving a helical track across the surface of the glass disc, which becomes the master to produce the injection-moulding formers for the mass production of the resultant CD. The playing surface of a CD is a thin layer of a special injection-moulded plastic material, which is indented from the non-playing side, with a series of 'pits' (Fig. 1.17).

The data stream shown to illustrate how a disc is manufactured 'conveniently' had no consecutive 1's in the data stream, but of course the digital data can comprise any combination of 0's and 1's, and instances will frequently occur when there will be long series of 0's or 1's (Figs 1.18 and 1.19).

When a series of 1's occurs (Fig. 1.18) the following problems can arise:

1 The frequency rate will increase.
2 The 'pit' length may become too short, and may be shorter than the track width.
3 The high 1's rate can be integrated within the servo circuits to create varying d.c. levels which can cause the servo circuits to become unstable.

When a series of 0's occurs (Fig. 1.19) the following problems can arise:

1 Without any 1's occurring at regular intervals, the lock of the phase lock loop of the voltage controlled oscillator within the CD player decoder circuit can become unstable.
2 If a long series of 0's causes long distances between the pits or bumps, the 'trackability' of the player can be affected due to long distances without any track being present.

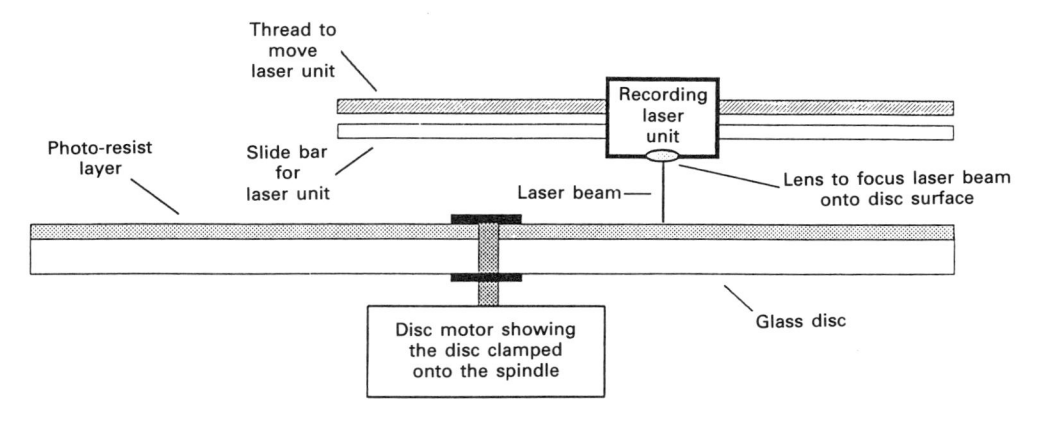

Figure 1.16 *Basic mechanical arrangement for disc recording*

Introduction

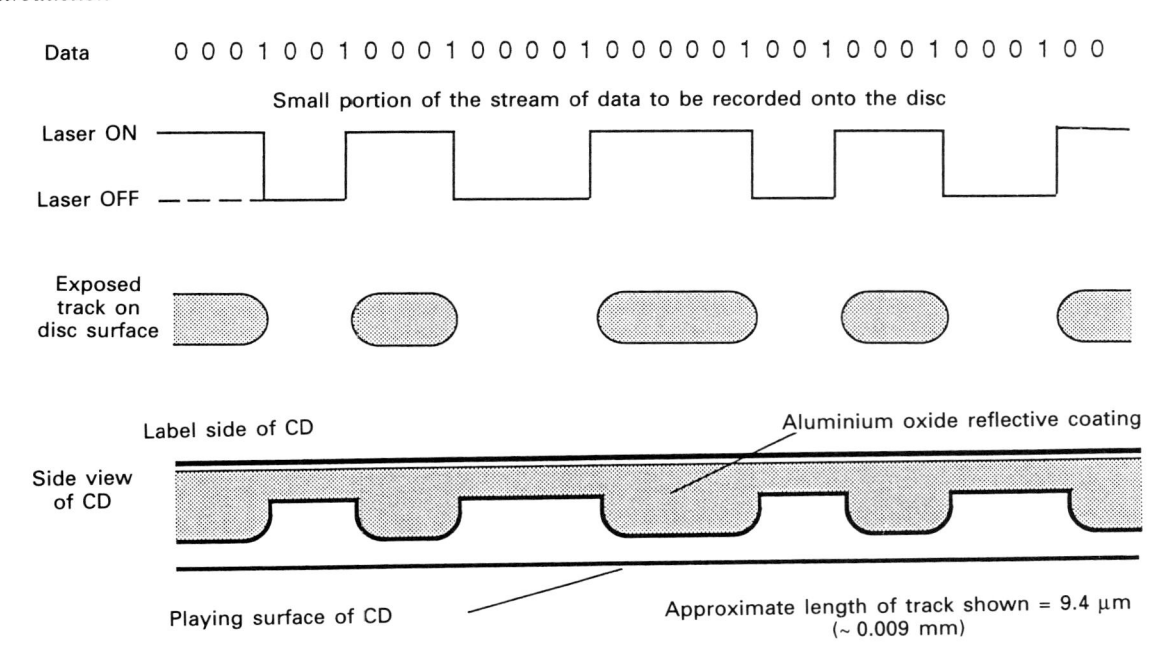

Figure 1.17 *Enlarged side section view of CD, showing the 'pits' indented into the back of the playing surface*

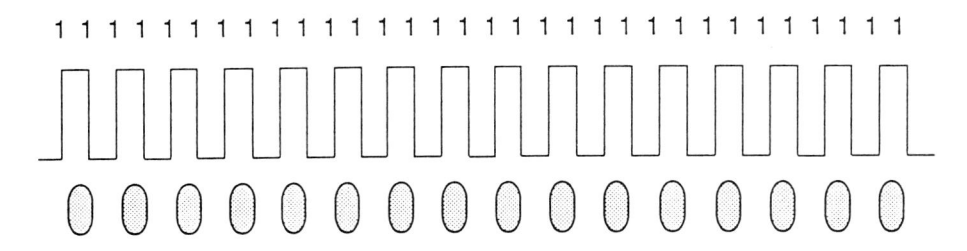

Figure 1.18 *A long series of 1's*

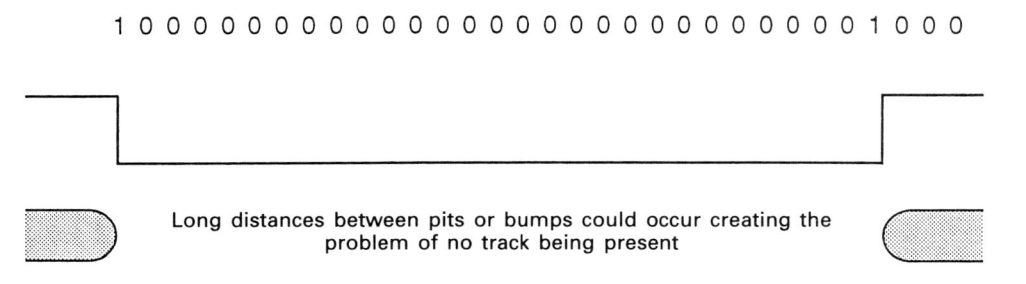

Figure 1.19 *A long series of 0's*

Introduction

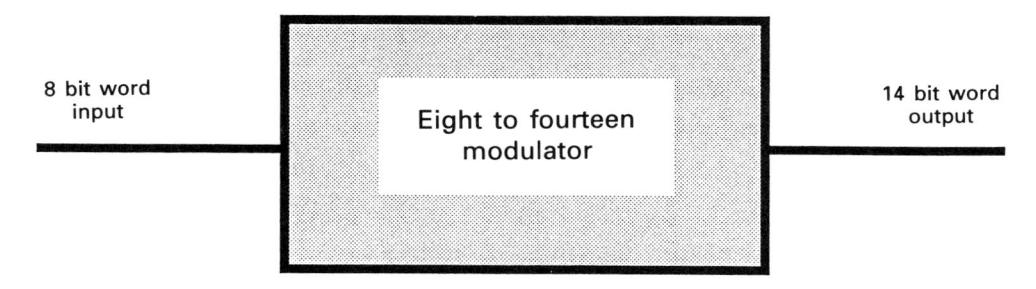

Figure 1.20 *The EFM process*

To overcome these problems the method of EFM was devised. It is purely a process of being able to effectively transfer the digital data onto the CD without any of the problems outlined above.

When a 16 bit word is prepared for the EFM process, the 16 bit word is split into two 8 bit words. This stage is easily achieved by taking the first 8 bits followed by the second 8 bits. Of the 8 bit sequence of data, there are 256 different combinations, from all the 0's

00000000

through to all the 1's.

11111111

With the 14 bit code there are 16 364 possible combinations from all the 0's

00000000000000

to all the 1's.

11111111111111

However, only 267 of these combinations satisfy certain criteria, which are: (1) that no two 1's are consecutive; (2) that a minimum of two 0's exist between two 1's; and (3) that a maximum of ten 0's exist between two 1's. Of the 267 combinations that fall into these criteria, 256 have been specifically selected, and put into a 'look-up' table. Each 8 bit combination is then 'paired' with a specific 14 bit code, of which three examples are:

00000010 = 10010000100000
01011001 = 10000000000100
11111100 = 01000000010010

The new 14 bit 'symbols' now represent the original data information when 'embodied' onto the CD, and when this information is retrieved from the disc within the CD player, another 'look-up' table will enable the 14 bit data to be transferred back into the original 8 bit code. Two 8 bit 'words' together will now comprise the original 16 bit word that was developed to begin with.

One trusts that this method of explanation will enable the 'light' to shine through reasonably brightly.

The eight to fourteen modulation process

Consider the following 16 bit word

0111101001111100

which is split into two 8 bit words

01111010   01111100

and fed into the eight to fourteen modulator where they are converted into two 14 bit symbols (Fig. 1.20):

10010000000010   01000000000010

Again consider the following 16 bit word

0111101001111100

split into two 8 bit words

01111010   01111100

and fed into the eight to fourteen modulator (Fig. 1.21).

The EFM criteria are as follows:

1 No two 1's will be consecutive.
2 The minimum spacing between two 1's will be two 0's.
3 The maximum spacing between two 1's will be ten 0's.

13

Introduction

Eight to fourteen modulation

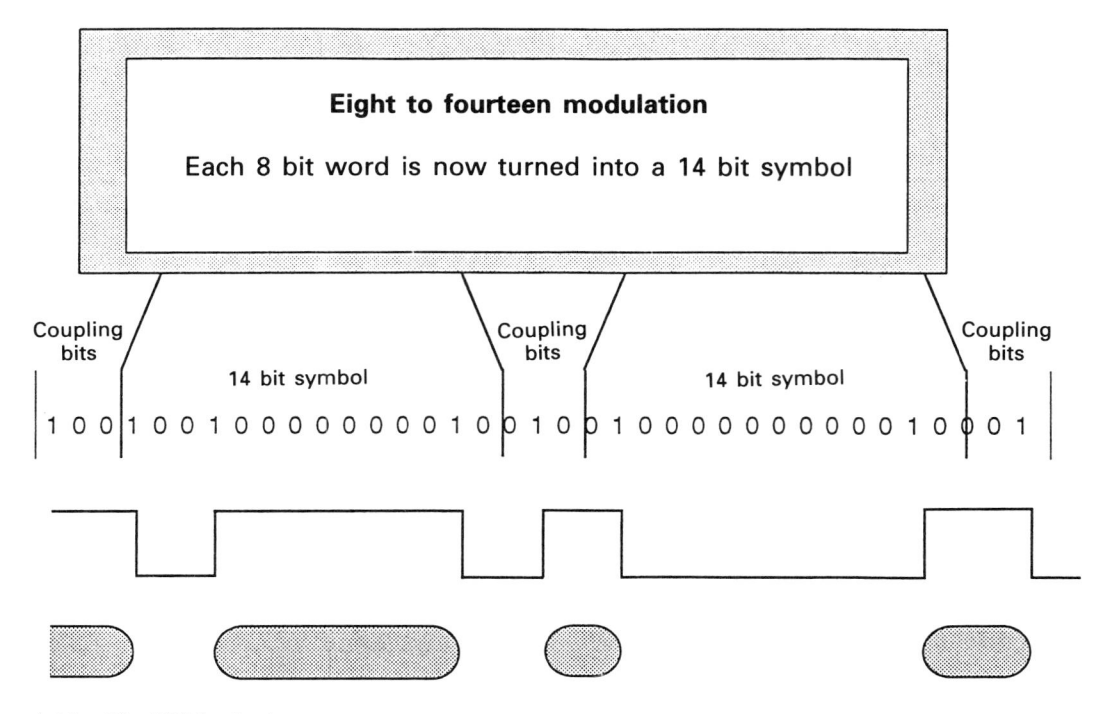

Each 8 bit word is now turned into a 14 bit symbol

Figure 1.21 *The EFM criteria*

These criteria will ensure the following:

1 The frequency bandwidth is reduced.
2 The d.c. content is reduced.
3 The pit length is now wider than the track width.
4 With regular appearance of 1's, i.e. no more than ten 0's, the phase locked loop of the VCO in the decoder will remain locked.
5 Trackability across the surface of the disc will be maintained.

Coupling bits

From the above process, it can be observed that three extra bits have been inserted between each 14 bit 'symbol'; these are referred to as coupling bits. The purpose of these extra bits is to allow for the fact that throughout the EFM process, it could not be guaranteed that when one 14 bit symbol ended in a 1, the next 14 bit symbol would not commence in a 1, therefore disrupting the achievements to be gained with this process. The extra bits are inserted under 'computer control' by analysing subsequent symbols in order that the relevant extra bits enable the required criteria to be maintained. In fact these extra bits are identified within the processing circuits of the CD player and then literally 'thrown away' as they fulfil no further function.

Pit lengths

As a result of the EFM process, there are only nine different 'pit' lengths to carry all the necessary digital data and information for effective CD player operation (Fig. 1.22). Reference is made in Fig. 1.23 to the final clock frequency of 4.3218 MHz, which is the final 'clock' rate for the data on CD, and will be referred to later. From this diagram it can be seen that the start and finish of any 'pit' length is related to the clock frequency of 4.3218 MHz, which is related to the VCO in the CD player, which normally operates at twice the clock frequency (i.e. $4.3218 \times 2 = 8.6436$ MHz). Whilst a CD is in operation, and whilst all the different 'pit' lengths are continually appearing in relation to the digital data information, each 'pit' length will develop a subharmonic of the clock frequency.

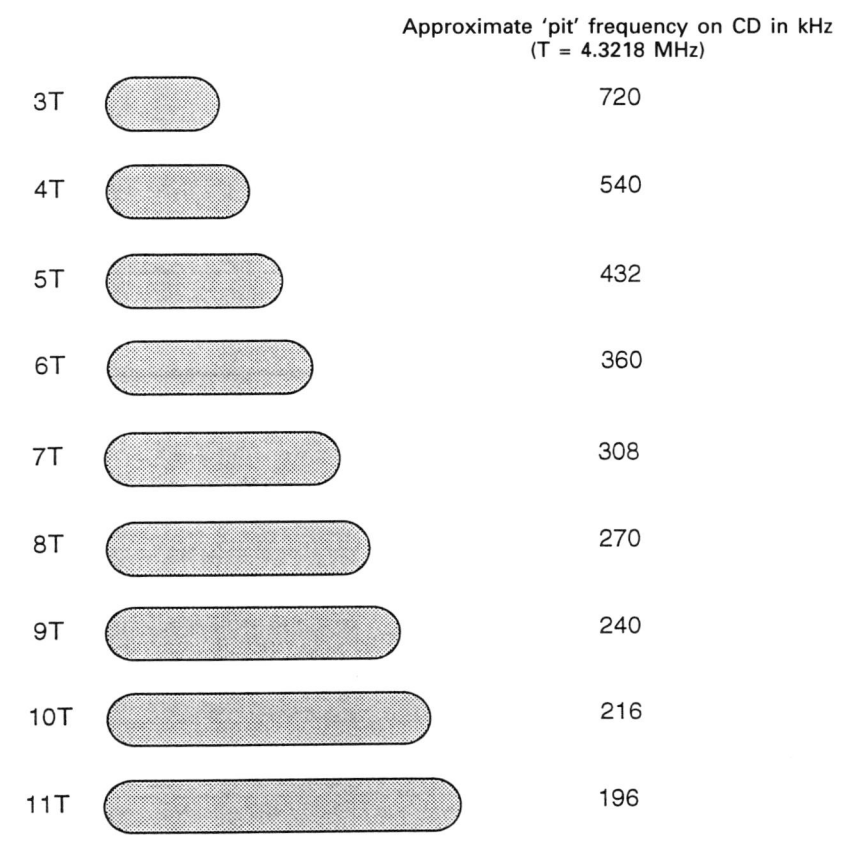

As a result of the EFM process, nine different 'pit' lengths will be present on a CD

Approximate 'pit' frequency on CD in kHz
(T = 4.3218 MHz)

Pit length	Frequency (kHz)
3T	720
4T	540
5T	432
6T	360
7T	308
8T	270
9T	240
10T	216
11T	196

Minimum 'pit' length (3T) 0.833–0.972 μm, maximum 'pit' length (11T) 3.054–3.56 μm depending upon disc velocity (1.2–1.4 m s^{-1}).

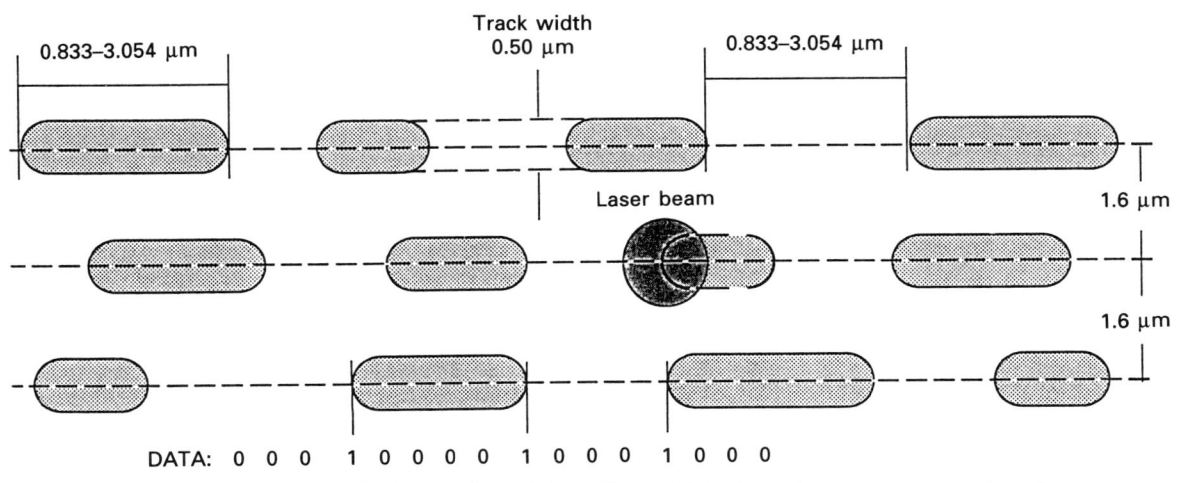

Track width 0.50 μm
0.833–3.054 μm
0.833–3.054 μm
Laser beam
1.6 μm
1.6 μm

DATA: 0 0 0 1 0 0 0 0 1 0 0 0 1 0 0 0

'Pits' are sequential along a helical track, and the reflected light from the laser beam varies when compared between a 'pit' and 'no pit' or space area. The start or finish of a 'pit' or 'space' indicates the 1's information, whilst the 0's are re-inserted in the decoder of the CD player

Figure 1.22 *Comparison of the various pit lengths*

Introduction

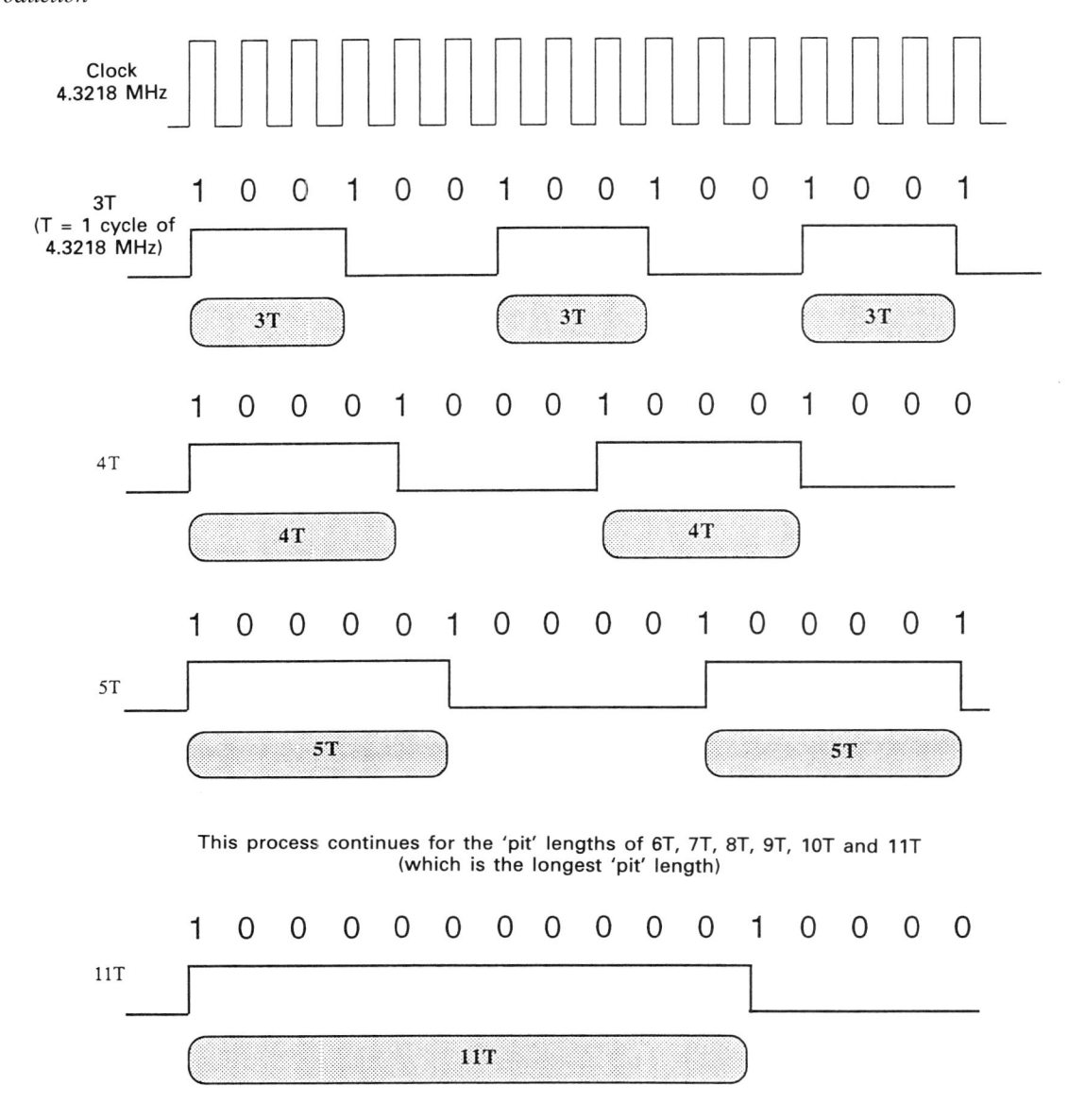

Figure 1.23 *Pit lengths*

Sync. word

The final information to be inserted in the recording sequence is the sync. word (Fig. 1.24), which is used in the CD player as the 'start point' for the data to commence the necessary processing, and also as a signal to be compared for the disc speed control circuits.

The sync. signal is uniquely different to all other data information on CD, and comprises a 24 bit word as follows: 100000000001000000000010.

Fig. 1.25, showing how the track is laid down, will indicate how the data are positioned within one frame.

The path of the original analogue signal from its humble beginnings through to becoming an almost insignificant part of a 'pit' or 'bump', depending upon which way you are looking at the disc, or even the space between, has now been completed.

CD technology is a very complex process, and it often a wonder that the technology works as well as it does, especially when installed in a motor vehicle bumping along some of the 'better' highways.

Many engineers have been known to almost run the other way when it comes to CD player servicing, but with a reasonable amount of knowledge CD players do not constitute a major problem. Having considered some of the basic concepts of the CD, it is now worthwhile considering a typical CD player and its basic operation.

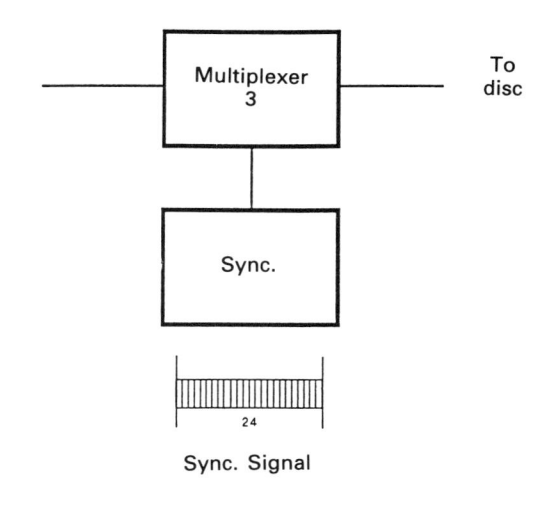

Figure 1.24 *Sync. signal data positioning*

Frame information
One frame = 588 bits
Bit rate or disc clock frequency 4.3218 MHz
Frame or sync. word frequency = 7.35 kHz
(7.35 kHz = 4.3218 MHz divided by 588)

Frame detail

Sync. word – This is used as the 'start point' for the processing stages within the decoder of the CD player to be able to identify individual data symbols for their relevant function or operation. The sync. word is also used for disc motor speed control
Control word – Used for track no. and timing, as well as table of contents, de-emphasis, mute control, etc.
Data symbols – 24 Symbols (12 left, 12 right) each frame, which is equal to 12 × 16 bit words (6 left, 6 right)
Error correction (CIRC) – 8 Symbols each frame (4 left, 4 right)

Track detail on the disc, showing the 'pits' or 'bumps'. The track is 0.5 µm wide, the start and the end of a pit or bump indicates the 1's information of the data stream; the 0's are 'fitted' back into the data within the decoder unit of the CD player

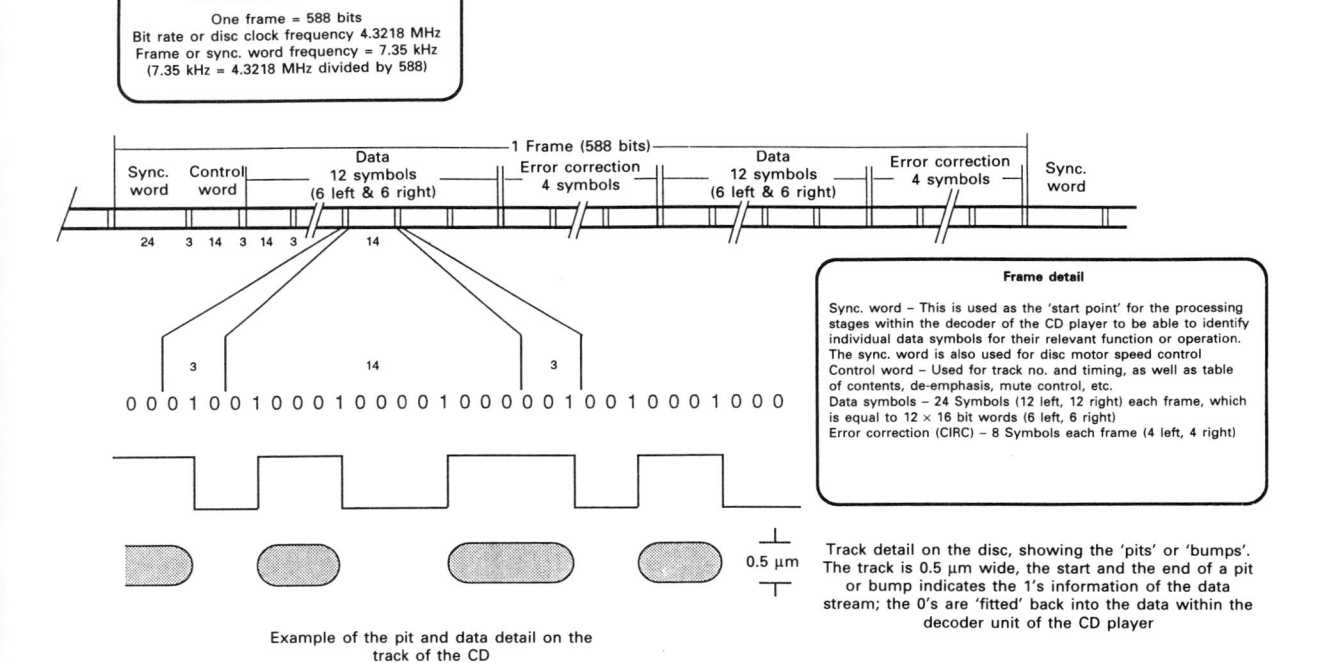

Figure 1.25 *Basic CD system track detail*

2

THE COMPACT DISC PLAYER

The main function of a CD player is to play the CD at the relevant speed and retrieve the digital data from the disc by means of a reflected laser beam, converting the digital data back into the original analogue form as accurately as possible (Fig. 2.1).

Much of a CD player will comprise various servo systems that will enable the laser beam to be accurately focused onto the surface of the disc and simultaneously track the laser beam across the surface of the disc, whilst it is rotated at the correct speed (Fig. 2.1). All CD players are required to operate in a specific manner, the sequence of which is controlled by some form of system control.

Most of the electronic operations take place within large-scale integrated circuits, which despite their complexity are usually extremely reliable.

Mechanical operations are relatively simple and are limited to motors to drive the disc, optical assembly and the loading and unloading mechanism, and a pair of coils which will enable the lens within the optical assembly to move vertically and laterally. An exception to this is the Philips radial system, where the lens moves only vertically, and where the complete optical assembly moves radially in a similar manner to the moving coil meter.

Optical assembly

The optical assembly is probably one of the most important blocks within the CD player, and there are numerous different designs. A fundamental description of some of the typical optical blocks that are generally available is given in Chapter 3.

Generally there are two main types of optical block:

1 The Philips or radial optical assembly. This unit will be found in the Philips and Marantz players, as well as Denon, occasionally Sharp and Technics, and some other players such as Binatone and Sentra.

2 The Japanese tangential tracking type, the alternative, appears in all other players, ranging from Aiwa, Akai, Hitachi, Kenwood, through to Pioneer and Sony, plus all other manufacturers in between, and possibly beyond.

Contained within this unit will be the laser diode to produce a laser beam that is reflected from the surface of the CD through certain components within the block onto a photo-diode array which will produce a range of signals that are in fact the effect of reflected light variations from the 'pit' or 'bump' information on the playing surface of the CD, which in turn become the current variations from the photo-diodes contained within the photo-diode array.

Suffice to say at this stage, these current variations will in due course produce a range of signals that will represent the music information or data from the disc, together with focus, tracking, and spindle or disc speed information that will enable the disc to be played in the correct manner.

Linked to the optical assembly is the automatic power control block which will enable the laser power to be controlled as required, as well as enabling it to be maintained at a constant power level.

RF amplifier

The current variation signals from the photo-diode array within the optical block are passed to the RF amplifier (Fig. 2.2), where the necessary processing takes place to produce the required information, such as data, focus and tracking which will be passed onto their relevant operating sections.

Other signals are also produced which are detailed in the signal processing circuits within the RF amplifier.

The compact disc player

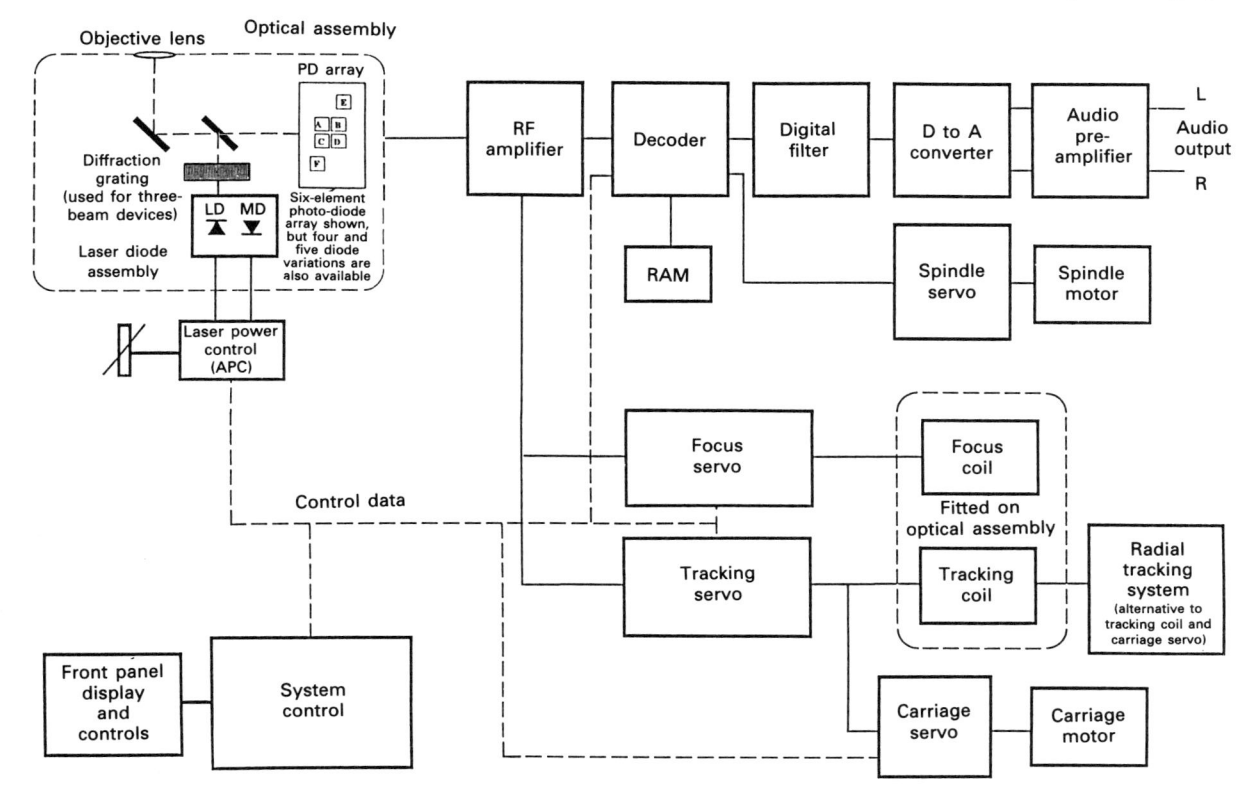

Figure 2.1 *Basic block diagram of a CD player*

RF amplifier: signal processing

The signals from the photo-diode (PD) array, which usually comprises either four or six photo-diodes, are processed to produce three basic signals: the data from the disc, the focus and the tracking signals. The basic circuit illustrated in Fig. 2.2 is normally contained within an integrated circuit, with the relevant outputs as indicated.

Focus error amplifier

This circuit amplifies the signals from four of the photo-diodes which will relate to focus information, and then the signals are passed to a differential amplifier to provide the focus error output, the value and polarity of which will be related to the amount and direction of focus error.

When the in-focus situation is achieved, the focus error signal will be zero, but when mis-focus occurs, then depending upon the direction in which the out of focus situation arises, the polarity of the signal will be either positive or negative with respect to zero, and the amount or level of the signal will relate to the extent of the mis-focus.

Tracking error amplifier

The outputs from either a special pair of photo-diodes, or possibly four photo-diodes, are used for tracking purposes only, and as the CD rotates, the helical track on the disc will cause different amounts of reflectivity from the surface of the disc. This in turn will develop an output from the differential amplifier which produces the tracking error signal, a signal or voltage, the value and polarity of which will be related to the amount and direction of tracking error, or in the case of the radial optical assembly, the radial error signal.

Summation amplifier

The input to this amplifier comprises the output from four photo-diodes. These signals are summed

19

The compact disc player

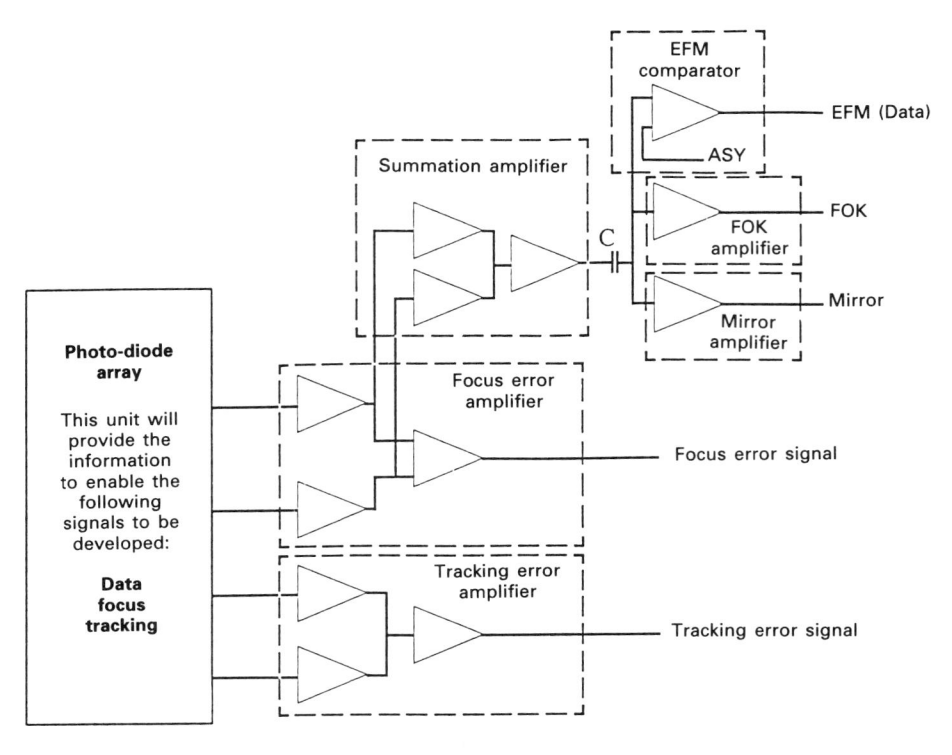

Figure 2.2 *RF amplifier*

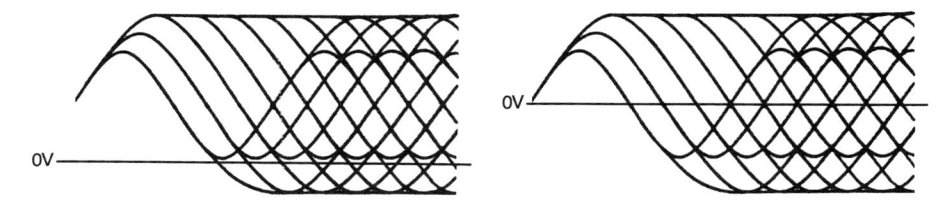

Figure 2.3 *RF eye pattern waveforms*

or added together within this amplifier to produce an output that will be passed on to the decoder. This signal is usually referred to as the RF signal or eye pattern waveform (Fig. 2.3) and comprises the data information from the CD that is to be further processed in order that the original analogue audio information may be obtained. The information relating to disc speed can also be obtained from this signal. The output from this amplifier is fed via a capacitor, C, to three more circuits. The purpose of the capacitor is twofold: firstly, to isolate any d.c. content from the summation amplifier and to produce an RF output signal, comprising frequencies within the range 196–720 kHz with the level varying either side of zero depending upon the reflectivity of the laser beam from the playing surface of the CD. Secondly, the waveforms resulting from the reflected signals from the disc do not tend to be symmetrical due to variations during disc manufacture, as well as the actual shape of the 'pits' or 'bumps' on the disc. Here the capacitor will tend to ensure that the waveform is reasonably symmetrical either side of the zero line, and therefore a waveform effectively crossing the zero line can indicate that a '1' is present within the data stream.

The RF or eye pattern waveform is characteristic

The compact disc player

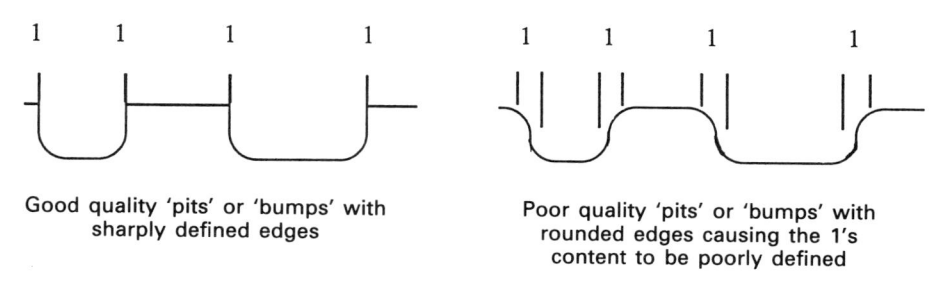

Figure 2.4 *Comparison of good and poor quality 'pits' or 'bumps'*

of any CD player and virtually any type of CD. The period lengths are related to the differing lengths of the 'pits' or 'bumps' on the playing surface of the disc, which in turn are related to the final clock frequency as a result of all the various processing of the data that takes place prior to the recording of the data onto the CD. The shortest period in the above waveforms is the result of the shortest 'pit' length or space between 'pits', and is equivalent to three cycles of the final clock frequency of 4.3218 MHz. Each period length thereafter is equal to the next 'pit' length and therefore a further cycle of the final clock frequency.

EFM comparator

The RF signal from the capacitor is now passed to the EFM comparator, where it is effectively amplified to provide a square wave output with clean or steep leading and trailing edges. It is extremely important to obtain as square an output as possible as the leading and trailing edges will indicate when a '1' occurs within the data stream.

In the injection mould or 'pressing' stage of the manufacturing process of the CD, variations in quality can occur during the lifetime of a specific mould as it begins to wear with each successive 'pressing', causing the the edges of the 'pits' or 'bumps' to become less defined. This can result in problems within the CD player of identifying where a 'pit' starts or commences, and thus impairs its ability to produce a '1' at the correct point in time (Fig. 2.4).

The EFM comparator within most CD players will have two inputs, one being the RF signal, and the other being the ASY (asymmetry) voltage or reference voltage. The latter is related to the timing of the EFM, or squared RF signal waveform, compared with a phase locked clock (PLCK) within the decoder, which after filtering, produces a reference voltage (ASY) for the EFM comparator to operate. The resultant output from the comparator is now passed to the decoder for further processing, to eventually enable the required 16 bit data words to be obtained for conversion back into their original analogue counterparts.

FOK amplifier

When a CD is inserted into the player, it is necessary for the focus servo to initially operate the objective lens to achieve focus onto the playing surface of the disc. This is achieved by causing the lens to elevate up and down usually two or three times. When focus has been achieved, which is related to a maximum RF signal from the disc, a HIGH output will be obtained from the FOK (focus OK) amplifier, which in turn communicates to the relevant control circuits that focus has been achieved.

In many players the FOK signal is also used to inform the control circuits of the player that a disc is present and ready for playing.

Mirror amplifier

The tracking servo enables the laser beam to gradually move across the playing surface of the CD by a combination of moving the objective sideways as well as physically moving the complete optical assembly.

It is possible for the servo to allow the laser beam to track between the tracks containing the digital information, and therefore it is essential for the tracking servo to 'know' that it is tracking correctly along the data track.

The mirror signal enables the control circuits to be aware that incorrect tracking may be occurring, and this is determined by the fact that this signal will be LOW when actually 'on track' and

The compact disc player

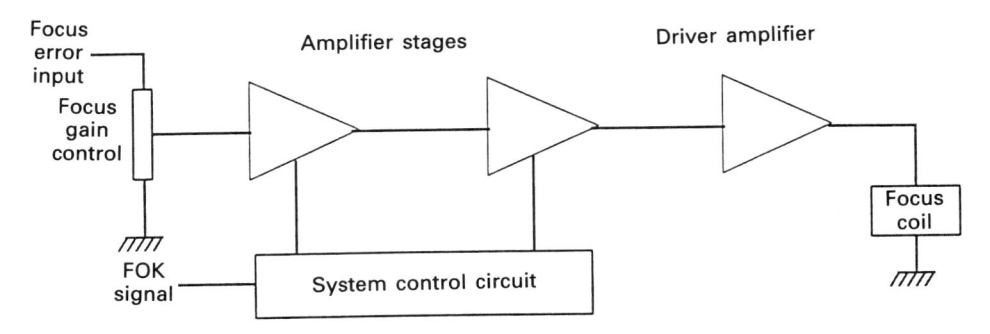

Figure 2.5 *Focus servo*

HIGH when 'off track'. If this signal stays HIGH beyond a defined period of time, the control circuit will effectively give the tracking servo a 'kick' until the mirror signal maintains a LOW level, thereby indicating an 'on-track' condition. The mirror signal is also used whilst track jumping takes place when a new music track on the disc is selected by the operator. The start of the required music track is computed within the player whilst the optical assembly traverses across the disc simultaneously counting individual tracks as they are crossed.

Servo circuits

Within most CD players there are usually four individual servo systems, but with the radial optical assembly only three servo systems will be present, i.e. the focus, radial and spindle servos:

Focus servo — to maintain the laser beam 'in focus' on the playing surface of the CD, by the vertical movement of the objective lens.
Tracking servo — to enable the laser beam to track across the playing surface of the CD by the sideways movement of the objective lens.
Carriage servo — to fractionally move the optical assembly when the objective lens begins to reach the limits of its operation. This servo works in conjunction with the tracking servo, and is often described as either the sled servo or slider servo.
Radial servo — with the radial optical assembly, this servo will take the place of the tracking and carriage servos. As the complete assembly is required to traverse radially across the surface of the disc in extremely fine movements measuring fractions of a micrometre, a wobble frequency is generated to overcome any 'stiction', a term relating to friction within servo systems, enabling a relatively large device to move in an extremely smooth manner.
Spindle servo — the speed of the disc is related to the position of the laser beam, and therefore the optical assembly, as it obtains the data information from the CD. The data rate from the CD is compared to an internal reference within the CD player, usually a crystal frequency source for stability, with a resultant signal to enable the CD to rotate at the correct speed, which is in the region of 500 rpm at the centre or start of the disc, decreasing to around 180 rpm at the end of a 12 cm CD. As the data are processed within the decoder, the 16 bit data words are briefly stored into a random access memory (RAM), and then taken out of the memory as required. This operation is linked to the crystal clock that drives the disc at the approximately correct speed. As the data are removed from the memory, the disc speed will increase and decrease to continually maintain a certain amount of data within the RAM. This servo is frequently referred to as the disc servo or occasionally as the turntable servo.

Focus servo

The focus servo (Fig. 2.5) amplifies the focus error signal and applies the signals to the focus coil which is attached to the objective lens to enable it to move vertically and maintain focus on the playing surface of the CD.

The system control circuit controls the operation of the servo, ensuring that it operates at the correct time, usually after the laser has been switched on. The first operation is to cause the lens to move vertically in order to search for the correct focal point. Having achieved this, and

The compact disc player

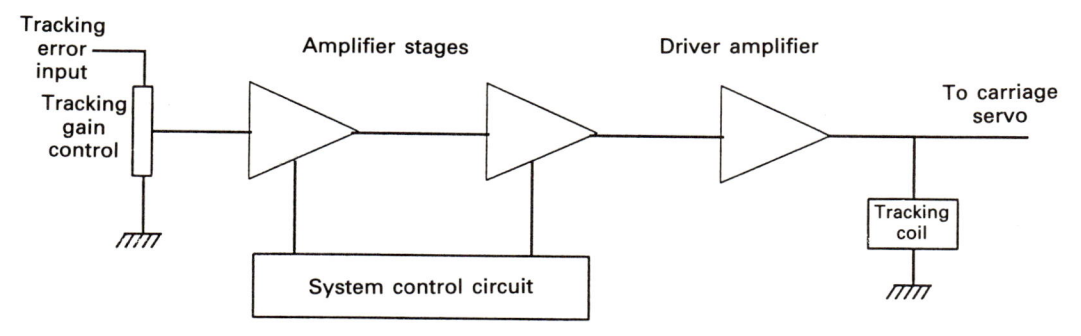

Figure 2.6 *Tracking servo*

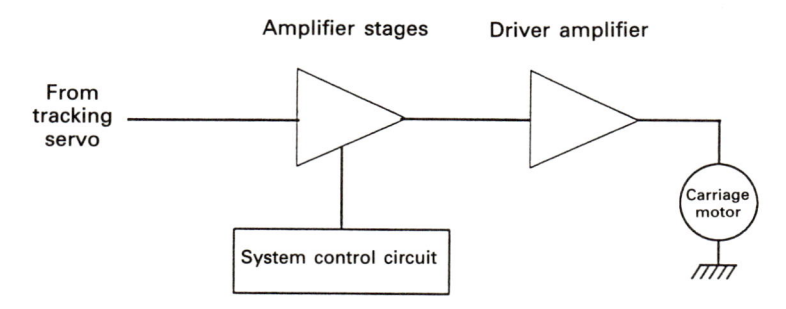

Figure 2.7 *Carriage servo*

when the FOK signal has been received, the servo will then be allowed to operate normally, maintaining the required 'in-focus' condition. During the focus search sequence the first amplifier stage is normally switched off to prevent the developing focus error signal from counteracting the search operation.

Tracking servo

The tracking servo (Fig. 2.6) has many similarities to the focus servo, but its function is to enable the tracking error signal to be amplified and passed to the tracking coil, which in turn is attached to the objective lens to enable the relevant sideways movement to maintain the tracking requirement across the playing surface of the CD.

The objective lens has only a limited amount of lateral or sideways movement (approximately 2 mm) but as the movement increases laterally this can be interpreted as an increasing output from the tracking circuit which can be used to drive the carriage servo when necessary.

Again, the system control circuits control the operation of the tracking servo, enabling the servo to commence the tracking function when the CD is rotating, which on many players commences after the laser has been switched on and the focus servo has in turn commenced its own operation.

Carriage servo

The function of the carriage servo (Fig. 2.7) is to gradually move the optical assembly across the surface of the CD by driving the carriage motor, which in turn is mechanically connected via a long threaded drive to the optical assembly.

The gearing ratio is extremely high, which means that it can take up to 75 min, depending on the playing time of the CD, to enable the optical assembly to cover the playing area of the disc which can be in the region of 4.5 cm.

As the objective lens moves laterally (up to approximately 1 mm of movement) in conjunction with the tracking servo, the increasing signal to the tracking coil is passed to the carriage servo where it is amplified. When the output voltage

23

The compact disc player

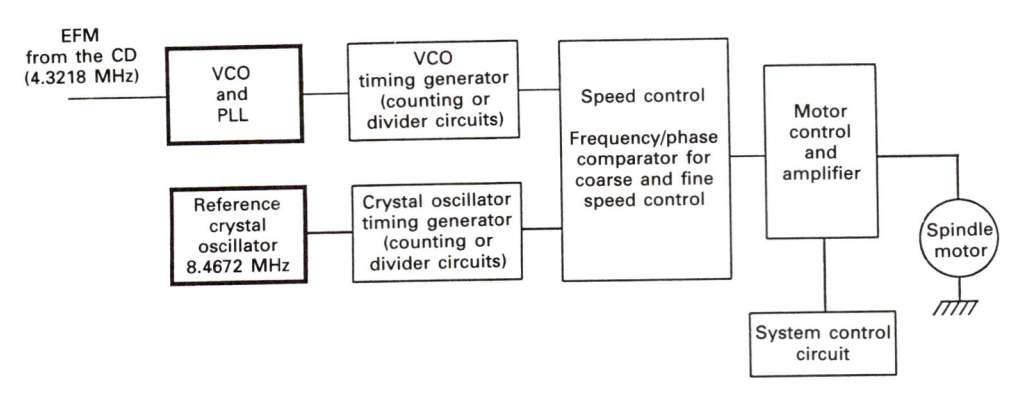

Figure 2.8 *Basic spindle or disc motor servo system*

has reached a level sufficient to operate the carriage motor, the motor will operate to move the optical assembly a very small amount. Simultaneously the objective lens will centralise its position again, with a resultant decrease in the signal to the tracking coil stopping the carriage motor, until the level has increased again to repeat the process. This procedure will continue until the disc has completed the playing sequence.

Some CD players utilise a linear drive motor system instead of a conventional motor, with a similar if not more complicated operating process. Again the servo is controlled by the system control circuit, which allows the carriage servo to come into operation when the tracking servo is functioning. Signals can also be developed from the control circuit which will drive the motor during track location sequences when searching for selected music tracks on the disc, as well as returning the optical assembly back to the centre or start position on completion of playing the CD, in readiness for playing another.

Spindle or disc motor servo

The spindle or disc motor ('spindle motor' will be used here for continuity) is required to rotate the disc at a speed that will maintain the surface speed of the disc passing the laser beam at a constant rate, which will be in the region of 1.2–1.4 m s^{-1}. In effect this will mean that for a 12 cm disc, the actual rotation speed of the disc will commence at the centre or start at around 500 rpm, slowing down to somewhere in the region of 200 rpm towards the outer edge of the disc.

During the sequence of enabling a player to actually play the disc, the spindle motor will 'kick start' usually after the laser and focus stages have operated correctly, with the system control circuit applying a control data signal to the motor control which enables the motor to start running up in speed.

A reference crystal frequency oscillator (8.4672 MHz) will provide, via the timing generator, a crystal locked frequency to the input of the frequency (coarse) and phase (fine) speed control stage (Fig. 2.8).

As the spindle motor increases in speed, the EFM signal will appear as an increasing frequency which is applied to the voltage controlled oscillator (VCO) and phase locked loop (PLL) stage, which will enable an output of 4.3218 MHz to be applied to the VCO timing generator stage.

When the disc rotation approaches the correct (coarse) speed, the two frequency inputs to the speed control stage will be within a certain tolerance, which will provide an output to the motor control to override the 'kick start' voltage. The complete circuit will now take overall control and maintain the speed of the compact disc to provide an EFM output that will be in the general region of 4.3218 MHz, and after further processing in the decoder stage, the resultant data will be stored in the RAM for further processing.

It will be apparent that the disc speed will not, and indeed cannot, be maintained at an absolute surface speed of 1.2–1.4 m s^{-1}, due to the requirement of continually reducing the disc speed as the disc is being played, and there is in fact fairly excessive 'wow' and 'flutter', which is corrected by the 'clocking' of the data through

The compact disc player

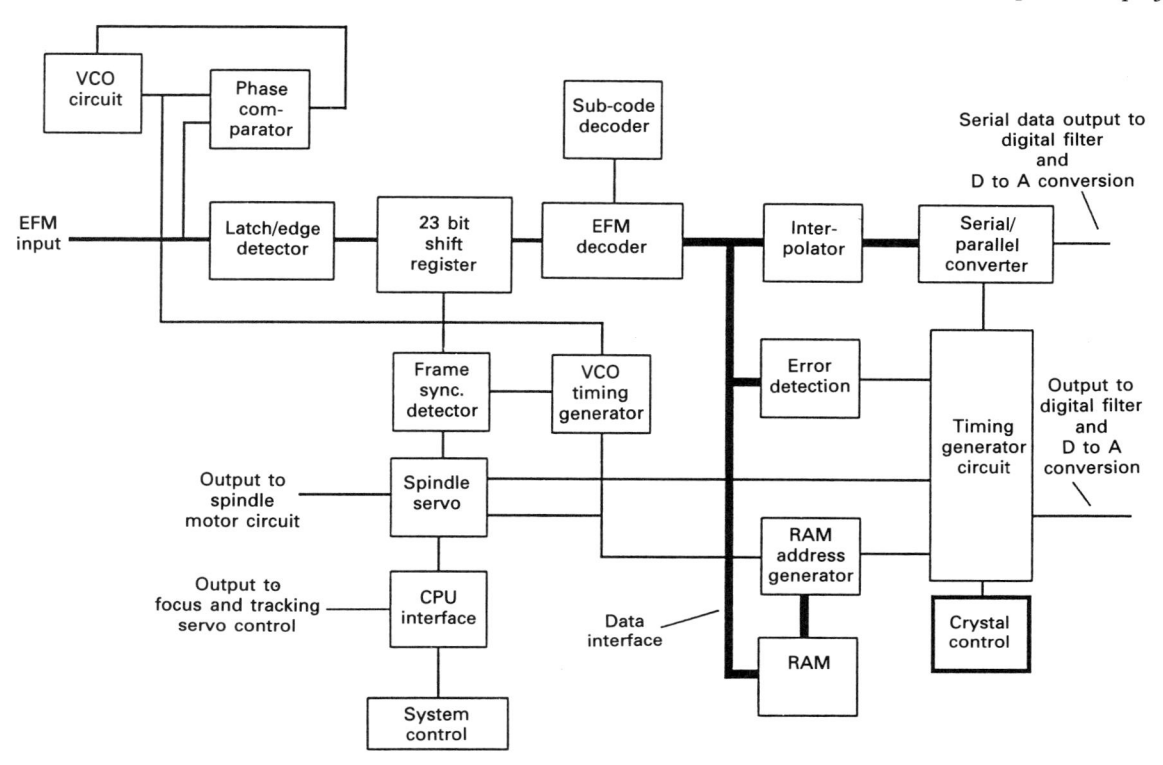

Figure 2.9 *Basic decoder*

The decoder

Much of the spindle servo circuitry is usually contained within the decoder, but despite its apparently small size as a component within the CD player (usually about the size of a standard postage stamp), it is fair to say that the decoder does quite a lot of work, with the support of a small amount of external circuitry. Fig. 2.9 illustrates the main features of the decoder section of a CD player.

Once the data from the disc have been initially processed from the original 'pits' or 'bumps', the data are now identified as the EFM signal and thus become the input to the decoder. Amongst the many processes that now take place, this signal is used to phase lock a VCO, usually operating at 8.64 MHz, twice the effective CD clock frequency, the sync. signal is removed within the 23 bit shift register, and passed to the frame sync. circuits, with the main bulk of the data now passing to the EFM demodulator. Within the EFM demodulator, the 14 bit symbols are now restored back into their original 8 bit counterparts, with the three coupling bits being virtually discarded. With the exception of the 8 bit control word, the remaining 8 bit data words are now paired together to form the original 16 bit data words that are in effect the digital samples of the original analogue information, but in a 'jumbled up' condition as a result of the interleaving process.

De-interleaving takes place with the use of the RAM, with error correction being applied using interpolation and relevant substitutions from the RAM as applicable, resulting from the error correction information contained within the cyclic redundancy check code (CRCC) data from the disc. The 8 bit control word is detected by the sub-code detector, with the P and Q bits being utilised accordingly.

System control plays its own part in the operation of the CD player by 'kick starting' the

25

The compact disc player

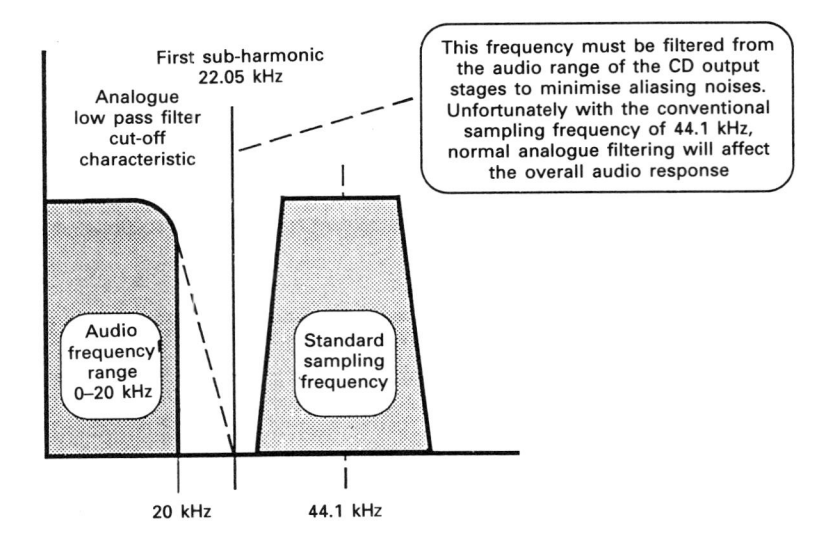

Figure 2.10 *Problems with conventional sampling at Fs = 4.41 kHz*

disc and analysing the fact that data are being retrieved from the disc. If any errors are being detected the system control will make decisions as to the type of correction to be applied, and if too many errors occur will, in due course, stop the operation of the player. An interface is also present to link up to the focus and tracking servos to ensure their correct operation.

The final digital data for processing into analogue can be either in serial or parallel form; serial data being the normal requirement for domestic CD players, usually fed in modern players to a digital filter for the next stage of processing.

The complete decoder operation is strictly controlled by the crystal control circuit which acts as a master reference for the majority of the decoder functions.

Digital filtering and digital to analogue conversion

When the digital data have been converted back into the original analogue form, it is necessary to pass the analogue signals through a filter network to remove any effects of the sampling frequency (44.1 kHz), which would appear as interference in the form of aliasing noise in the analogue output circuits. In order that the effects of the interference are minimised it is essential to ensure that the first subharmonic of the sampling frequency (22.05 kHz) is removed, which will ensure that any effects from 44.1 kHz will be minimal. However, the filter cut-off characteristics must operate from 20 kHz to 22.05 kHz, a cut-off rate which is not possible with conventional inductive, capacitive, resistive types of filters without degrading the analogue top end frequency response, which was a problem with the earlier CD players (Fig. 2.10).

Many developments have occurred within CD players with respect to improvements in the digital to analogue conversion and filtering stages, and whilst filtering is an area to be improved, it is worthwhile considering the restructuring of the analogue information from the digital data from the CD.

As seen in Fig. 2.11, another problem occurs with respect to quantisation noise, whereby noise pulses can occur at the sampling points, which though at 44.1 kHz can still cause effects to decrease the signal to noise ratio of the CD player. A method of overcoming both of these problems, i.e. the reduction in the high frequency response in order to overcome aliasing noise and the effect of quantisation noise, is to introduce the technique of oversampling. But it must be remembered that the digital data on the CD are effectively 16 bit data words at the sampling frequency of 44.1 kHz and therefore cannot be altered, whereas techniques within the CD player can.

The compact disc player

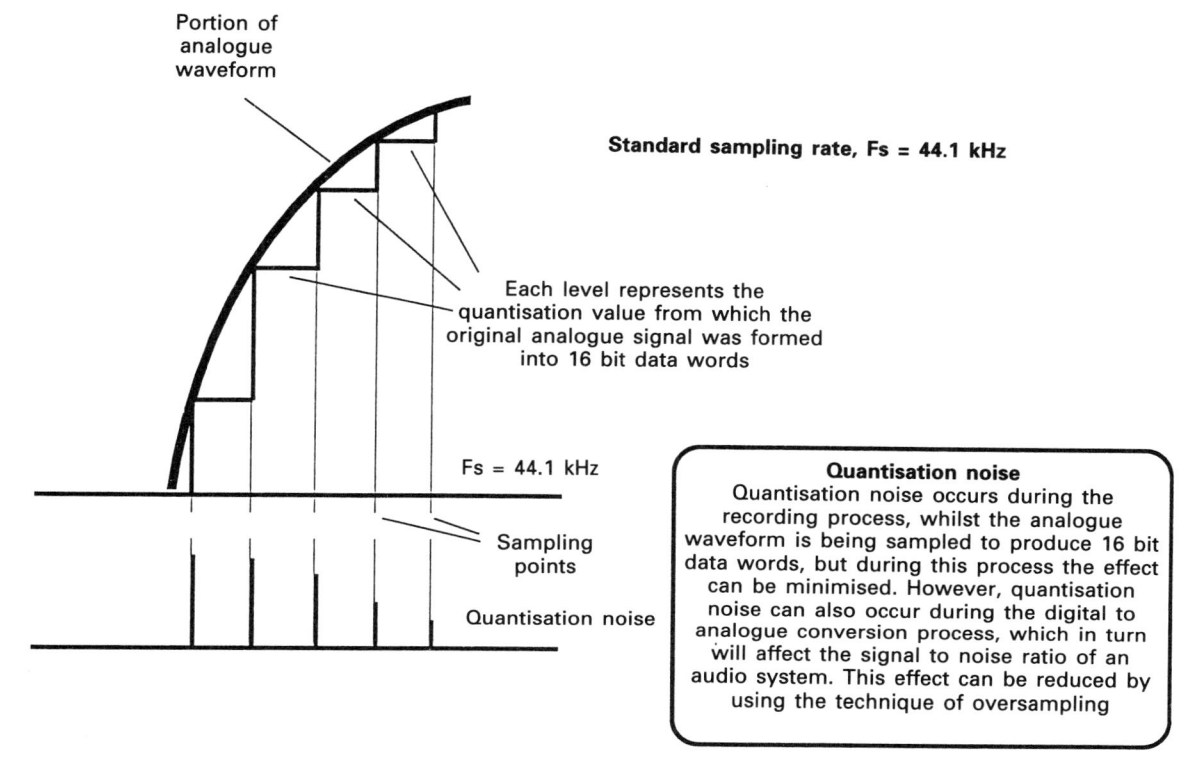

Figure 2.11 *Restructuring the original analogue information*

[Diagram labels: Portion of analogue waveform; Standard sampling rate, Fs = 44.1 kHz; Each level represents the quantisation value from which the original analogue signal was formed into 16 bit data words; Fs = 44.1 kHz; Sampling points; Quantisation noise]

Quantisation noise
Quantisation noise occurs during the recording process, whilst the analogue waveform is being sampled to produce 16 bit data words, but during this process the effect can be minimised. However, quantisation noise can also occur during the digital to analogue conversion process, which in turn will affect the signal to noise ratio of an audio system. This effect can be reduced by using the technique of oversampling

By the use of additional memory, delaying and multiplication techniques, it is possible to make acceptable predictions of a data word that can be included between any two of the sampling points described in Fig. 2.11.

A simple method of describing the achievement of an additional data word which can be effectively 'fitted in between' two original samples is by interpolation, i.e. adding two consecutive data words together, dividing by two and fitting the result in between the two original samples. Two important factors emerge from this concept: (1) the sampling frequency must double in frequency to 88.2 kHz, and (2) quantisation noise will reduce, thereby improving the signal to noise ratio. The effective doubling of the sampling frequency ensures that a conventional filter can be used to remove the first subharmonic, which is now 44.1 kHz, with the effect of improving the high frequency response of the CD player.

Figs 2.12 and 2.13 illustrate the effect of two times and four times oversampling techniques, though many players utilise even higher sampling frequencies, especially in the realms of bit stream digital to analogue methods which will be highlighted later.

The frequency response of the filtering circuits in the output stages is illustrated in Fig. 2.14 where a new sampling frequency of four times the standard used on the CD, i.e. 176.4 kHz, enables a more conventional filter to be designed, with less steep characteristics.

Digital filtering

The increases in the sampling frequencies can be achieved by multiplication and delaying techniques, as illustrated in Fig. 2.15. From this figure it may be appreciated that with the concept of four times oversampling it is possible to produce three digitally produced extra data words, which are in fact predictions, to be inserted between any two actual samples from the CD.

With the technique of delaying the original 16 bit data words, and then multiplying each resultant word four times, each multiplication being a different coefficient, a resultant series of 28 bit

The compact disc player

Sampling rate, Fs = 88.2 kHz

> **Interpolation**
> Interpolation occurs when the average of two successive data words is inserted between those two data words. In practice interpolation takes place within the digital filter of a CD player, where improved interpolation is achieved by analysing a series of data words

Portion of analogue waveform

Interpolation points

2 Fs = 88.2 kHz

Original sampling points

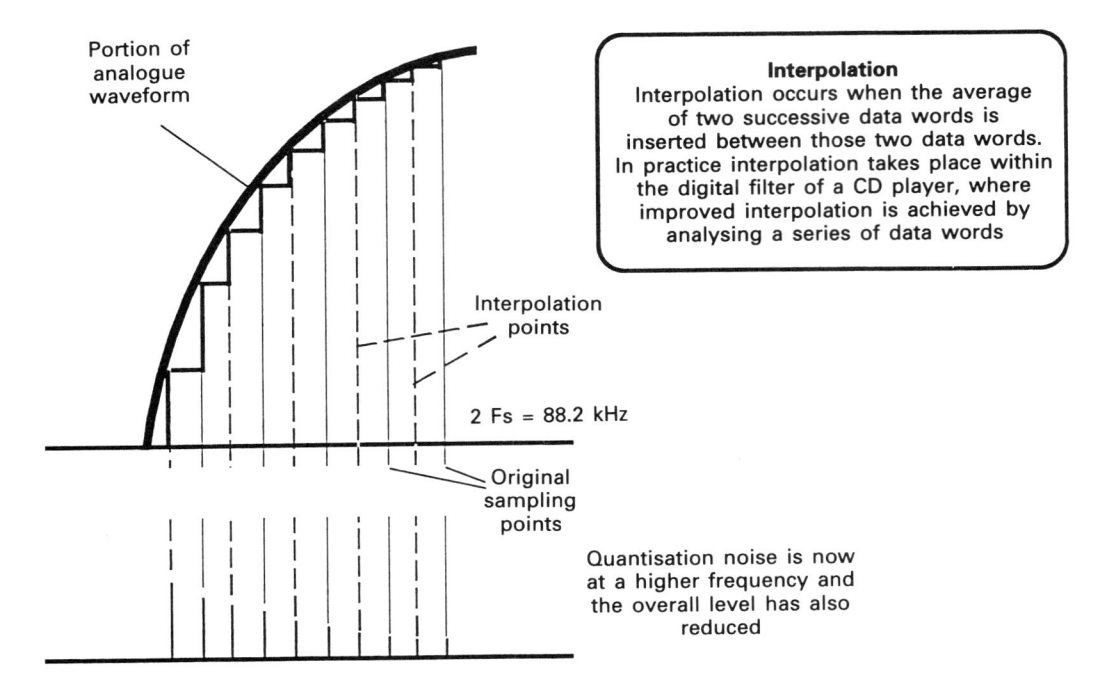

Quantisation noise is now at a higher frequency and the overall level has also reduced

Figure 2.12 *Two times oversampling*

words is produced. Summing a complete sequence or group of these resultant 28 bit words produces a weighted average of a large number of samples. Whilst this description relates to four times oversampling, further improvements can be achieved using eight times oversampling, where the sampling frequency becomes 352.8 kHz. The resultant 28 bit data words are now passed on to noise shaping circuits where further improvements in the signal to noise ratio can be achieved.

Noise shaping

The circuit shown in Fig. 2.16 can be used to improve the signal to noise ratio by utilising the fact that the noise level of the signal will be contained within the least significant bit area. By extracting the 12 least significant bits, delaying them by one sampling period and then subtracting the resultant from the input signal, the output will now comprise the 16 most significant bits. Fig. 2.16 provides an example of this concept, and can be identified as a single integration type of noise shaping circuit. More complex types are available comprising double integration or even multi-stage noise shaping (MASH) circuits, all of which are progressions in the evolution, by various manufacturers, of the processing of the original 16 bit data words coming from the CD, the basic concept of which will covered at a later stage.

The 16 bit data words from the circuit shown in Fig. 2.16 can now be passed to the digital to analogue converter for conversion back into the original analogue information.

Digital to analogue conversion

The 16 bit data words are passed to the D to A converter where the data information in terms of 0's and 1's will control internal switches to enable a current to be produced that ideally relates to the original analogue value of the signal when it was first recorded onto CD.

The simplified diagram in Fig. 2.17 illustrates a series of current generators, each of which provides a current that is equal to half of the current provided by its left-hand counterpart. The

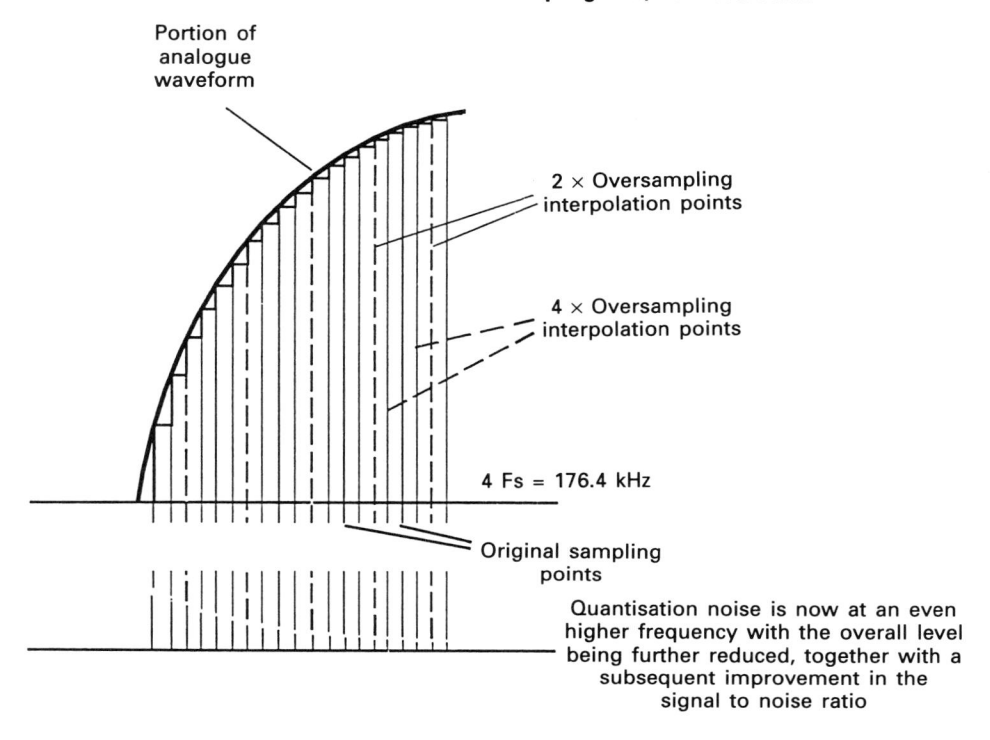

Figure 2.13 *Four times oversampling*

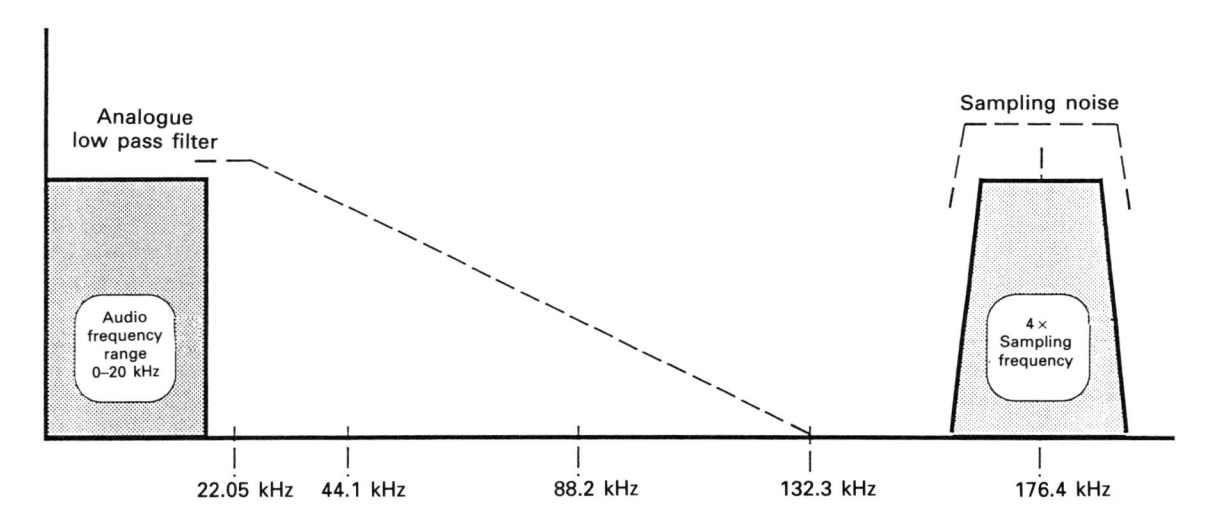

Figure 2.14 *The improved filtering effects with four times oversampling*

The compact disc player

[Figure 2.15: Digital filtering diagram showing:
- DIGITAL INPUT 16 bits 44.1 kHz feeding into a series of D (delay) blocks
- "Series of delays, each delay being equal to one sampling period of 44.1 kHz. 24 Individual delays are typical of this arrangement"
- Row of 4× M (multiplier) blocks
- "Each 16 bit data word is now multiplied four times, with different coefficients, adding 12 extra bits to provide outputs of 28 (16 + 12) bit data words at 176.4 kHz (4 × 44.1 kHz)"
- SUMMATION SECTION: "All the multiplied data words are now summed together to give a 'weighted average' or 'long term interpolated' output"
- Clock input 64 Fs (2.8224 MHz)
- DIGITAL OUTPUT 28 bits 176.4 kHz]

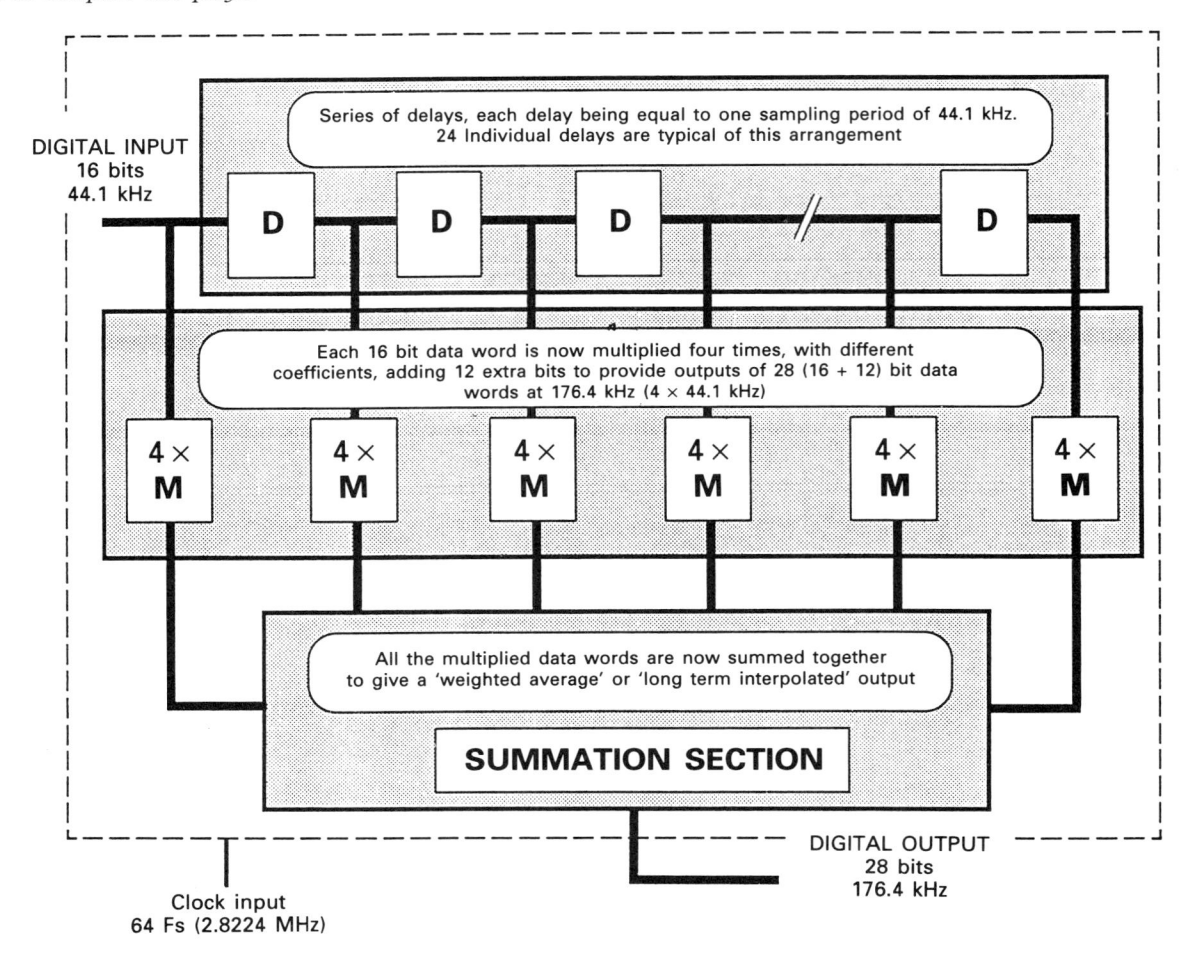

Figure 2.15 *Digital filtering*

left-hand current generator will provide half of the total value of current available.

The 16 bit data word is passed into the circuit, and whenever a 1 occurs within the word, the relevant switch S1–S16 will close. The sum of the currents related to when the 1's occur will pass to the current to voltage converter to provide an output that should be equivalent to the original analogue value at the moment of sampling or quantisation during the recording process.

Each subsequent data word will therefore enable the original analogue signal to be sequentially restructured, and therefore the speed of operation is related to the sampling frequency of the data words being applied. However, D to A converters suffer from inherent problems which can cause discrepancies in the output signal in the form of non-linearity errors, zero cross distortion and glitches.

Non-linearity errors

Each current generator is usually formed by a resistive network to enable a specific current to flow, and it is essential for each generator to be a specific fraction of the one next to it. Thus if the MSB current generator is designed to provide half of the total current required, then the next generator must provide half of that value and so forth. Any variations of these tolerances will not enable a faithful reproduction of the original analogue signal to be achieved.

30

The compact disc player

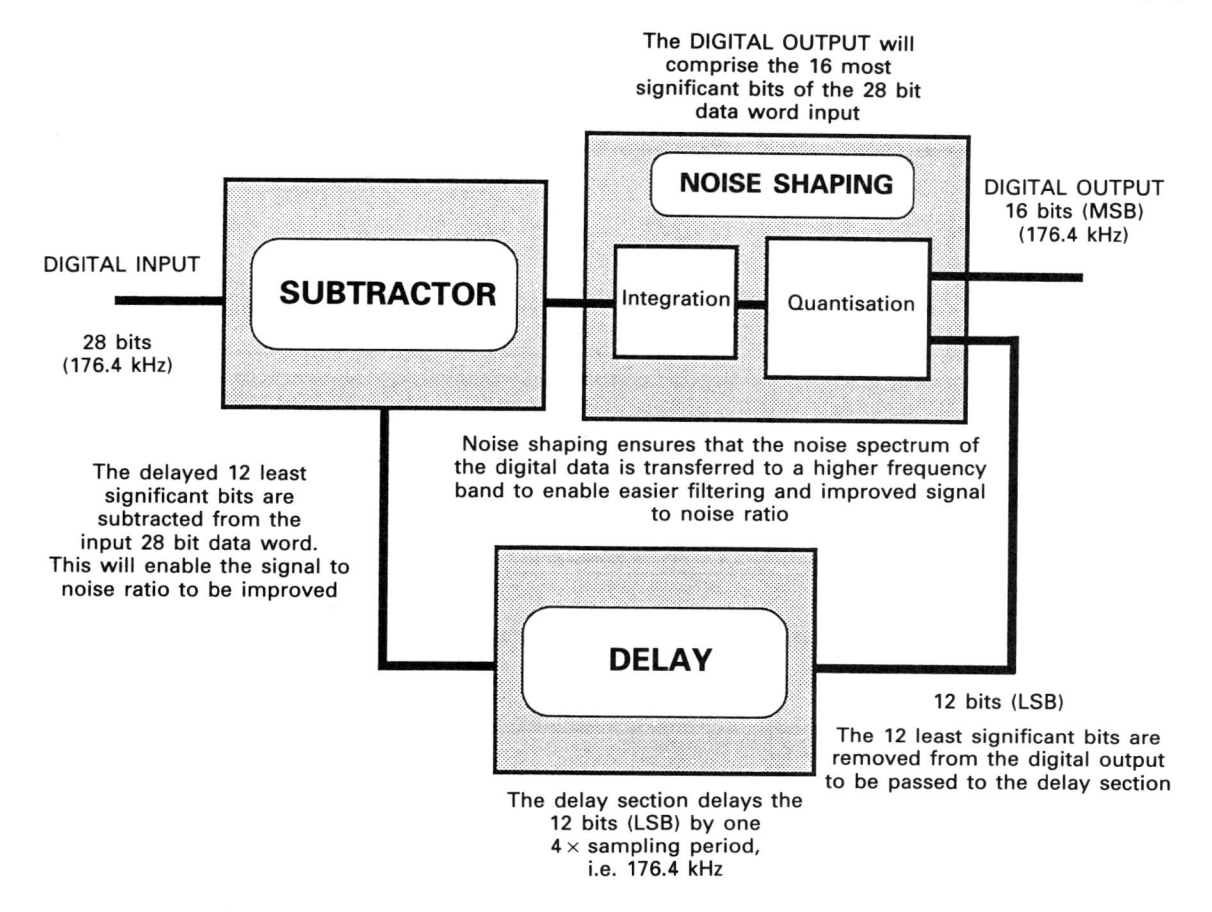

Figure 2.16 *Noise shaping*

Zero cross distortion

When the analogue signal is operating within its postive half cycle the 16 bit digital data will be the complement of its opposite negative value, so as the analogue signal traverses the zero cross point the sequence of bits will be reversed. The MSB during the positive half cycle is always a 0, whilst during the negative half cycle the MSB will always be a 1. Therefore with low level signals and any non-linearity within the LSB area of the D to A converter, distortion of the signal output can be caused.

Glitches

Glitches are caused when any of the switches do not operate at the precise moment of the occurrence of each relevant bit, with the result that an incorrect output can be momentarily obtained.

To overcome some of these problems, 18 bit and 20 bit D to A converters have been designed which enable a more faithful replication of the original signal to be achieved. The 18 or 20 bit data words can be derived from the digital filtering or oversampling process, but whether it be a 16, 18 or even 20 bit D to A converter, accurate manufacture of these items for the domestic CD player can prove extremely expensive.

Another method of digital to analogue conversion proving to be extremely popular with modern players is the bit stream or one bit system.

As previously mentioned, there has been a gradual evolution, by various manufacturers, regarding the processing of the digital data back into their original analogue form as accurately as possible, and noise shaping is an essential prerequisite.

The compact disc player

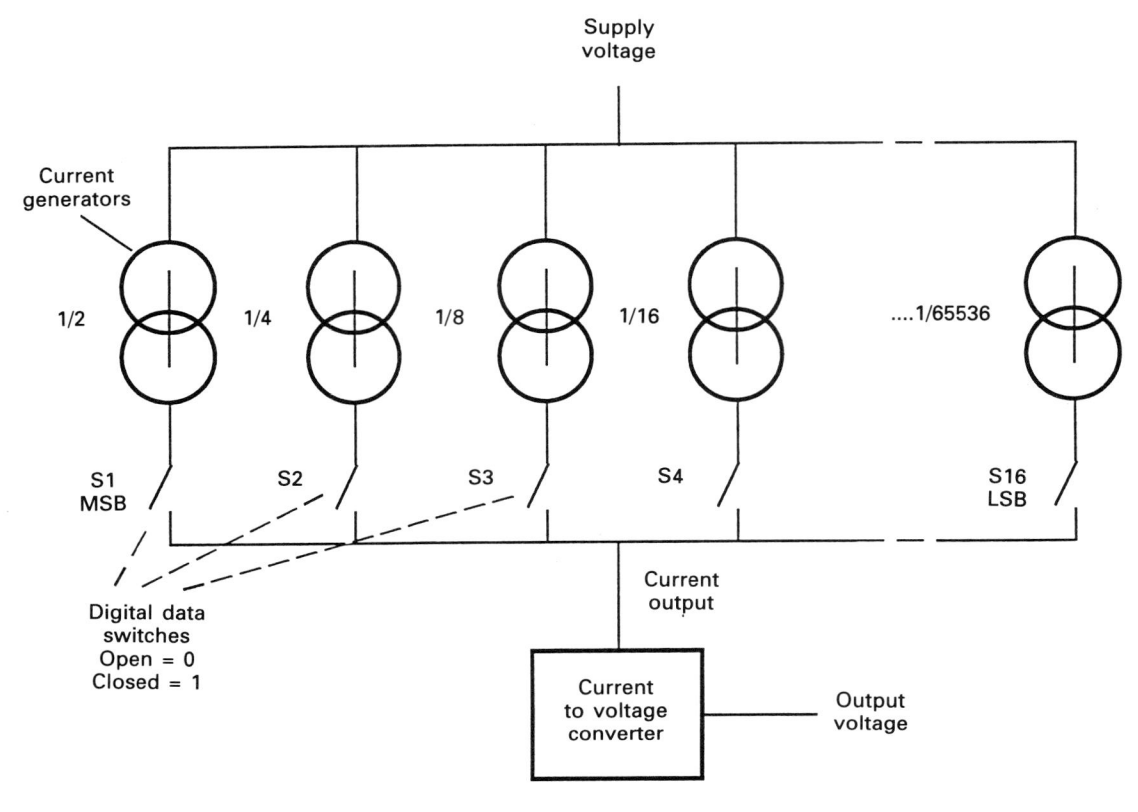

Figure 2.17 *A series of current generators*

Multi-stage noise shaping

As an example, the digital input from the digital filter can be in the form of a series of 28 bit data words at a sampling frequency of four or even eight times the original sampling frequency (i.e. 8Fs = 8 × 44.1 kHz = 352.8 kHz); this implies that there are either three or seven interpolated samples between any two data words from the CD. The MASH (Fig. 2.18) will now process the digital input at a much higher rate, in the order of 32 times the original sampling frequency (i.e. 32Fs = 44.1 kHz × 32 = 1.411 MHz), though some players use higher rates, and it would appear at this stage that the sampling frequencies are becoming inordinately high when compared to the original sampling frequency. In order that these increased sampling frequencies may be processed, it is necessary to use high clock frequencies which are obtained from a crystal oscillator.

But the implication now is that at 32Fs there must be 31 interpolated samples between any two data words from the disc. As a result, it is now not necessary to maintain such a high number of bits in each data word. In fact the output from the MASH circuit above is in the form of four bits, which will provide sufficient detail to effectively indicate the analogue content in due course.

The 4 bit output is now passed to a pulse width modulator; 11 of the possible 4 bit combinations are used to provide 11 different pulse widths, and the width of these pulses is related to the analogue information. By passing these varying pulse widths to a low pass filter, or integrator, the required analogue output will be obtained.

The resultant 4 bit data words basically represent the 'trend' of the signal between two original sampling points, and therefore a series of 16 bit or more data words becomes unnecessary. Only 11 different 4 bit data words are passed to a pulse

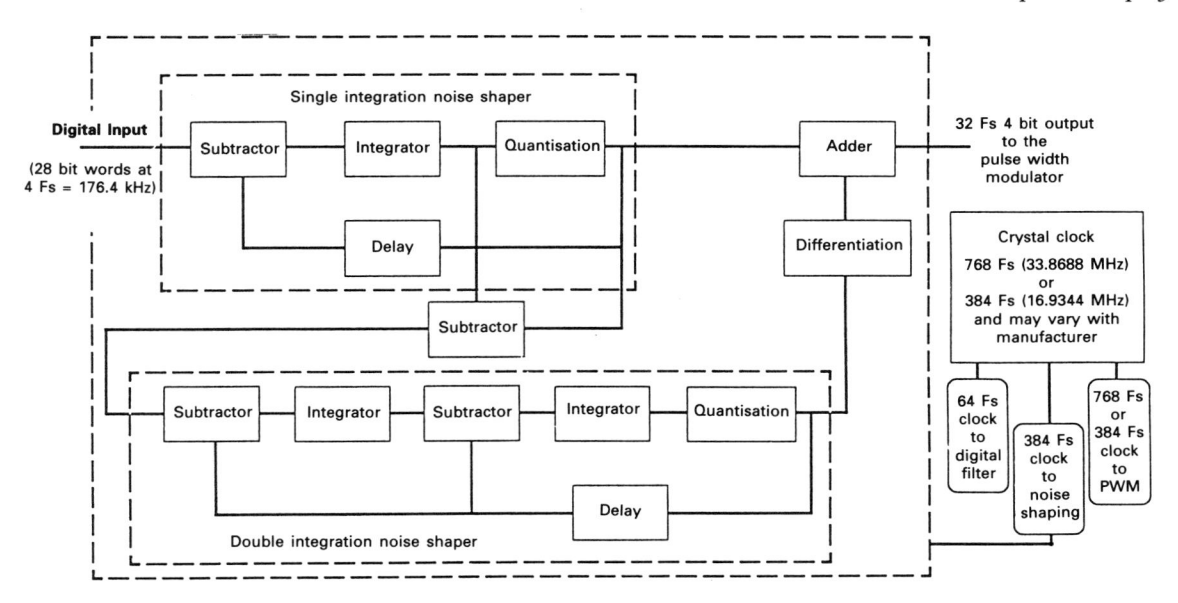

Figure 2.18 *Multi-stage noise shaping*

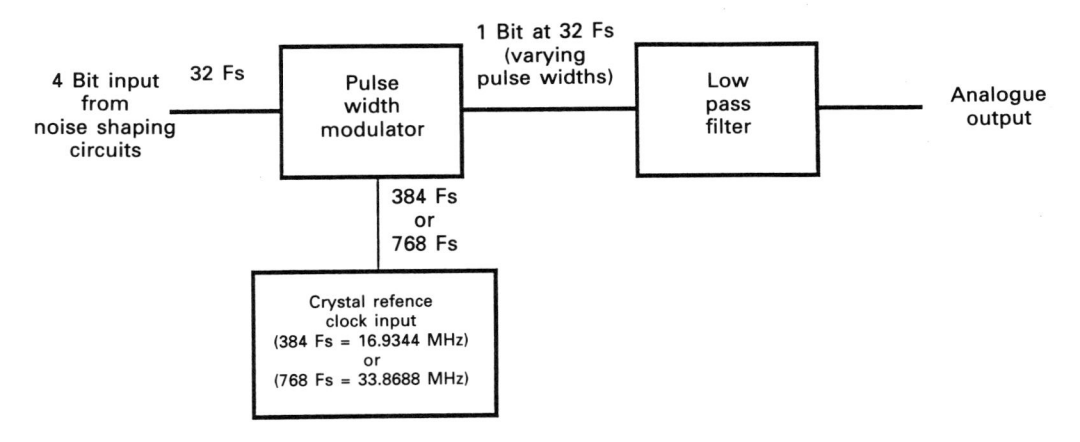

Figure 2.19 *Pulse width modulation*

width modulator (Fig. 2.19) where each 4 bit word is converted into a single pulse each of a different width.

This is where the term '1 bit digital to analogue conversion' has caused confusion, as the question that has often been raised is... 'How can a 16 bit word become a 1 bit word?' Possibly the best way to accept this concept is to consider any one of the 11 different pulse widths, which now represents a small sample of the extra interpolated information between two original 16 bit data words from the disc, as a bit of information, and not necessarily as 1 bit of data.

Whilst a high clock frequency is used, timing circuits will provide the relevant clock frequencies for the different related circuits, which must be 'locked' to the same reference source. The output from the pulse width modulator is described as 1 bit information insomuch that it is a series of single pulses, the width of which is varying (up to 11 different widths), and it is therefore no longer a stream of data.

The compact disc player

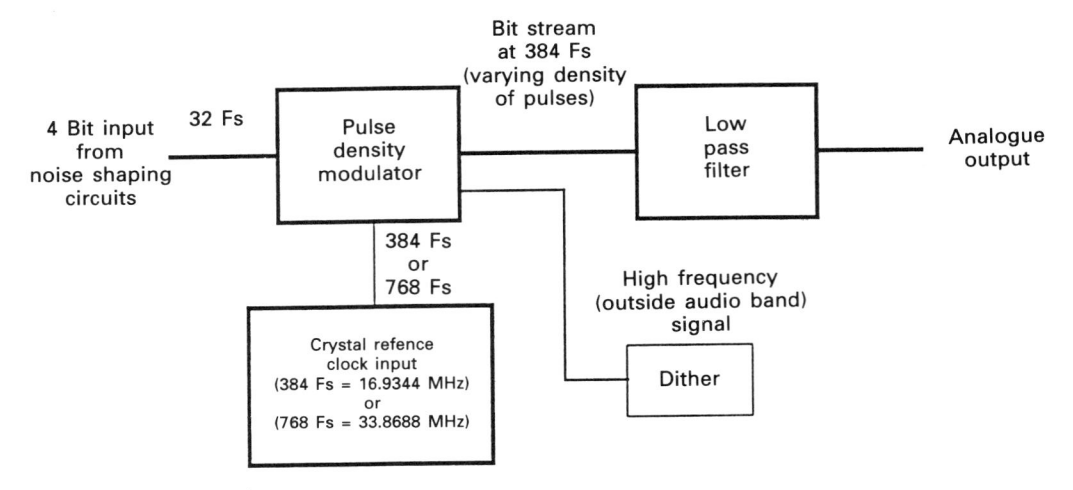

Figure 2.20 *Pulse density modulation*

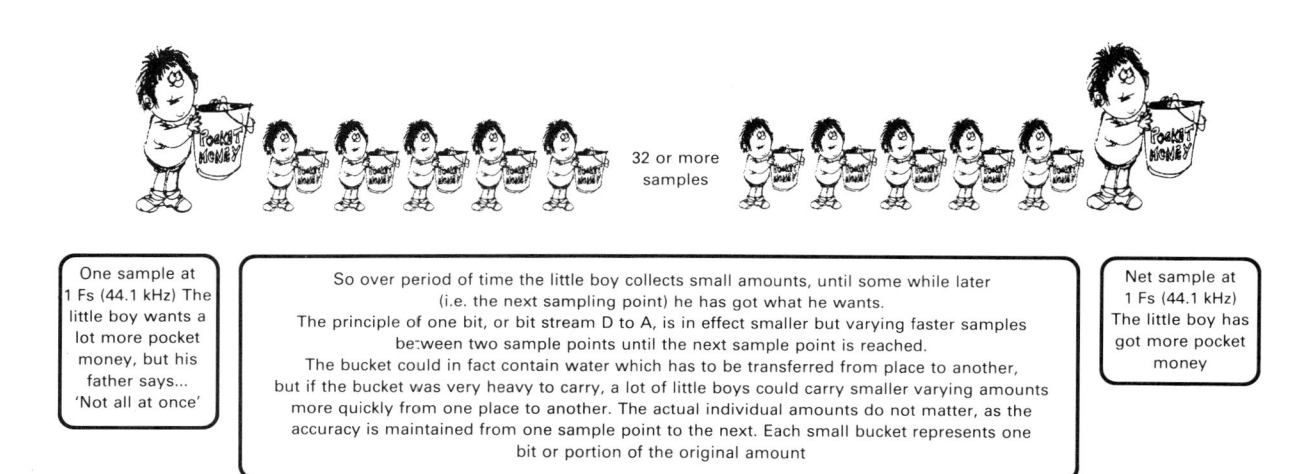

Figure 2.21 *One bit or bit stream D to A – a pictorial analogy*

Passing the information through a low pass filter will enable the original analogue information to be retrieved with much greater accuracy and simplicity compared to the more conventional method of D to A conversion, and therefore the problems of non-linearity, zero cross distortion and glitches are resolved. There is, however, one main problem with this method and that is 'jitter'. Jitter is related to the accuracy of the actual start and finishing points of each pulse width, which can affect the final sound quality. This effect may be more apparent by using the higher clock frequency of 786 Fs, and to endeavour to overcome this problem some players may use the lower clock frequency of 384 Fs.

Another method of resolving the problem of jitter is to fill the varying pulse widths with the high clock frequency pulses, with the final result becoming a continuous stream of bits (hence the term 'bit stream'). Passing these bits through the low pass filter produces the same analogue output. This method was originally used by Philips, but has now become a standard feature of many other types of CD player. However, this method can create other problems, and when zero or low level signal levels occur, it is possible to hear spurious noises resulting from the 'idling' of the data that will occur at these levels. This can be overcome by applying a high out of audio range 'dither' signal to the circuit. This will remove the spurious noises, without affecting the final integration of the bit stream information in the low pass filter.

With the pulse density concept (Fig. 2.20) the circuit arrangement is virtually identical, the main difference being that instead of producing varying pulse widths, suffering from possible jitter, a stream of 384 Fs pulses is used, the number or density of which will relate to the original analogue output from the low pass filter. The pulse density modulation or bit stream principle is virtually the latest technology being utilised within CD technology and the foregoing descriptions are intended to provide an overall appreciation of some of the technologies involved.

3

OPTICAL ASSEMBLIES

The optical assembly, or pick-up unit, is one of the most important components within a CD player, and is essential to being able to determine the data or music information that has been recorded on to the CD.

Though optical assemblies from different manufacturers will vary, they will contain similar elements which will basically include:

1 *a low power laser* to illuminate the disc track;
2 *a lens and prism system* to direct the laser beam towards the disc surface, and the reflected beam towards the photo-diode array;
3 *a photo-diode array* to provide the data, focus and tracking signals obtained from the disc, and which can comprise four or six photo-diodes; and
4 *focus and tracking coils* to enable the laser beam to be focused onto the disc, as well as enabling it to track across the disc surface. Some types of optical units, such as those usually found in Philips CD players, will not contain the tracking coil.

Throughout the wide range of CD players that service engineers may encounter, the optical assembly will usually be identified as one of three types:

1 a single-beam device with radial tracking,
2 a single-beam device with linear/straight line tracking,
3 a three-beam device with linear/straight line tracking.

Again, it is possible to find additional variations from one manufacturer to another; Sharp and Technics to name but two can be found to have a radial optical system in some players, and a linear/straight optical system in others. When the need arises to replace an optical assembly, it may be either a situation of a quick replacement with little or no adjustments, or a major operation which involves essential mechanical adjustments that may prove critical to the final operation of the CD player.

Single-beam device with radial tracking

Single-beam optical assemblies with radial tracking (Fig. 3.1) are usually found in Philips/Marantz CD players, some of the Panasonic/Technics models, as well as Denon, Binatone and Sentra, to name but a few.

The basic internal arrangement of the optical assembly used in the radial system is shown in Fig. 3.2. With this type of device the objective lens will move vertically for focusing, but will have no lateral movement for tracking, this being achieved by gradually moving the complete assembly radially across the playing area of the CD. The laser beam is directed onto the CD and reflected back via a beam-splitting prism to produce two beams which are directed towards the photo-diode array, which comprises four photo-diodes from which must be obtained the necessary signals for data, focus and tracking.

Figure 3.1 *Radial optical assembly*

Optical assemblies

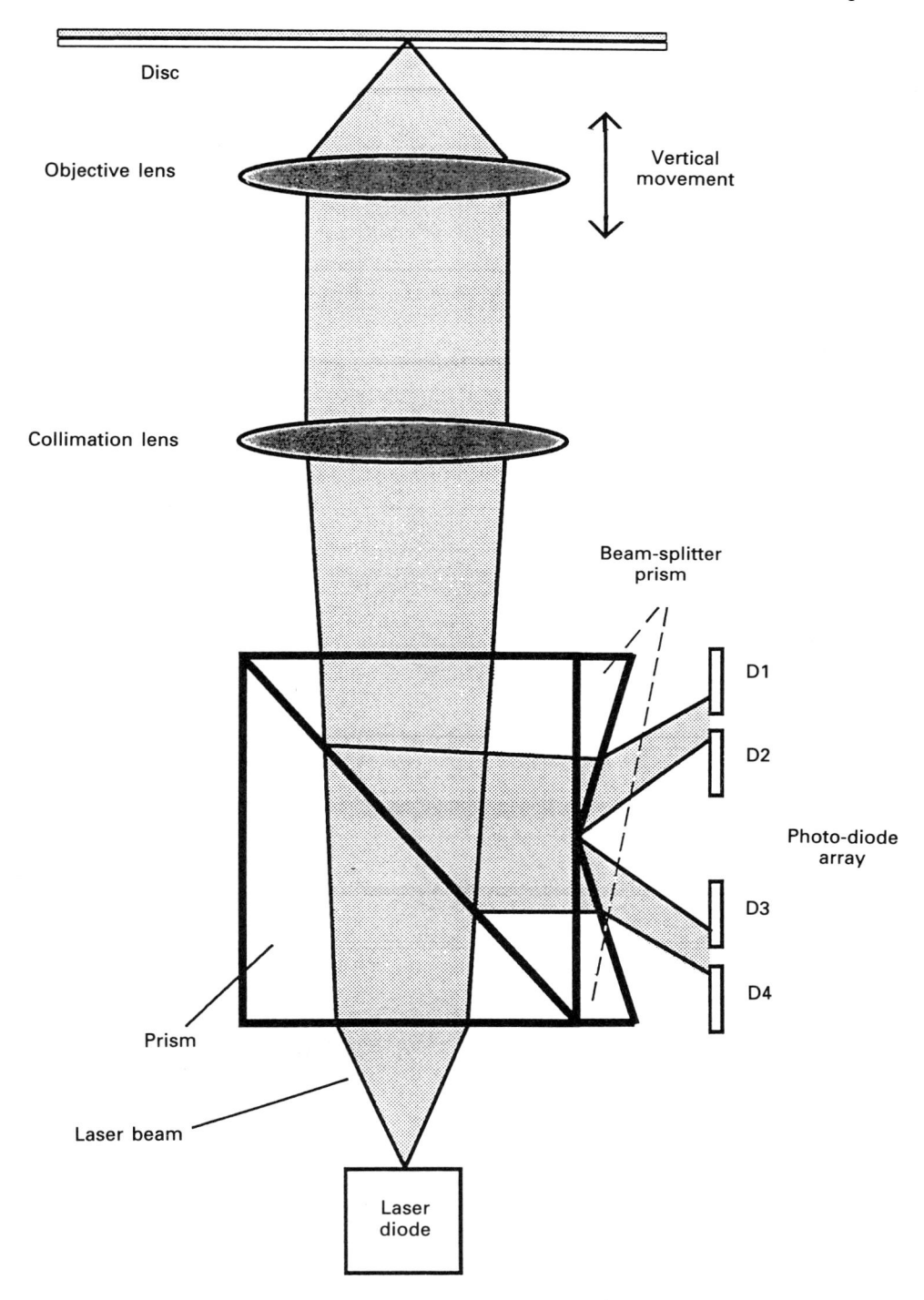

Figure 3.2 *Philips optical assembly*

Single-beam device with linear/straight line tracking

Single-beam devices with linear tracking (Fig. 3.3) are usually found in the Panasonic/Technics models, though this type did appear in the first Pioneer in-car CD players.

In this type the objective lens has the capability of moving vertically to achieve focus, and laterally for tracking, but as there is only a limited amount of lateral movement of the lens (~ 2 mm), a method of moving the optical assembly across the surface of the disc, at right angles to the track, has to be achieved.

As with the previous device, the photo-diode array comprises four photo-diodes to develop the data, focus and tracking signals. Later versions may contain six photo-diodes for improved operational capabilities with scratched or marked discs.

Three-beam device with linear/straight line tracking

It could be said that this type of optical assembly will appear in the majority of the players originating from the Far East, e.g. Akai, JVC, Kenwood, Panasonic/Technics, Pioneer, Sanyo, Sansui, Sony, Toshiba and Yamaha.

With these units, three beams are provided, where the main or centre beam provides data retrieval as well as focus information and two side beams provide tracking signals. The objective lens is able to move vertically and laterally in a similar fashion to the previously described device. In order that the two extra beams can be obtained from the single laser beam a diffraction grating is necessary. Again the lens has only a limited amount of lateral movement, and therefore a method of moving the complete assembly gradually across the surface of the disc has to be found (Fig. 3.4).

Referring to any of any of the illustrated optical blocks, similarities will occur with respect to the main components. Occasionally it can be observed that different manufacturers choose to include certain component arrangements or concepts that will suit a specific design criterion, but overall the operation of all optical blocks will be generally similar regardless of their manufacturing origin. Exceptions to this rule are those optical blocks that do not contain tracking coils (which is typical of Philips) and those that have a diffraction grating (which is typical of the Pioneer CD players). The following descriptions apply to any or all of the illustrated optical blocks, with additional information as applicable, commencing from the laser diode and effectively progressing through the optical block, until the reflected light reaches the photo-diode array.

Components of optical blocks

Laser diode

The laser beam is essentially the controlling feature of the CD system insomuch that this type of light operates at a specific frequency (~ 790 nm), whereas white light covers a wide range of frequencies. Because of this specific frequency of the laser beam, it is possible to achieve a beam width in the region of 0.6 µm within a CD player, thus enabling the reading of the data from a disc containing a track width of 0.5 µm – a feature that is not possible with white light. The power of the laser is extremely small; in the region of 0.23 mW in the early CD players, to around 0.12 mW in the majority of the players currently available.

It is essential to maintain the light output level of the laser diode at a consistent level, which is achieved by monitoring a portion of the laser output via a photo or monitor diode, and feeding the registered level back through an automatic power control circuit where it is compared to a predetermined level which can be adjusted by the laser power control, thereby stabilising the light intensity output at the setting determined by the relevant manufacturer.

The laser diode is switched On and Off via the LDON line which is originally controlled from the system control (Fig. 3.5).

Diffraction grating

Diffraction grating is an extremely small special lens that will produce three beams from the single beam of the laser diode. It can be compared to the special 'add on' lenses that can provide multiple images of a single scene for photographic or video cameras, though the actual size of the lens or grating is extremely small.

This element will only appear in the three-beam

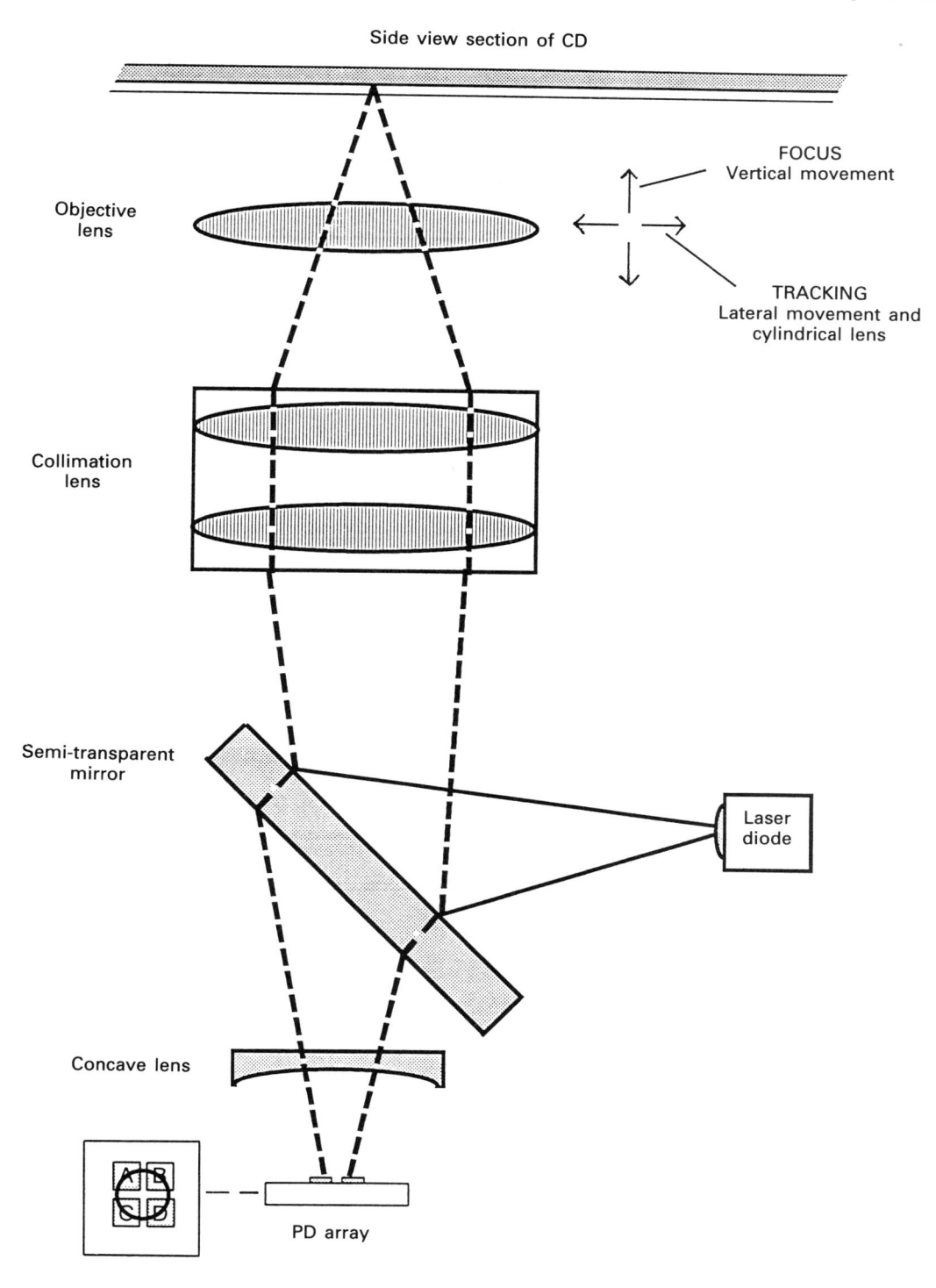

Figure 3.3 *Single-beam optical assembly*

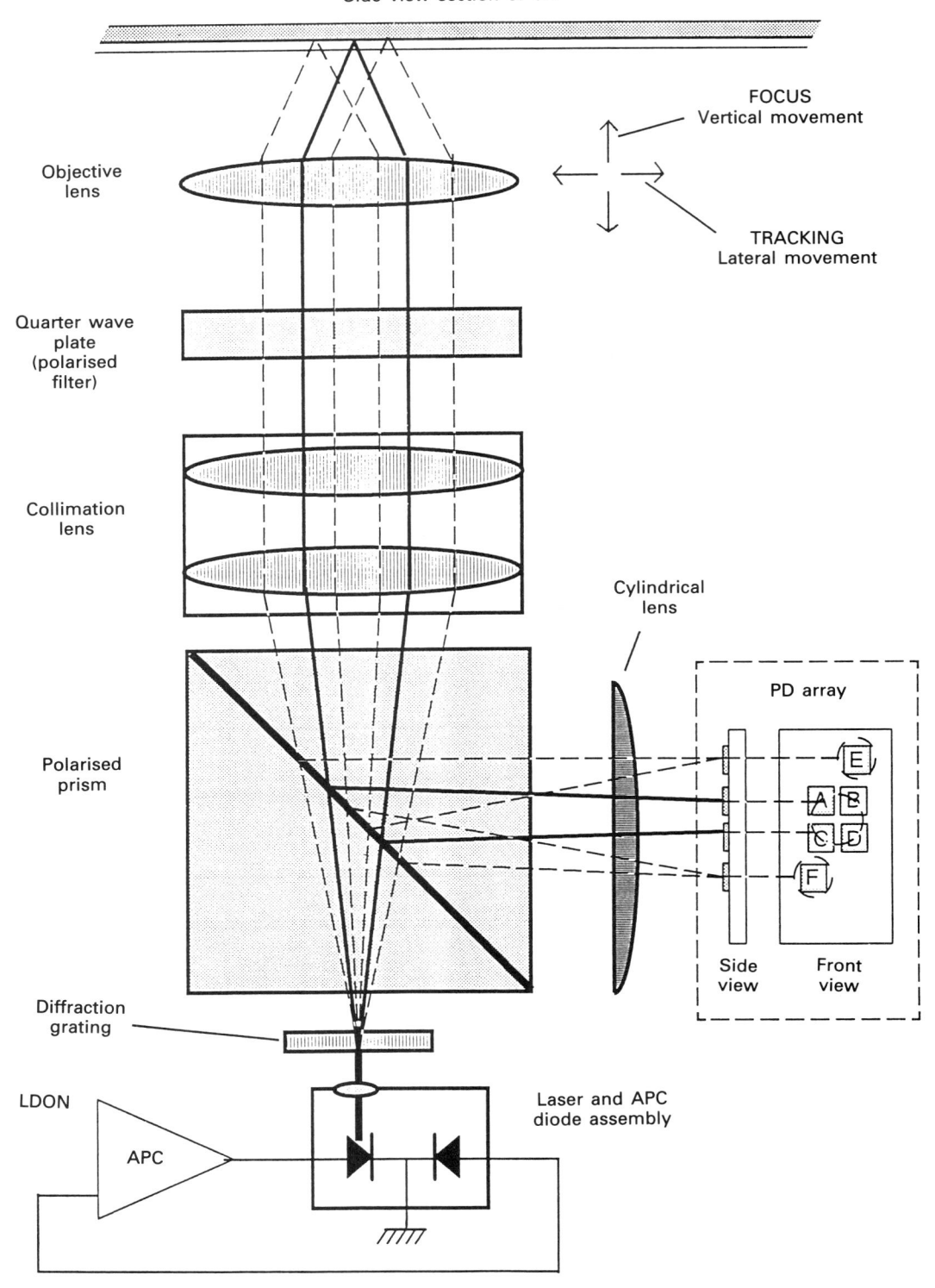

Figure 3.4 *Three-beam optical assembly*

Optical assemblies

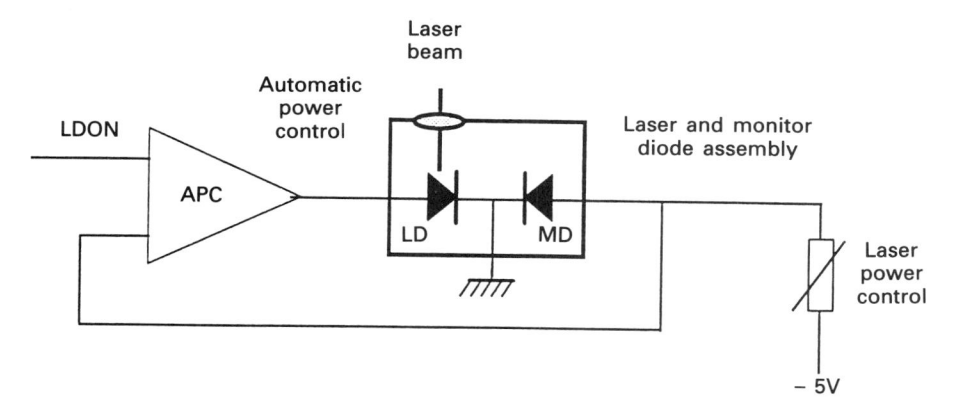

Figure 3.5 *Laser diode power control*

devices, where the extra two beams, frequently referred to as side beams, will enable the signal for the tracking servo to be achieved. With some manufacturers, typically Pioneer, there is a requirement to adjust the grating whenever it proves necessary to replace a faulty optical block. The method for adjustment of the diffraction grating is fully described under the adjustment procedures in Chapter 6. However, the latest Pioneer optical block (the 92 series) does not require the grating to be adjusted as it is factory set during manufacture.

Prism – Three-beam optical block

Depending upon the type of optical block, there may be more than one prism. Essentially it is required to separate the direct laser beam going towards the disc from the reflected beam from the disc for processing by the photo-diode array.

The reflective surface of the prism is frequently covered with polarising material which will enable the reflected beam to be efficiently separated from the main or direct laser beam. Typical of this arrangement is the illustrated three-beam device.

Semi-transparent mirror – single-beam optical block

This reflects the laser beam towards the collimation and objective lenses and onto the surface of the disc.

Collimation lens

The collimation lens together with the objective lens ensures that the laser beam achieves focus on the playing surface of the CD.

Quarter wave plate – three-beam optical block

The direct laser beam passing through this plate towards the CD is optically altered by 45°, and the returning beam is altered by a further 45°, therefore causing the optical polarisation between direct and reflected beams to differ by 90°. The reflected beam is now of the correct polarisation to enable efficient deflection within the prism towards the photo-diode array.

Objective lens

The objective lens is capable of vertical movement in all optical assemblies to achieve focus of the laser beam onto the CD, and in many players is capable of lateral movement to enable the optical assembly to follow the track of the disc. In players that use the radial method of track following, as in the majority of Philips players, the objective lens does not move laterally.

Optical assemblies

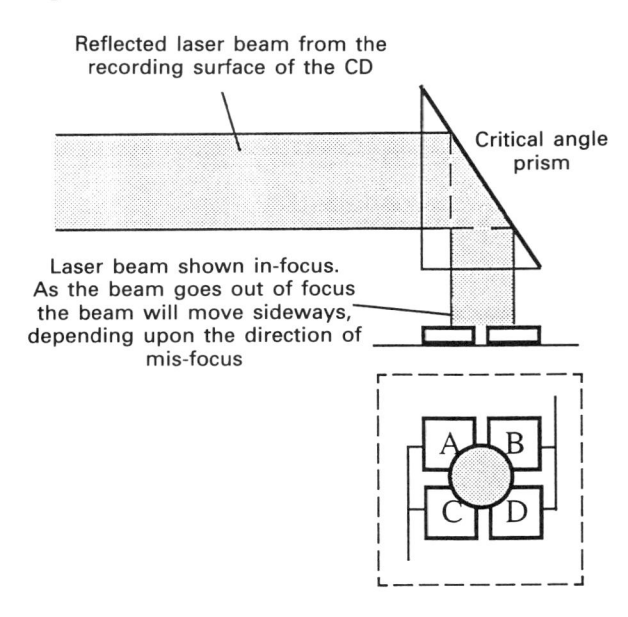

Figure 3.6 *Critical angle prism*

Cylindrical lens – three-beam optical block

The function of this lens is to enable the reflected beam from the CD to assist in creating the necessary signal to ensure that focus of the laser beam onto the playing surface of the disc can be maintained.

When the beam is correctly focused, a round beam of light will land on the four photo-diode elements, as shown in Fig. 3.12. However, if the beam becomes out of focus, the cylindrical lens will in effect distort the beam elliptically (frequently referred to as 'astigmatism') as illustrated, depending upon the direction of mis-focus. Further description of how this achieves the focus error signal is given under the section on Photo-Diode Arrays.

Critical angle prism

An alternative method of determining focus can be achieved with the critical angle prism system, as shown in Fig. 3.6.

The critical angle method utilizes an optical phenomenon when mis-focus occurs, where the reflected beam moves fractionally sideways as the result of passing through a prism at a specific angle, with the amount of deflection being related to the direction amd amount of mis-focus.

Beam-splitting prisms – Philips optical block

This method of determining mis-focus and mis-tracking is usually found in the Philips players and some of the Sharp and Panasonic/Technics players, and is further described under the section on photo-diode arrays.

Concave lens – single-beam optical block

This concentrates the reflected beam onto the photo-diode array, and improves sensitivity.

Photo-diode arrays

Depending upon the type of optical assembly involved, the photo-diode array will usually comprise either four or six diodes, with the prime function of developing the data, focus and tracking signals, which are the result of the reflected laser beam from the playing surface of the CD (Fig. 3.7). However, the latest Pioneer optical block has five photo-diodes, which together with a hologram prism provide a somewhat different method of processing the reflected beam (see Fig. 3.17).

It is the photo-diode array, whatever its form, that is required to produce the three main signals to enable a CD player to operate effectively:

1 data information: to enable the music or other information to be produced;
2 focus error: to ensure that the laser beam remains in focus on the disc whilst it is being played; and
3 tracking error: to enable the laser beam to track across the playing surface of the disc.

To produce the data information from the diodes is relatively straightforward by adding the sum of the four diodes (A + B + C + D) together and together with the necessary amplification, an RF signal will be obtained, which will be later referred to as the 'RF eye pattern waveform', and will be described in later sections.

The focus and tracking error signals are developed by various methods of combining the diodes to produce the required signals which will operate the focus and tracking servos, and whilst these

Optical assemblies

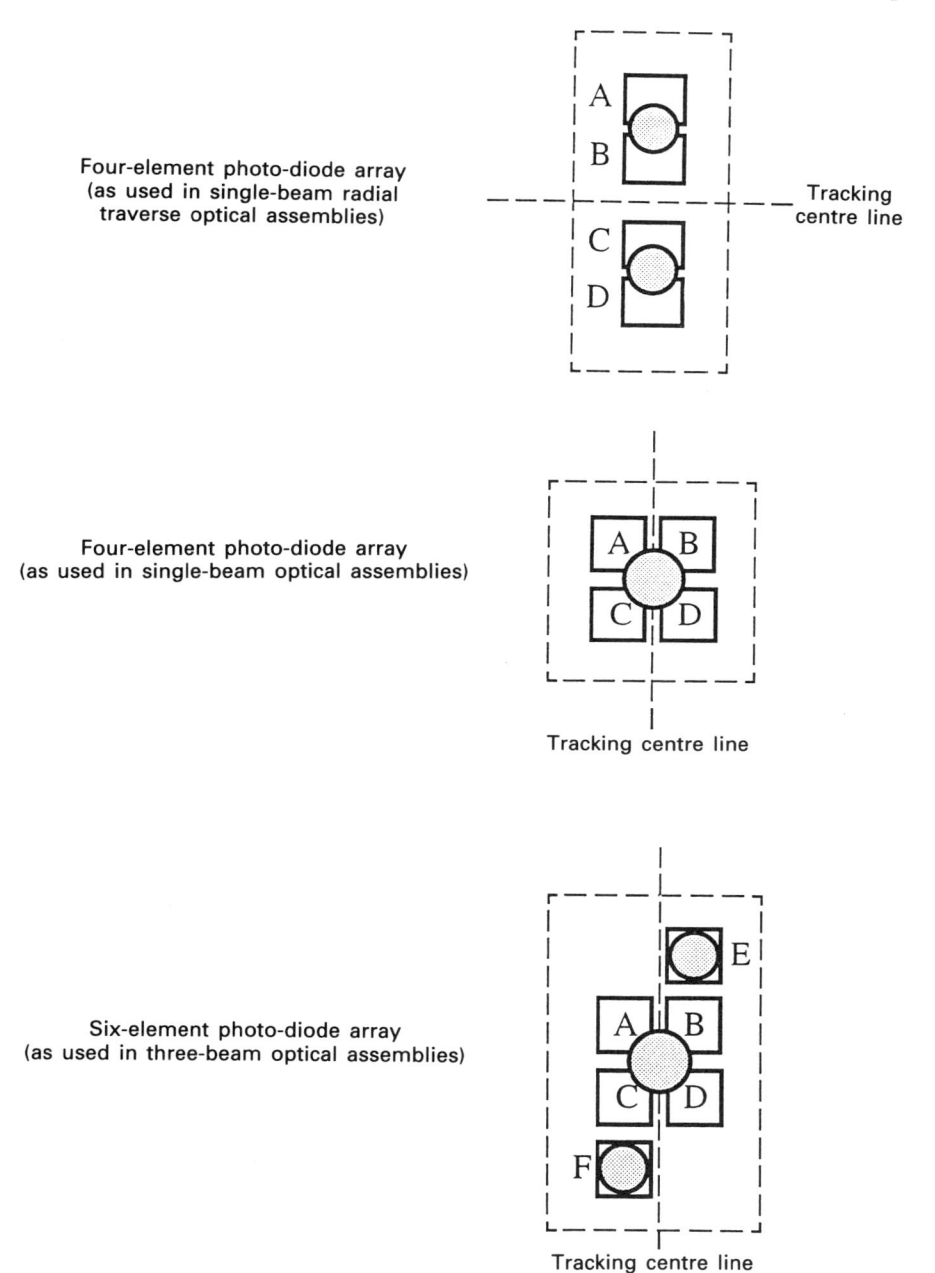

Four-element photo-diode array
(as used in single-beam radial
traverse optical assemblies)

Tracking centre line

Four-element photo-diode array
(as used in single-beam optical assemblies)

Tracking centre line

Six-element photo-diode array
(as used in three-beam optical assemblies)

Tracking centre line

Figure 3.7 *Comparison of photo-diode arrays*

Optical assemblies

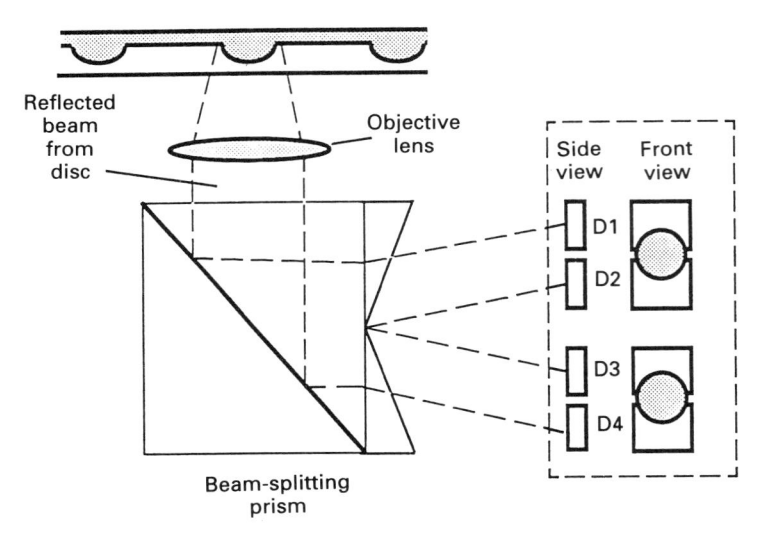

Figure 3.8 *Beam-splitting prism: in-focus condition with the photo-diodes receiving equal amounts of reflected light*

will be covered in Chapter 4, the various methods of developing these signals will now be described.

Beam-splitting prism

Developing the focus signal

This method is used in the Philips radial tracking optical assembly which is a single laser beam unit. It is the shape of the prism that causes the reflected beam to be split into a pair of equal beams (Fig. 3.8).

Each beam will land equally onto a pair of diodes when the in-focus condition has been achieved, but as mis-focus occurs, the two beams will effectively move outwards or inwards depending upon the direction of focus error (as shown in Fig. 3.9).

Developing the tracking signal

When the laser beam is directly on the track of the disc, the reflected light will appear as an equal pair of beams which each land onto a pair of photo-diodes (Fig. 3.10). As the disc rotates, the reflectivity from the track will alter, causing the light intensity monitored by each pair of beams to vary (as shown in Fig. 3.11).

As there is a difference of light intensity in relation to the amount and direction of tracking error, it is now possible to develop a tracking error signal for the tracking servo.

Four-element photo-diode array

Developing the focus signal

Two methods can be described, with one type using a cylindrical lens which produces astigmatism, a form of distortion of the reflected laser beam as mis-focus occurs (Fig. 3.12). The other method is the critical angle prism which produces a slight sideways movement of the reflected laser beam when mis-focus occurs (Fig. 3.13).

When the laser beam is in focus, the cylindrical lens will produce a round beam of reflected light spanning the surface of the four photo-diodes, but when mis-focus occurs, the cylindrical lens will distort the reflected beam virtually into an ellipse, which will span the diodes (B + C) or (A + D), depending upon the direction of the

Optical assemblies

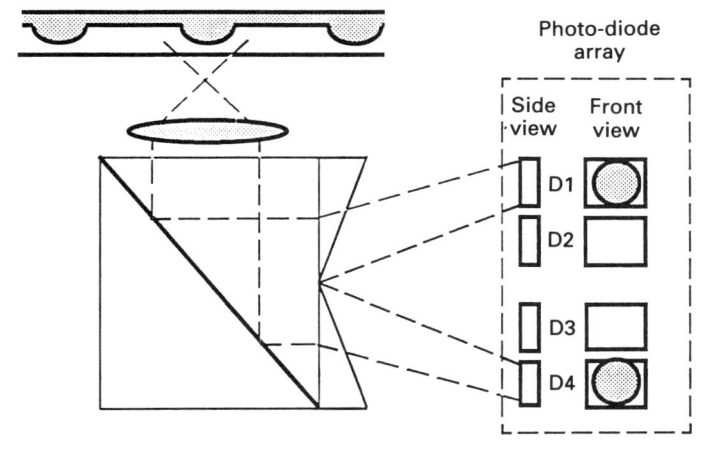

Out of focus condition: too close

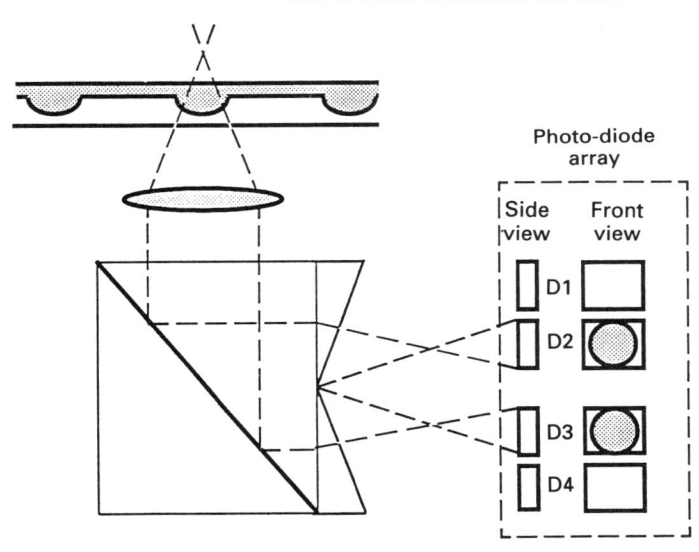

Out of focus condition: too far

Figure 3.9 *Beam-splitting prism: method of achieving focus error*

Optical assemblies

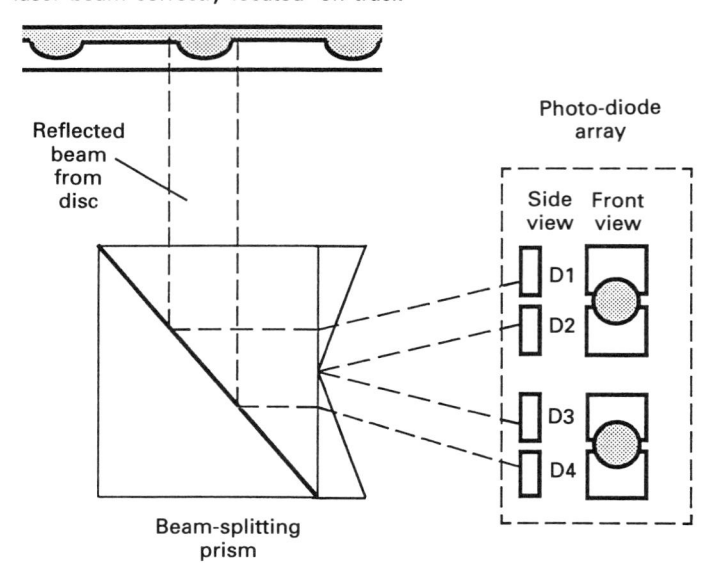

Figure 3.10 *Beam-splitting prism: on-track condition. The pairs of photo-diodes will receive equal amounts of reflected light*

focus error. By cross connecting the diode pairs (B + C) and (A + D), it is possible to develop a focus error signal.

Critical angle prism

The reflected beam is directed onto the four photo-diodes via the critical angle prism, which is a reflective surface set at approximately 42°. As mis-focus occurs, the reflected beam will move fractionally sideways to land either on the (A + C) or (B + D) diode pairs (Fig. 3.13). Therefore, depending upon the direction of mis-focus, it is possible to develop a focus error signal to maintain the laser beam in focus on the disc.

Developing the tracking error signal

With the four photo-diodes it is possible to connect them in two ways to enable a tracking error signal to be developed, either by connecting them in pairs (A + C) and (B + D), or diagonally as (A + D) and (B + C).

Push–pull method

With a push–pull arrangement the photo-diodes are connected in pairs (A + C) and (B + D) (Fig. 3.14). When the laser beam is on track the reflected light intensity on the two pairs is identical and when applied to the tracking error amplifier, will produce a zero tracking error output.

As the track moves to one side or the other, the light intensity received by the two pairs will vary, resulting in a difference between the two pairs, which will provide a tracking error output from the amplifier that will relate to the direction of the tracking error.

This is probably one of the easier methods of obtaining a tracking error signal (compared to, for example, the phase difference method).

Phase difference method

With the phase difference method (Fig. 3.15) the photo-diodes are cross connected in pairs (A + D) and (B + C). Each pair is fed to a summing amplifier and also a differential amplifier, where

Optical assemblies

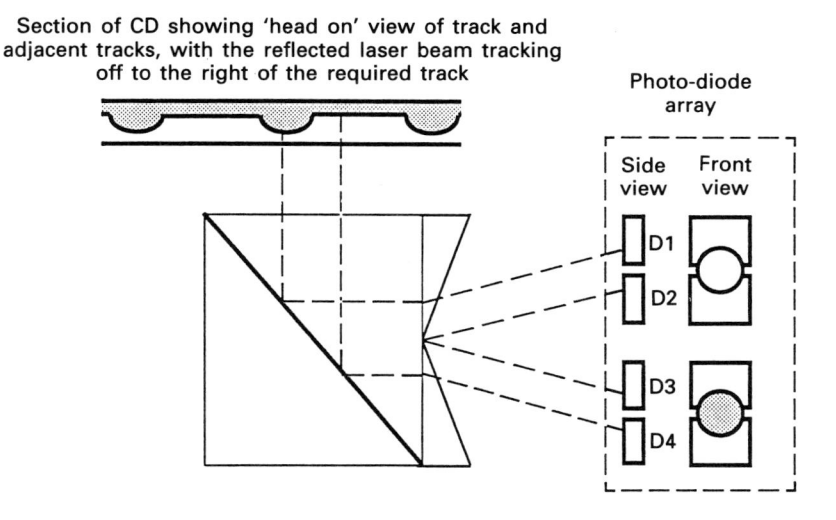

Section of CD showing 'head on' view of track and adjacent tracks, with the reflected laser beam tracking off to the right of the required track

Photo-diode array

Side view Front view

D1
D2
D3
D4

Section of CD showing 'head on' view of track and adjacent tracks, with the reflected laser beam tracking off to the left of the required track

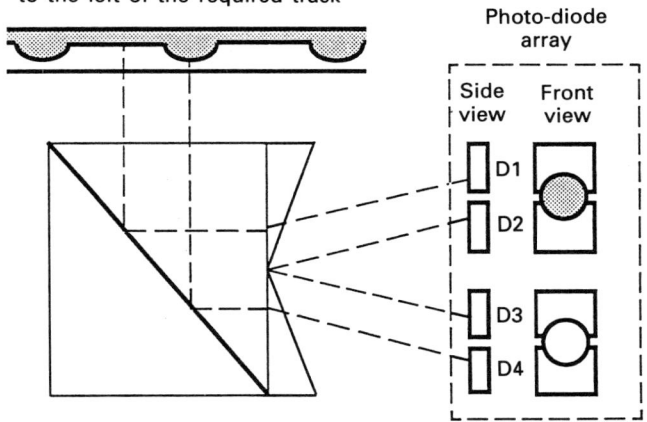

Photo-diode array

Side view Front view

D1
D2
D3
D4

Off track condition: as the reflected laser beam tracks either side of the track, the light intensity monitored by the pairs of photo-diodes will vary depending upon to which side the mis-tracking occurs

Figure 3.11 *Beam-splitting prism: method of achieving tracking error*

Optical assemblies

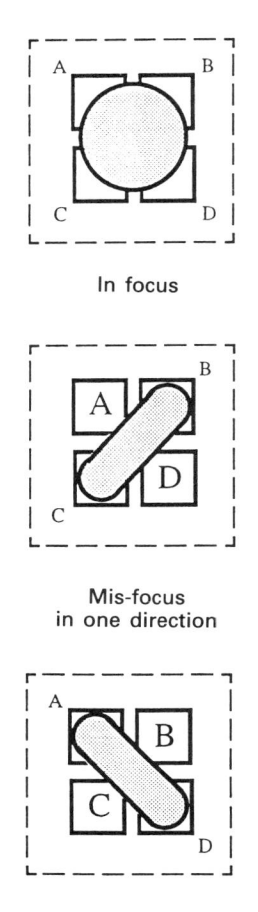

Figure 3.12 *Four-element PD array: method of achieving focus error 1*

further processing takes place as described under the relevant section in Chapter 4. It is possible to produce a tracking error signal that is related to differences in phase which are the result of the amount of tracking error signal.

Three-beam optical block method of achieving tracking error

The three-beam optical block (Fig. 3.16) is probably one of the most common types of optical blocks in recent years. Whilst the block has a single laser beam, it is possible to produce three beams by passing the beam through a diffraction grating lens (not too dissimilar to the type of lens that can be attached to the front lens of a camera to be able to produce multiple images of the same scene).

The centre beam is used for data retrieval from the disc, and also to enable the focus error signal to be produced. The two outer beams are used for developing the tracking error signal, and are arranged to land upon the (E) and (F) photo-diodes.

It is possible with some types of optical blocks, especially the Pioneer types, to adjust the diffraction grating to align correctly onto the (E) and (F) photo-diodes, but many types are factory set, and therefore require no further adjustment.

As the disc rotates, and assuming that the tracking is correct, the reflected light beams from the disc will land onto the photo-diode array as shown in Fig. 3.16, with the beams for the (E) and (F) photo-diodes receiving reflected light from either side of the track, which will relate to zero tracking error.

When a tracking error occurs on one side or the other of the track, the reflected light monitored by the (E) and (F) photo-diodes will now vary with respect to each other, depending on which side of the track the error occurs, and this will result in a signal the value and polarity of which will be related to the amount and direction of the tracking error.

Pioneer '92 optical block

Compared to previous types of optical blocks, the most recent Pioneer type (Fig. 3.17) has three major variants: a five-element photo-diode array, a hologram prism, and the laser diode and photo-diode array contained within the same component (these are usually individual components placed in different areas of the complete optical block).

The single laser beam passes through a diffraction grating to develop the three beams, in common with many modern optical blocks, which pass directly through the hologram prism area towards the surface of the disc via a polarised objective lens.

Previous optical blocks have usually contained a polarising prism to ensure that the reflected beams are optically correct to be deflected towards

Optical assemblies

Reflected laser beam from the recording surface of the CD

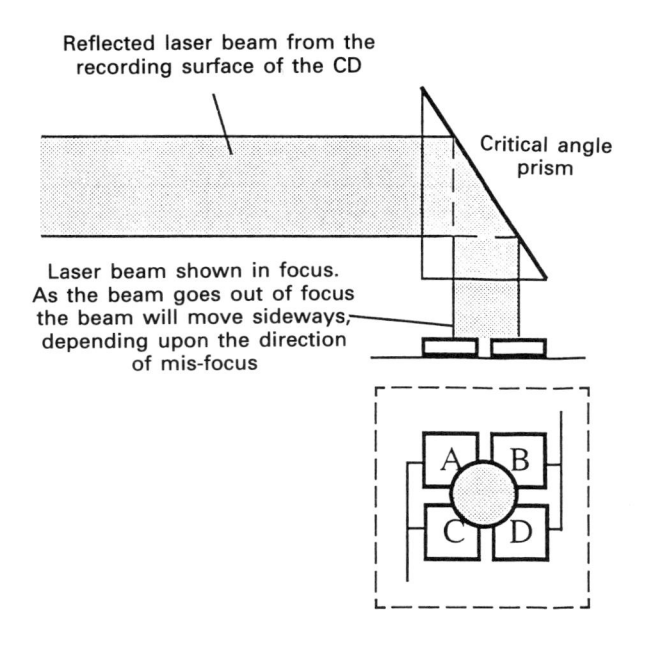

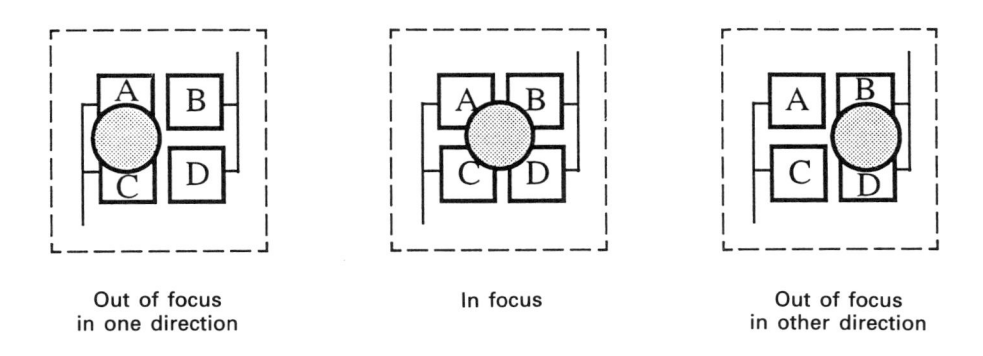

Critical angle prism

Laser beam shown in focus. As the beam goes out of focus the beam will move sideways, depending upon the direction of mis-focus

Out of focus in one direction

In focus

Out of focus in other direction

Figure 3.13 *Four-element PD array: method of achieving focus error 2*

Optical assemblies

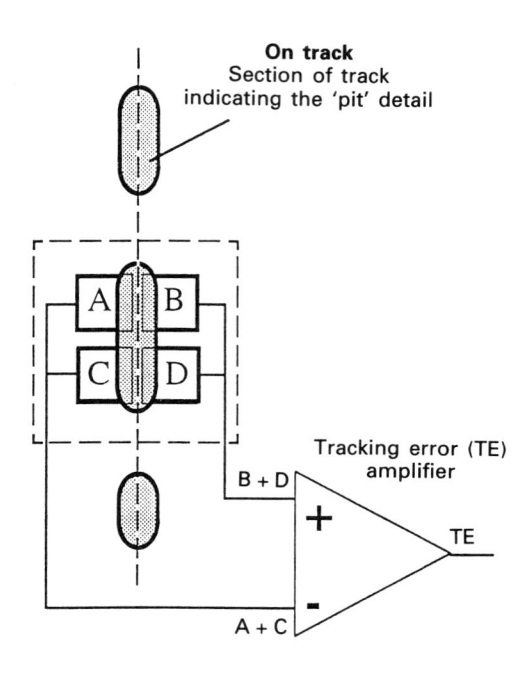

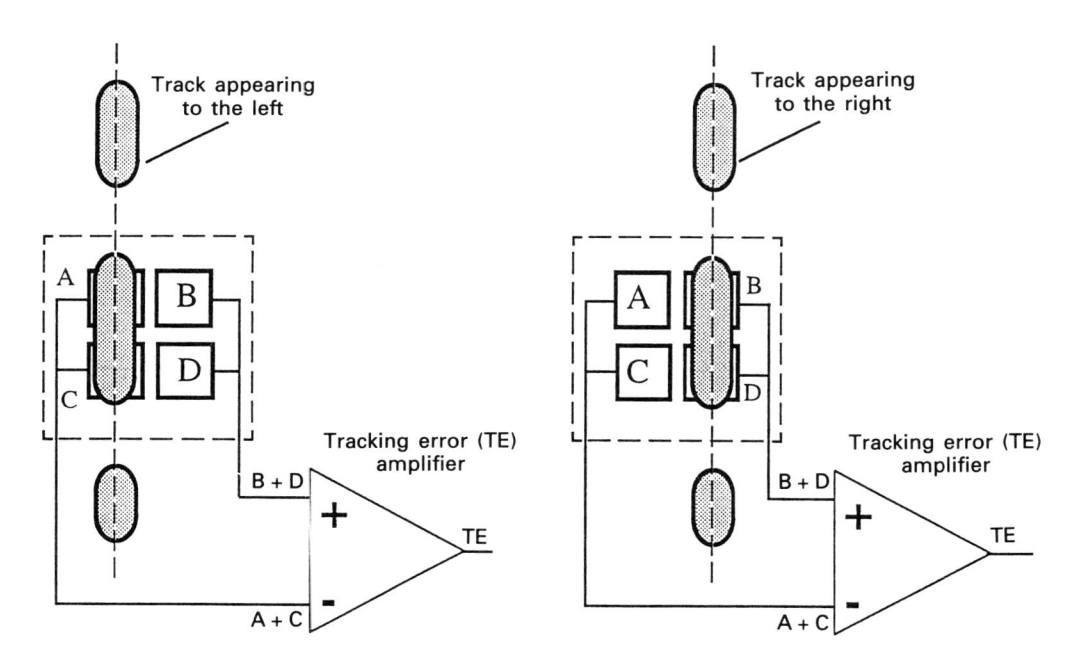

Figure 3.14 *Four-element PD array: push–pull method*

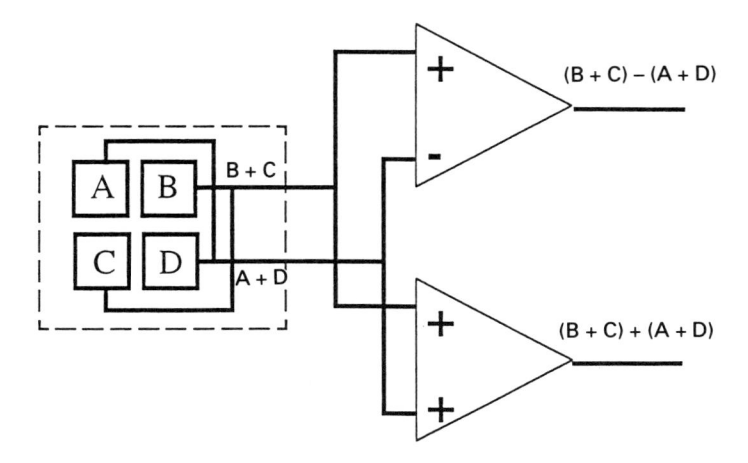

Figure 3.15 *Four-element PD array: phase difference method*

the photo-diode array by the relevant prism. Polarising the objective lens fulfils the same requirement whilst reducing the number of components within the block.

The reflected beams will now be in the correct optical phase or alignment to ensure that the beams are now deflected sideways by the hologram prism towards the photo-diode array which also splits each beam into two, making six beams land onto the five-element photo-diode array (Fig. 3.18).

The five diodes shown in Fig. 3.18 provide the necessary signals as follows:

1 (D1) and (D5) are used to produce the tracking error signal;
2 (D2) and (D3) are used to produce the focus error signal; and
3 the sum of (D2), (D3) and (D4) is used to produce the RF eye pattern waveform.

With the tracking passing through the centre of the array as illustrated in Fig. 3.18, any sideways deviation or error will be monitored by (D1) and (D5) to produce the tracking error signal as described above for the three-beam optical block method.

The focus error signal is detected by (D2) and (D3), which will monitor a slight sideways movement as mis-focus occurs – a technique which is similar to the critical angle prism method.

With the sum of (D2), (D3) and (D4), the RF signal produced will have improved benefits compared to some previous types. When scratched discs are played there is a tendency for skipping and jumping to occur due to the interference the scratches cause to the reflected beams, as well as sound glitches being caused by the RF signal being occasionally disrupted.

Ideally, to maintain effective focus and tracking, the size of the photo-diodes needs to be small in order that the focus and tracking servos do not respond to large noise spikes due to scratches, whilst it would be beneficial if they were larger to be able to respond more effectively to the 'pit' information to produce the RF signal, and meeting these ideals would possibly enable discs with reasonably small scratches to be played more effectively.

With the five-element array, the (D1), (D2), (D3) and (D5) diodes are small, whilst the (D4) diode is significantly larger. As the hologram prism produces six beams, these will also be small to be accommodated onto the photo-diodes, and this will result in the diodes and beams being small for effective focus and tracking, whilst the area of the (D2), (D3) and (D4) diodes will be larger to enable an improved RF signal to be developed.

Optical assemblies

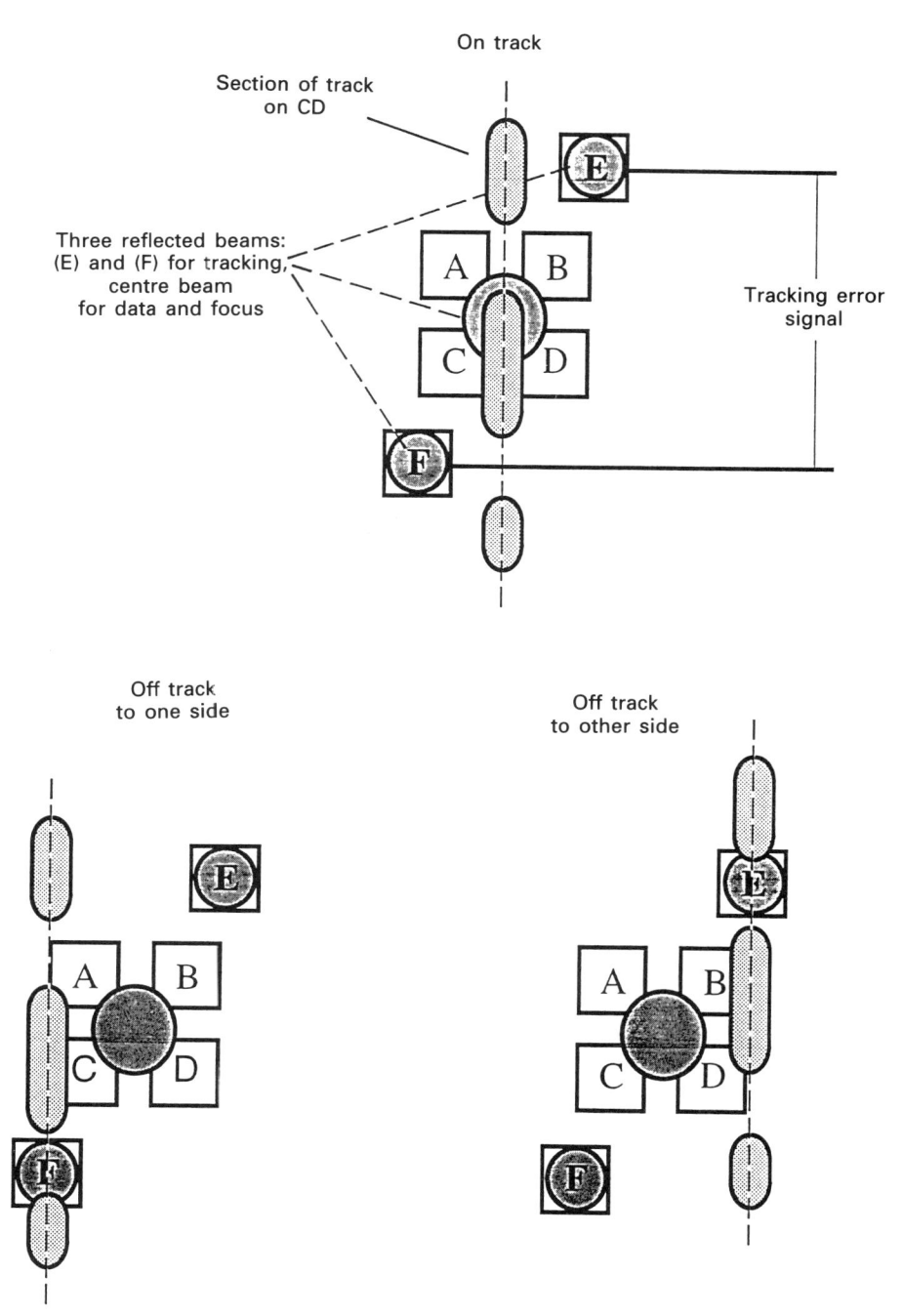

Figure 3.16 *Six-element PD array: method of achieving tracking error*

Optical assemblies

Optical system end-on view

Optical system side view

Compact disc

Objective lens

Collimation lens

Hologram prism

Reflected beams

Diffraction grating
Produces two side beams from the single laser beam. The side beams are used for tracking, but are omitted here for clarity

The hologram prism will deflect the three reflected beams (one for data and focus, the other two for tracking) sideways

The hologram prism will also act as a beam splitter, causing the reflected beams to be split as shown on the right. This will in effect appear to create six beams as illustrated below

Laser diode

Photo-diode array
End-on view and top view

Hologram prism develops 6 laser spots from reflected beam

Photo-diode array
Side view and top view

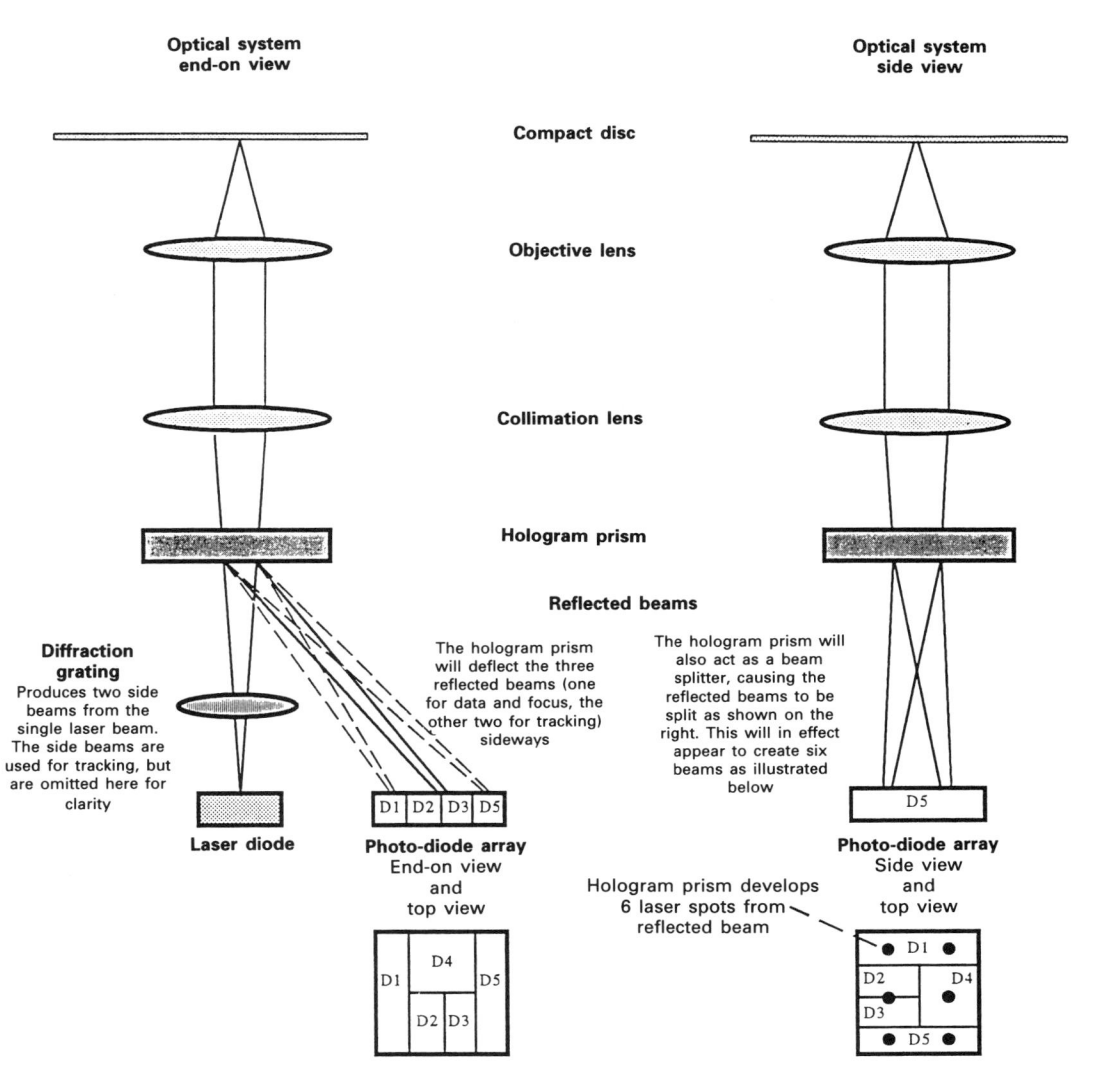

Figure 3.17 *New Pioneer '92 optical block using a hologram prism*

Optical assemblies

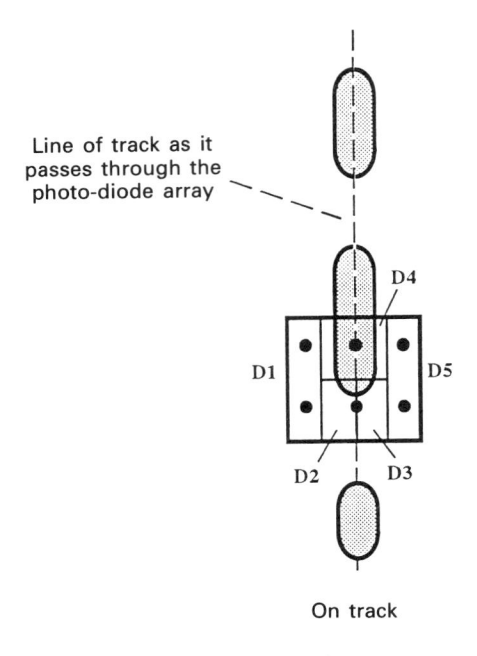

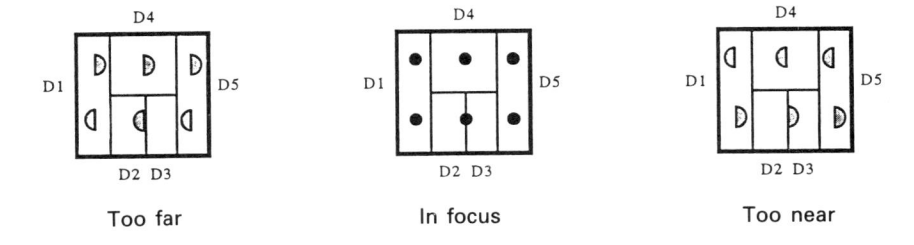

On track

Too far In focus Too near

Figure 3.18 *New Pioneer '92 optical block, five-element PD array: method of achieving focus error*

54

4

SERVO SYSTEMS

In order that the data can be retrieved from the CD, it is necessary to maintain the laser beam in focus on the playing surface of the disc, as well as being able to gradually track across the playing area. This is achieved via a range of servo systems which provide the requirements of focus, tracking and disc rotation.

Whilst players may differ from one manufacturer to another, the basic function of any of the servo systems will be similar. The basic servo (Fig. 4.1) can be considered to comprise:

1. a sensing element,
2. an amplifier system, and
3. a controlled element.

Because the various stages of a servo system are d.c. coupled, it will be necessary to include some form of offset adjustment to neutralise undesirable d.c. potentials within the system, as well as a gain control to enable optimum operating efficiency to be achieved.

Some players may well have some form of automatic method of correcting for any undesirable d.c. offset potentials, as well automatic gain adjustment.

The sensing element will comprise components of the photo-diode array referred to in Chapter 3. It will provide a signal that will be in relation to any errors that may have been sensed. These signals are processed within an amplifier system which will normally include some form of frequency shaping to optimise the operation of a specific servo system, before passing on to the controlled element. This will normally be in the form of a coil to operate a lens or motor to drive the CD or move the optical assembly across the surface of the disc, continuously neutralising the errors as they are being sensed.

As described in Chapter 5, various switching operations will take place within any of the servo systems, which are controlled by the system control circuit to ensure that the correct operational sequence of the player takes place.

Focus servo

The photo-diode array will provide the initial focus information, via an amplifier and a control system, to develop a signal to be applied to the focus coil which is attached to the objective lens (Fig. 4.2).

Depending upon the CD player manufacturer, the photo-diode array will comprise either a combination of four photo-diodes (see Fig. 3.7), or three of the five photo-diode arrays used in the Pioneer '92 optical block (Figs 3.17 and 3.18).

When the CD player is first switched on and prior to the commencement of playing a CD, it is necessary to ensure that the laser beam is focused correctly on to the playing surface of the disc. To achieve this it is necessary for the focus lens to 'search' for the correct focal point, by applying a specific search signal to the focus servo system, to cause the lens to physically move vertically, until focus is achieved.

At the point of correct focus, a maximum reflected signal will be received from the disc. This will result in a focus OK (FOK) signal, which is passed onto the system control circuit to enable the next stage of player operation, and in the case of many CD players, indicate that a CD is present.

Various methods exist for determining the focus error signals, as indicated in Chapter 3, but it is now relevant to consider some of the applications within some of types of focus servo systems that have appeared in some types of CD players.

Beam-splitting photo-diode array method

A beam-splitting photo-diode array (Fig. 4.3) is consistent with the radial traverse optical systems that have been extensively used in Philips and Marantz CD players, as well as some Technics and players of other manufacture.

Servo systems

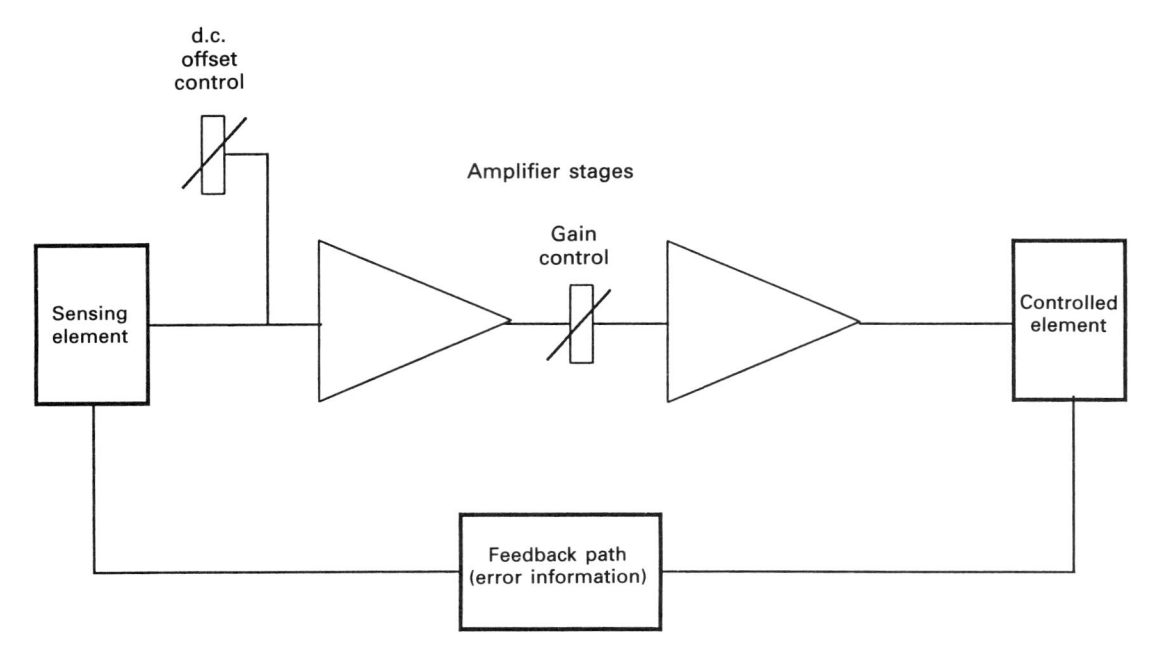

Figure 4.1 *Basic servo system*

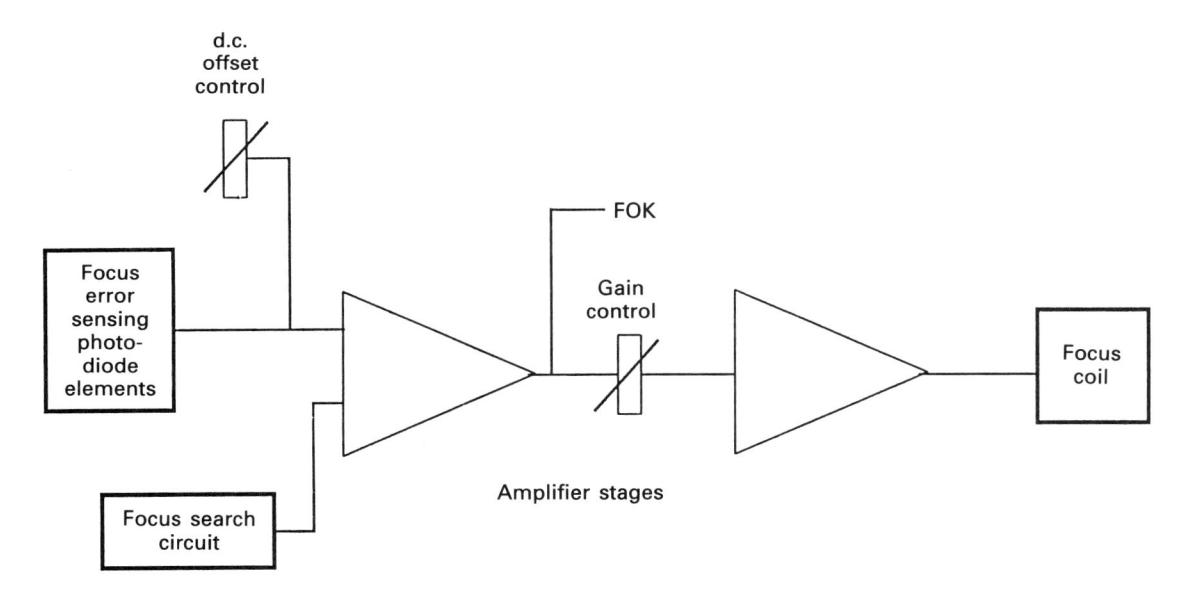

Figure 4.2 *Basic focus servo system*

Servo systems

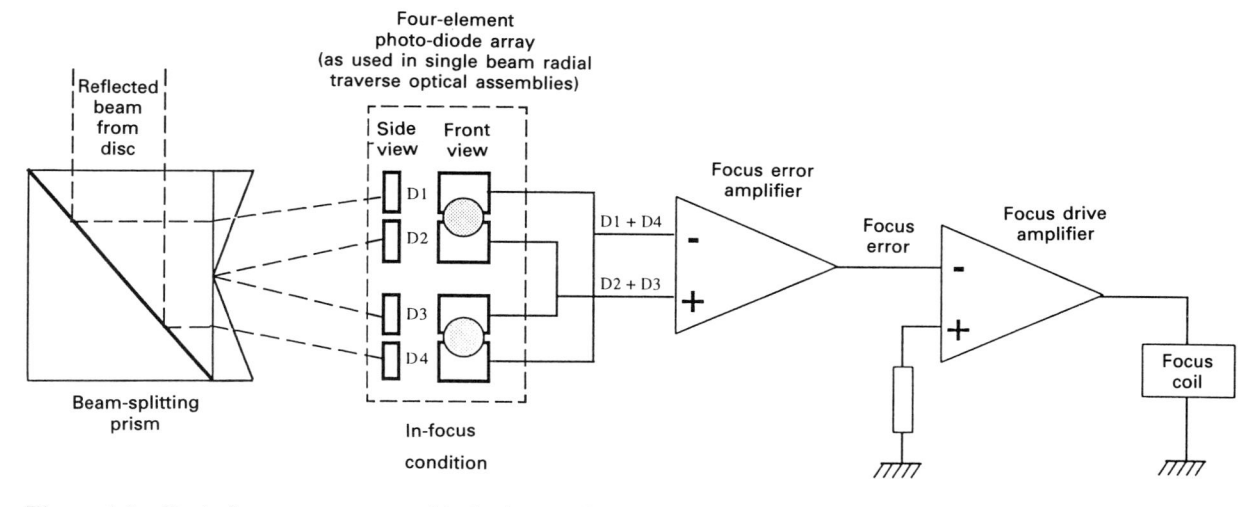

Figure 4.3 *Basic focus servo system with the beam-splitting photo-diode array*

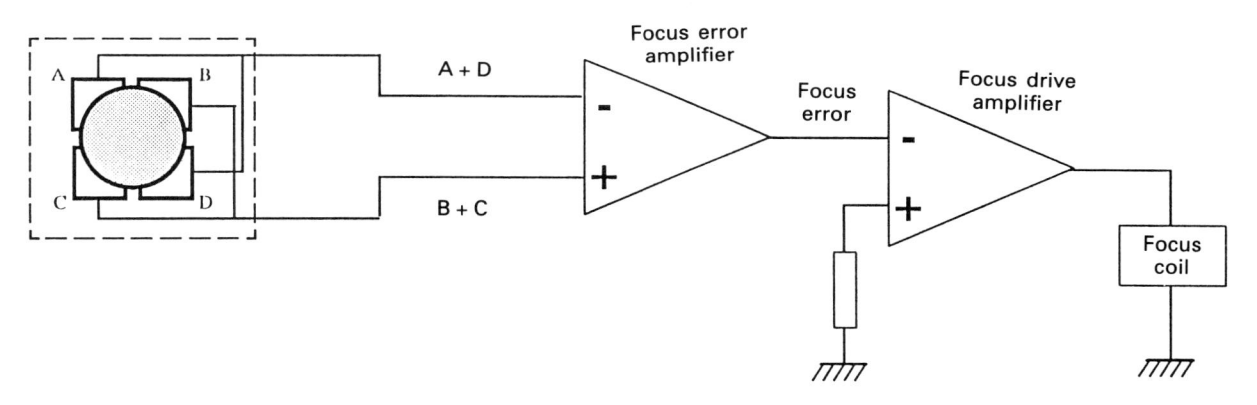

Figure 4.4 *Basic focus servo system with the four-element photo-diode array (as used in single-beam optical assemblies, using the cylindrical lens or astigmatism method)*

When the reflected beam from the surface of the disc is in focus, the two beams of light from the beam-splitting prism will land on the two pairs of photo-diodes, as shown in Fig. 4.3. The resultant inputs to the focus error amplifier will comprise (D1 + D4) and (D2 + D3), with the output being the difference between the two inputs, which for the in-focus condition will be a zero error signal.

Ideally the focus lens will be in a neutral or mid-point position.

As mis-focus occurs, the two beams will move outwards towards either the D1 and D4 diodes, or inwards towards the D2 and D3 diodes. The resultant output from the focus error amplifier will be a potential, the value and polarity of which will be related to the amount and direction of focus error.

Applying this signal to the focus coil via the focus drive amplifier will enable the focus lens to move in the direction to correct the focus error and maintain the laser beam correctly focused on the playing surface of the CD.

Cylindrical lens – four-element photo-diode array method

The cylindrical lens method (Fig. 4.4) is possibly the most frequently used system of focus servo

Servo systems

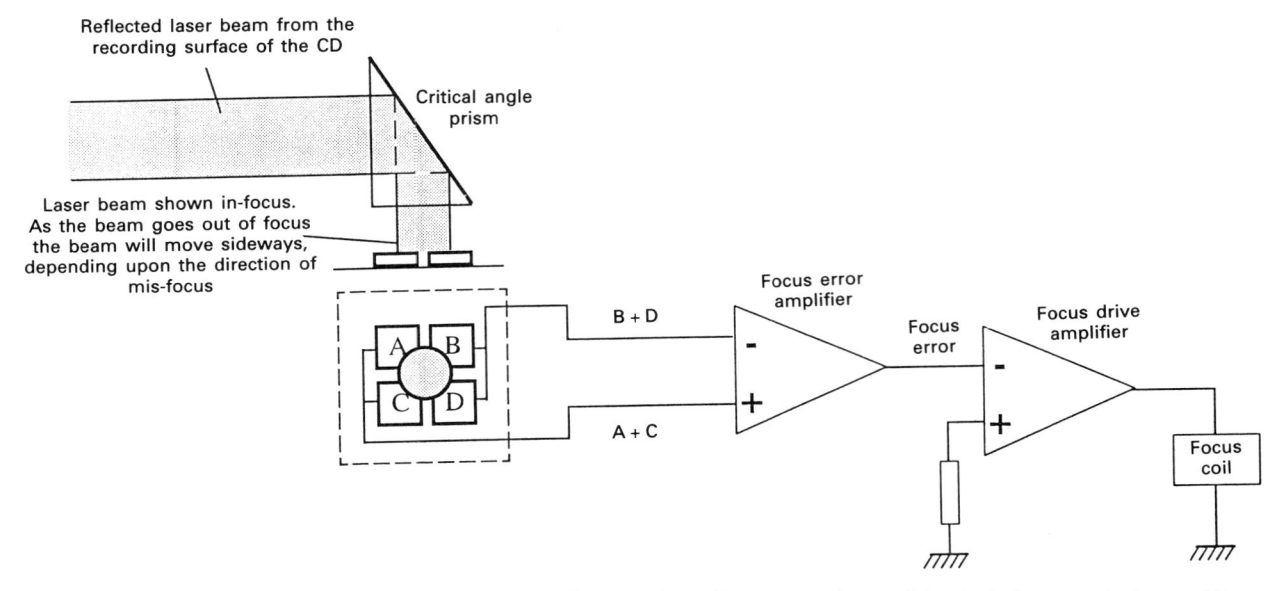

Figure 4.5 *Basic focus servo system with the four-element photo-diode array (as used in single-beam optical assemblies, using the critical angle prism method of focus control)*

control, and is typical of single-beam (four photo-diodes) or three-beam (six photo-diodes) types of optical block.

With the in-focus condition, the reflected beam will equally cover the four photo-diodes. As the opposite diodes are connected together, i.e. (A + D) and (B + C), the resultant potentials fed to the focus error amplifier will be the same, providing a zero error focus signal output which is applied to the focus coil via the focus drive amplifier, once again enabling the focus lens to achieve a neutral or mid-point position.

As mis-focus occurs, the reflected beam will become elliptical due to astigmatism caused by the cylindrical lens. This will cause the potentials (A + D) or (B + C) to differ with respect to each other, thereby creating a focus error signal which will enable the focus lens to move in the required direction to correct any mis-focus that may occur.

Critical angle prism – four-element photo-diode array method

The critical angle prism is an alternative method of utilising a four-element photo-diode array, and was used in earlier types of single-beam optical blocks. It makes use of a phenomenon where the reflected beam is directed via a prism, which is set at a 'critical angle' of 42°. As mis-focus occurs the beam will move fractionally sideways across the surface of the photo-diodes (Fig. 4.5).

With this system the photo-diodes are connected in pairs (A + C) and (B + D), which will result in identical inputs being applied to the focus error amplifier when the beam is correctly focused onto the playing surface of the disc.

Mis-focus will cause the reflected beam to move fractionally sideways, resulting in the potentials (A + C) and (B + D) differing with respect to each other, which in due course will enable the focus lens to move in the direction to correct any focus error.

Hologram method of focus servo

With the hologram method (Fig. 4.6), whilst the basic circuit concept of the focus servo still remains similar to previous circuits, the error signal, which is developed from only two of the photo-diodes, D2 and D3, is applied to the focus error amplifier. When the laser beam is on-focus, minimum output will be achieved from the focus error amplifier, whilst mis-focus will enable a potential to be produced which will relate to the amount and direction of the focus error.

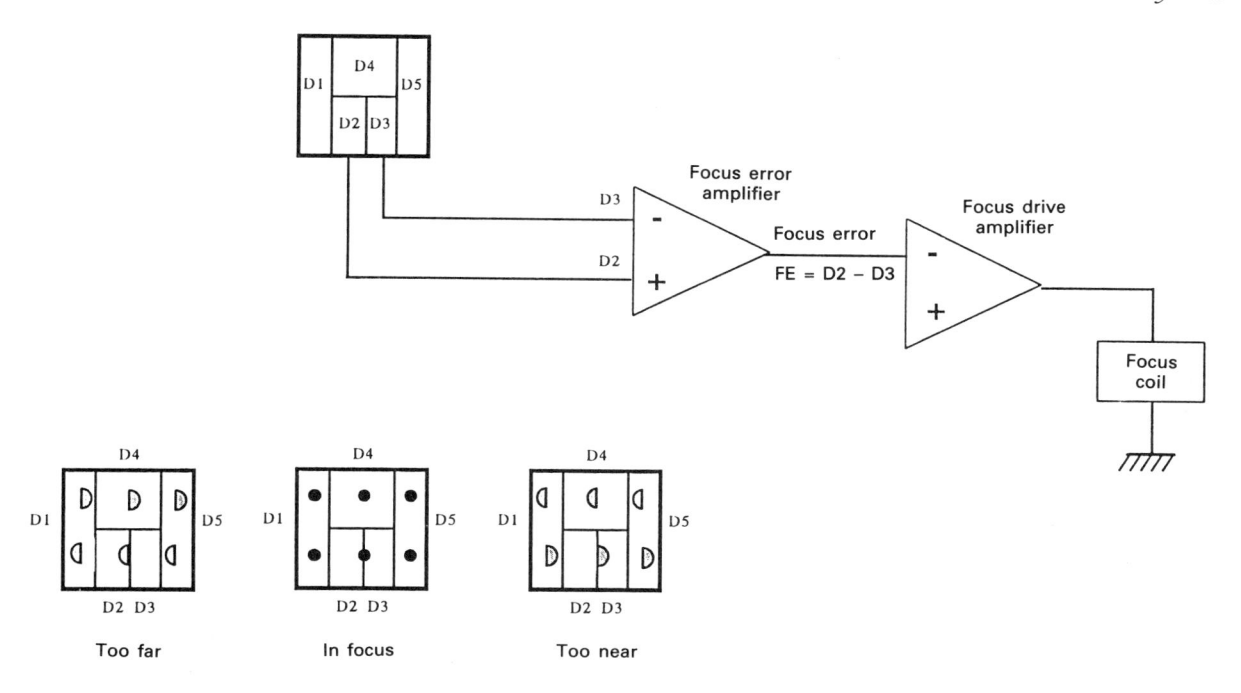

Figure 4.6 *Basic hologram method of focus servo system*

Tracking servo

Whilst the various focus servo systems are basically similar, the various types of tracking servos that have appeared in CD players vary in their complexity. Probably the three-beam optical assembly, where the two extra beams are used specifically for developing the tracking error signal, has proved to be the most popular and least complex.

However, developing the tracking error from a four-element photo-diode array or single-beam optical block can require some specialised signal processing circuits.

Whilst various types of tracking servo exist, two particular groups of optical block have been used throughout the numerous ranges of CD players that have been produced. Either the objective lens is fixed laterally, such as in the Philips radial tracking system, or the lens has a small amount of lateral movement which enables the lens to track gradually across a small surface area of the disc, which in turn is co-ordinated with a motor drive system to gradually move the complete optical assembly across the surface of the disc.

Basic tracking servo with lateral movement of the lens

The tracking error signal is derived from the relevant photo-diodes, which can comprise two diodes from either a four photo-diode array used with single-beam optical blocks, or from a five or six photo-diode array, normally associated with a three-beam device (Fig. 4.7).

A range of adjustments may be included within the tracking servo circuit in order to optimise the operation (see Chapter 6).

Because a tracking error signal has to be developed that will be related to the amount of deviation of the laser beam either side of the track on the CD, many CD players will include a tracking balance control to ensure that equal amounts of deviation of the laser beam either side of the track will result in equal and opposite polarity error signals.

As the disc rotates, a continual tracking error signal will be developed which is applied to the tracking coil via the necessary amplifier stages to continually neutralise the tracking error signal.

The tracking coil will have a lateral movement

Servo systems

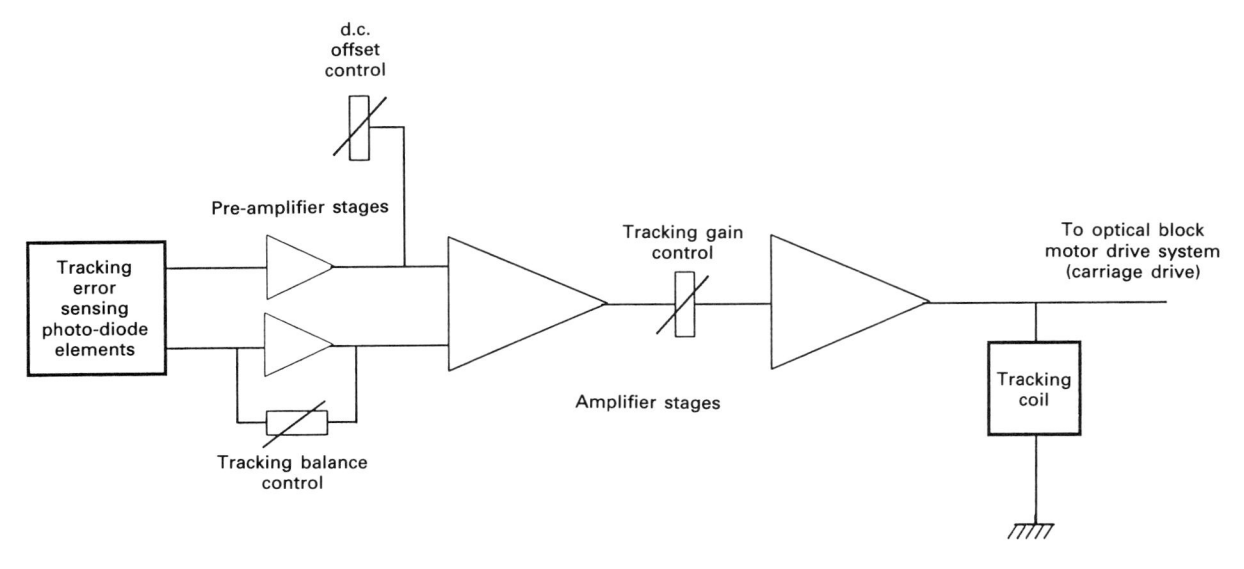

Figure 4.7 *Basic tracking servo system – lateral lens movement type*

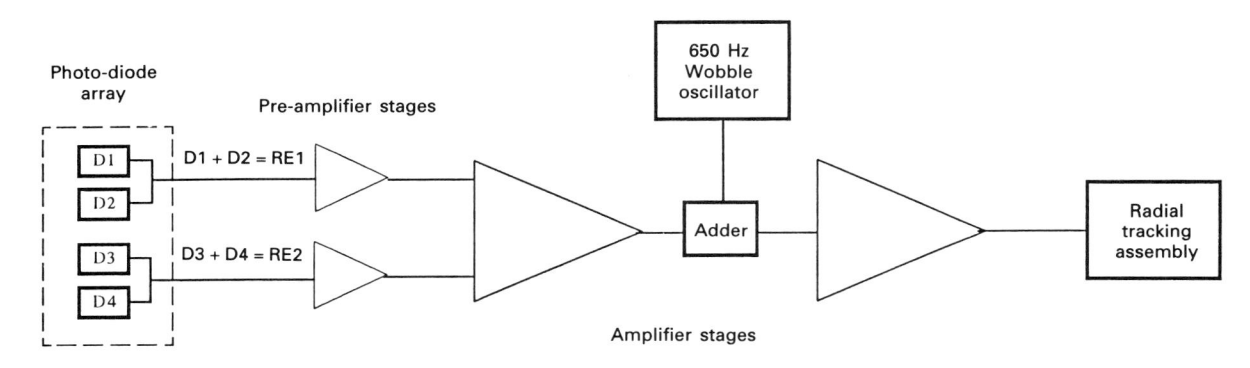

Figure 4.8 *Basic radial tracking servo system – fixed lens system*

of only a few millimetres, and therefore it is necessary to monitor the signal being applied to the tracking coil circuit. This signal will be in the form of an increasing potential, and will be in proportion to the lateral movement of the tracking coil and lens combination. As the signal reaches a certain level, a motor drive system will enable the optical block to gradually 'creep' across the surface of the disc. Therefore there is a continual compromise situation between the tracking coil and the motor drive system to effectively maintain efficient tracking across the surface of the disc.

The motor drive system has various identifications depending upon the manufacturer of the CD player: carriage, sled, sledge and slider, and no doubt other options, are in existence.

Basic tracking servo: radial tracking system

The radial tracking type of tracking servo (Fig. 4.8) is usually associated with the Philips and Marantz players, as well as some Sharp and Technics players. Other manufacturers have also utilised this method.

The major difference with this concept is that the objective lens is fixed, and that the complete

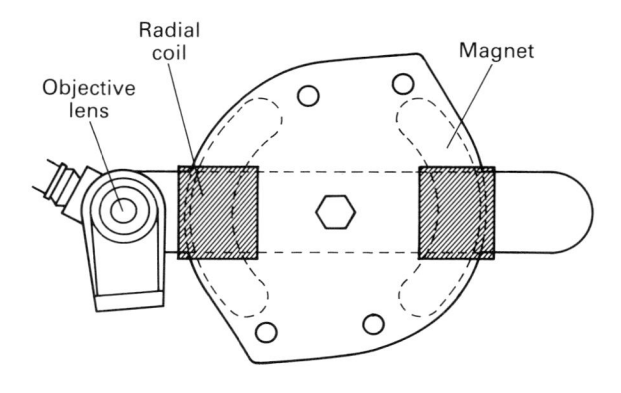

Figure 4.9 *Diagram of radial tracking assembly*

optical assembly moves laterally across the surface of the CD, in a manner similar to a moving coil meter (Fig. 4.9).

As this arrangement requires the complete optical assembly to move minutely in fractions of a micrometre, problems can occur with respect to tracking and friction. A term which some engineers may be familiar with relating to this type of friction in servo systems is static friction or 'stiction', which was quite common in early mechanical servo data transmission systems.

These early systems were used very much in the armed forces to relay such information as compass bearings from one central source to indicators situated in remote areas within a specific environment such as a ship or aircraft. To overcome the 'stiction' problem it was necessary to inject a high audible frequency source. This was referred to as a 'dither' frequency, and enabled the indicators to respond to small movements of information such as a gradual change in compass bearing as the ship or aircraft changed direction.

A similar method is introduced in the radial tracking system with the insertion of a 650 Hz wobble frequency, which causes the optical assembly to continually 'wobble' or oscillate either side of the track on the disc.

The radial optical assembly moves in an arc (as indicated in Fig. 4.10). When the optical unit is at the start or centre of the disc, the angle of the photo-diode array will be at right angles to the track, whilst towards the outer edge of the disc the angle will alter to approximately 45° – a situation which could cause problems with respect to discerning the tracking error signal.

The four-element photo-diode array is arranged as two pairs of photo-diodes side by side, and the radial tracking error is developed from the signals D1 + D2 = RE1 and D3 + D4 = RE2.

The 'on-track' condition will enable (D1 + D2) and (D3 + D4) to receive equal amounts of the reflected laser beam from the disc, causing equal potentials of RE1 and RE2 to be produced, which will result in a zero radial tracking error signal.

As the disc rotates and tracking errors are developed, the resulting levels of the RE1 and RE2 potentials will vary according to the direction and amount of deviation of the reflected beam from the centre of the track, which will result in a radial tracking. The error signal will vary in level and polarity, depending upon the amount and direction of tracking error.

Because of the changing angle of the photo-diode array with respect to the track as the optical block travels across the surface of the disc, the tracking error signal will not vary in proportion to the actual radial movement of the optical block, and can cause possible asymmetry of the reflected signal.

If the optical block, and consequently the photo-diode array, were to traverse gradually across the surface of the disc, a tracking error signal would be produced that would vary depending upon the position of the photo-diode array in relation to the individual tracks. Optical blocks that move linearly at right angles to the track will develop a consistent quality of tracking error signal across the disc that is proportional to the amount of movement of the optical assembly (Fig. 4.11). But with the radial tracking method, the quality of of the tracking error signal will deteriorate towards the outer edge of the disc as the angle of the photo-diode array alters with respect to the track. To minimise this undesirable effect, it is possible to make use of the effect of the 650 Hz wobble frequency that is applied into the radial tracking circuit. Fig. 4.12(a) illustrates the variation in level of the reflected light output or eye pattern waveform, which is related to the position of the photo-diode array compared to track position. When the array is effectively 'on track' the reflected signal will be at a minimum due to deflection of the laser beam onto the 'pits' or 'bumps' resulting in a reduction of actual reflected light, and consequently the resultant signal.

When the photo-diode array is between tracks, maximum reflection occurs, due to the mirror surface between tracks, and therefore results in a maximum signal level developing.

Servo systems

Arc of travel of the radial optical assembly

The relative angle of the photo-diode array will be at 90° with respect to the track, whilst at the outer edge it will be in the region of 45°

D1–D4

D1–D4

Section of track at the outer edge of the disc

Change of angle of the photo-diode array, as radial optical assembly moves across the disc

Section of track at the centre of the disc

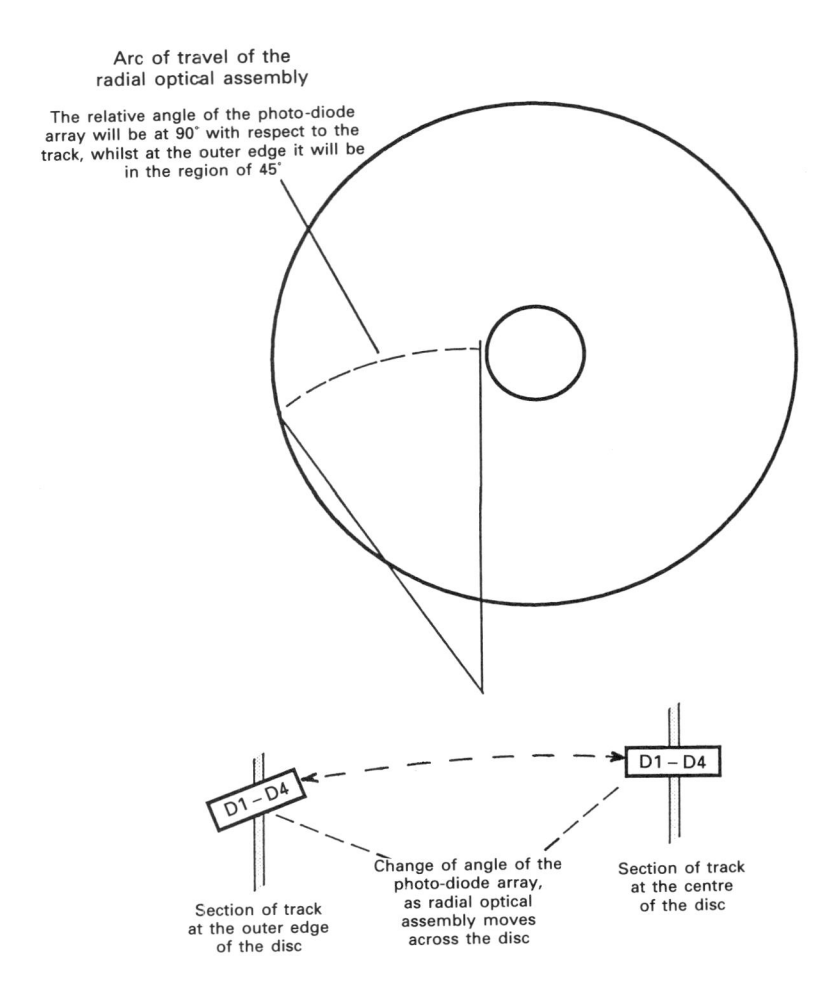

Figure 4.10 *Radial optical assembly*

On-track condition

By enabling the radial assembly to wobble or oscillate at a frequency of 650 Hz either side of the track, and assuming an 'on-track' situation, the basic minimum reflected signal which will increase in level as the radial assembly oscillates either side of the track centre to produce the waveform (a), which will modulate slightly the eye pattern waveform.

The waveform (a) is applied to the input of the synchronous detector circuit shown in Fig. 4.12(b), as signal X. The synchronous detector switch, which is operated by the 650 Hz + 90° wobble frequency, will alternately provide at the output as signal Z, either the direct signal X or the inverted signal Y.

Therefore during the on-track condition, the output signal Z will comprise alternate positive and negative half cycles which when applied to an integrating circuit will result in zero output.

Off-track to the right

Assuming an off-track condition, say to the right, will result in waveform (b) being applied to the synchronous detector input, which will result in

Servo systems

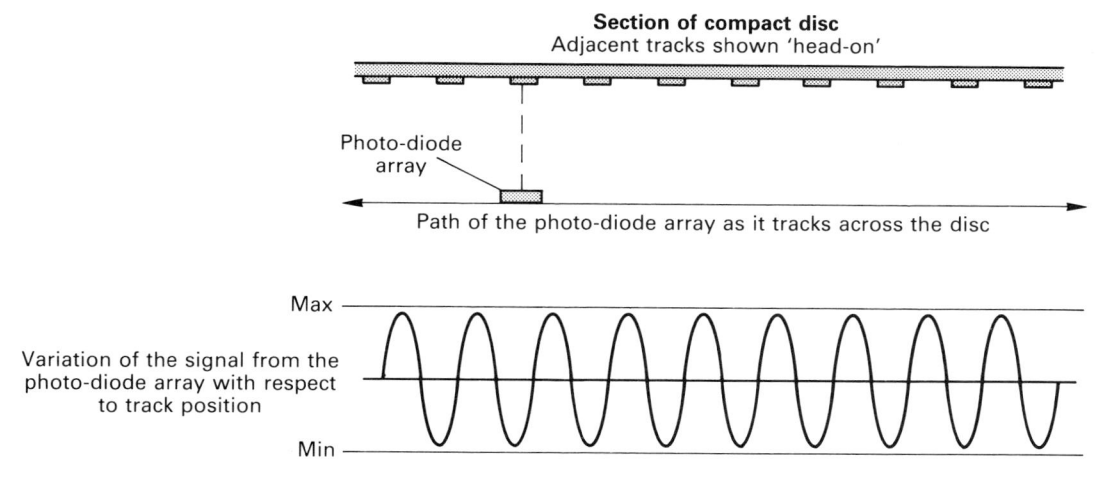

Figure 4.11 *Photo-diode/track relationship*

the output as waveform (d), and in turn provide a positive signal after integration.

Off-track to the left

For this situation, waveform (c) is now applied to the input of the synchronous detector, which in turn will result in waveform (e), i.e. a negative signal when integrated.

Therefore this method will enable a generally consistent quality of radial tracking error signal to be developed as the radial optical assembly gradually tracks towards the outer edge of the disc.

However, it is possible for other factors to affect the efficient operation of this tracking method, and these are related to the laser power, the reflective qualities of the disc itself, and possible phase differences of the wobble signal applied to the radial assembly when they are compared to the resulting wobble frequency that must be present on the radial tracking error signals at either the beginning or the end of the CD as the photo-diode angle alters.

These factors can influence the overall gain of the radial tracking servo system, and can result in inefficient and unstable operation. Therefore it is necessary to maintain a consistent gain throughout the overall operation of the radial tracking servo.

To achieve an effective gain control it is necessary to compare the phase of the wobble frequency applied to the radial assembly to that present on the radial tracking error signal, and as the phase alters the gain is varied accordingly.

The 650 Hz wobble oscillator is fed via adder (5) into the radial servo system which will cause the radial motor to oscillate by the extremely small amount of 0.1 μm, and will result in radial error signals which are developed via the four photo-diodes, D1–D4, to be 'modulated' with the 650 Hz wobble frequency (Fig. 4.13).

The signal outputs from the four photo-diodes, D1–D4, are connected after amplification and via the adders (1) and (2) to provide the two radial error signals (D1 + D2) = RE1 and (D3 + D4) = RE2.

The amplifiers for each of the diodes are gain controlled in relation to the sum of the radial error signal, and the level of the RF signal determined from the sum of the four diodes.

From Fig. 4.13 it can be determined that the sum of D1–D4 is passed to the RF amplifier stages, which includes a high frequency stage and an equalisation stage to provide the standard RF eye pattern waveform, for further processing in the decoder circuits, but it is also fed via the HF slice stage to maintain specific limits, to the gain control stage.

The sum of RE1 and RE2, which is derived via adder (3), is passed through a low pass filter (LPF) to remove the 650 Hz content, and fed to the gain control stage. The resulting sum from these two signals will provide a d.c. level to maintain the gain of the first amplifier stages within defined limits.

The complete radial servo signal path comprises

Servo systems

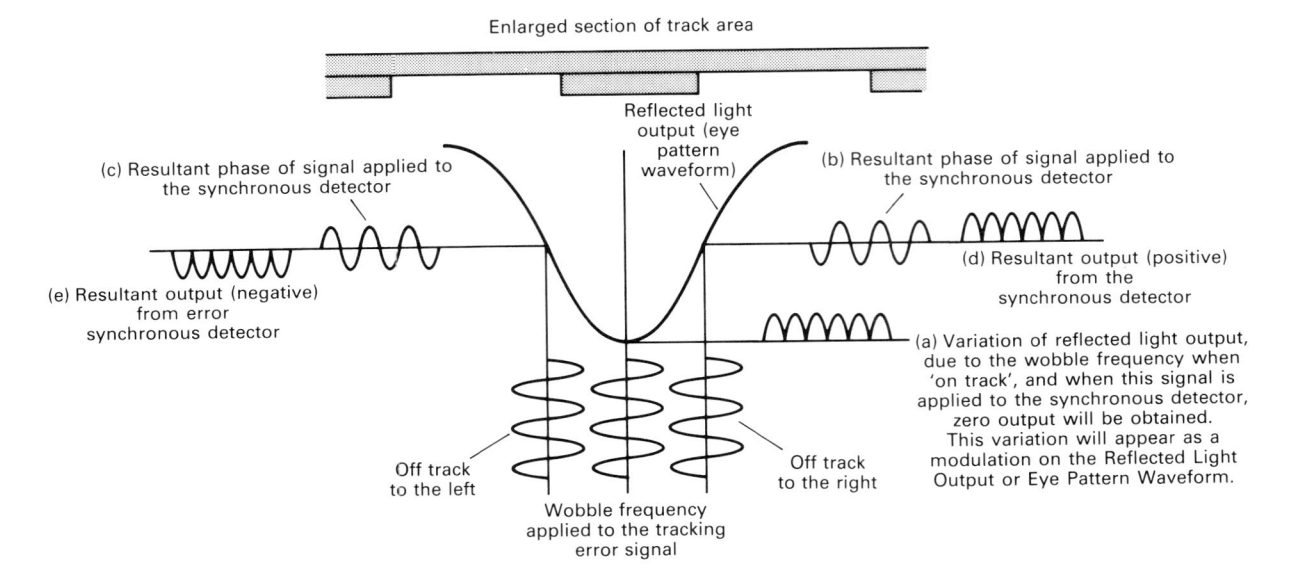

Figure 4.12(a) *Relative phase of the 650 Hz wobble frequency, in relation to track position*

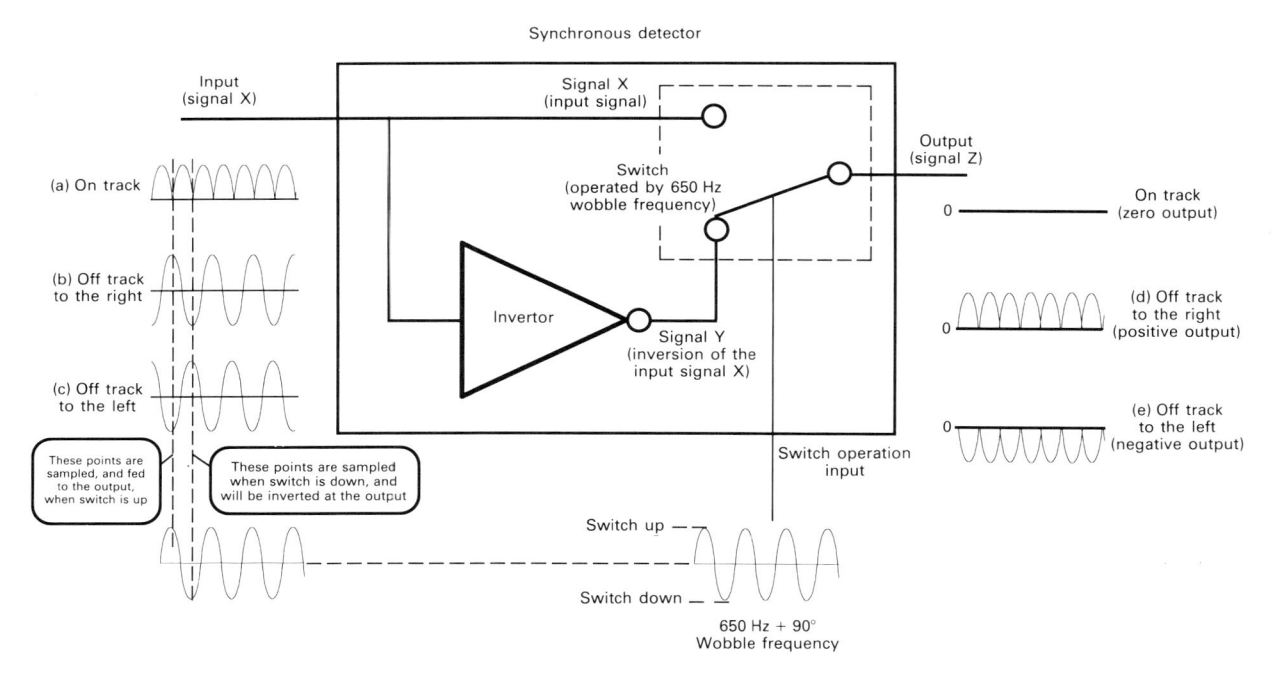

Figure 4.12(b) *Operation of the synchronous detector*

Servo systems

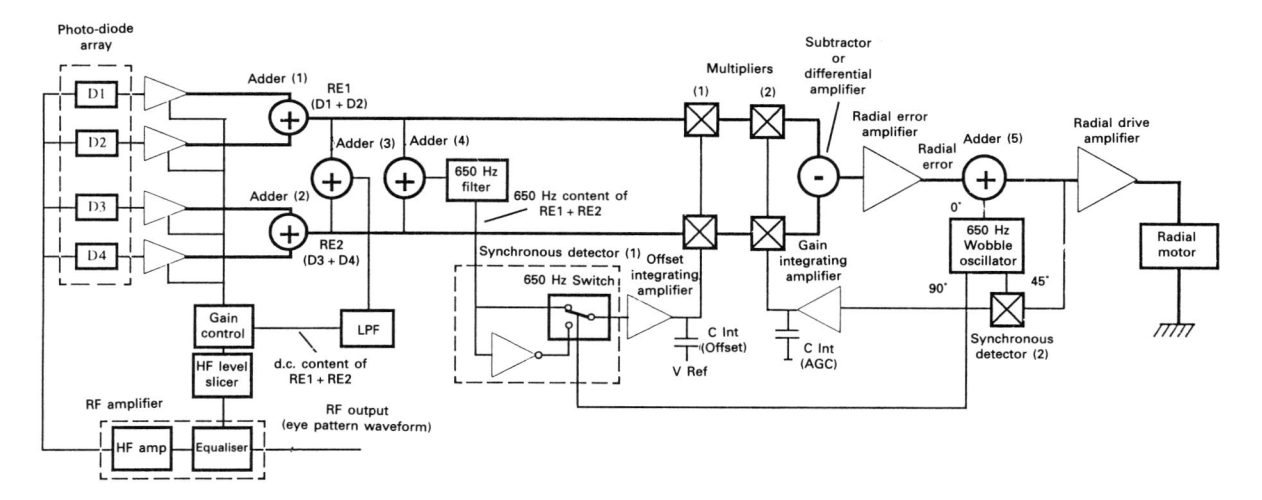

Figure 4.13 *Radial tracking servo*

the error signals RE1 and RE2 passing through multipliers (1) and (2), into a subtractor or differential amplifier to provide a radial tracking error signal which continues via the radial error amplifier and adder (5) to the radial motor drive circuit.

Fundamentally, the radial tracking error signal will be a d.c. level which will relate to the amount and direction of the tracking error. However, because of the various problems caused by the movement of the optical assembly (as described above), it is necessary to provide additional processing to maintain a consistent and efficient radial tracking operation.

Radial error signals RE1 and RE2 are combined via adder (4) and filtered via the 650 Hz filter to provide the wobble frequency content of the radial tracking error signals, which is now fed to the synchronous detector (1) to provide outputs in relation to the amount of radial tracking error (as illustrated in Fig. 4.12).

An output from the wobble oscillator is fed to the synchronous detector to operate the 650 Hz switch, which will enable the required error signal to be developed. The detector output is passed to the offset integrating amplifier to provide a level, which is related to radial tracking error, and when combined at multipliers (1), together with the RE1 and RE2 signals, will enable an effective error signal to be achieved.

It was previously mentioned that it was also necessary to maintain a consistent gain which is related to the phase of the original wobble frequency, and the wobble frequency content of the radial tracking error signal being applied to the radial drive circuit. This is achieved at the synchronous detector (2) resulting in an output which after processing in the gain integration circuit provides a level which is passed to the multipliers (2).

The following tracking servo descriptions are now related to the method of moving the optical assembly laterally across the surface of the CD, where the photo-diode array is consistently maintained at right angles to the track. With the exception of the phase difference method (Fig. 4.15), these alternative tracking servo methods prove to be a simpler method of operation than the radial tracking system.

The four photo-diode arrays always relate to single-beam optical assemblies, whereas the five and six photo-diode arrays are used in the three-beam optical assemblies, which would generally appear to be more widely used types of optical assemblies.

With the push–pull tracking servo (Fig. 4.14) the photo-diodes are connected in pairs (A + C) and (B + D). When the laser beam is on track, there will be equal amounts of reflected light received by the two pairs of diodes, which when applied to the first tracking, or differential amplifier, will result in zero output.

As the track moves to one side as the disc rotates, the proportions of reflected light on the two pairs of diodes will differ, providing a tracking error (TE) output from the differential amplifier

Servo systems

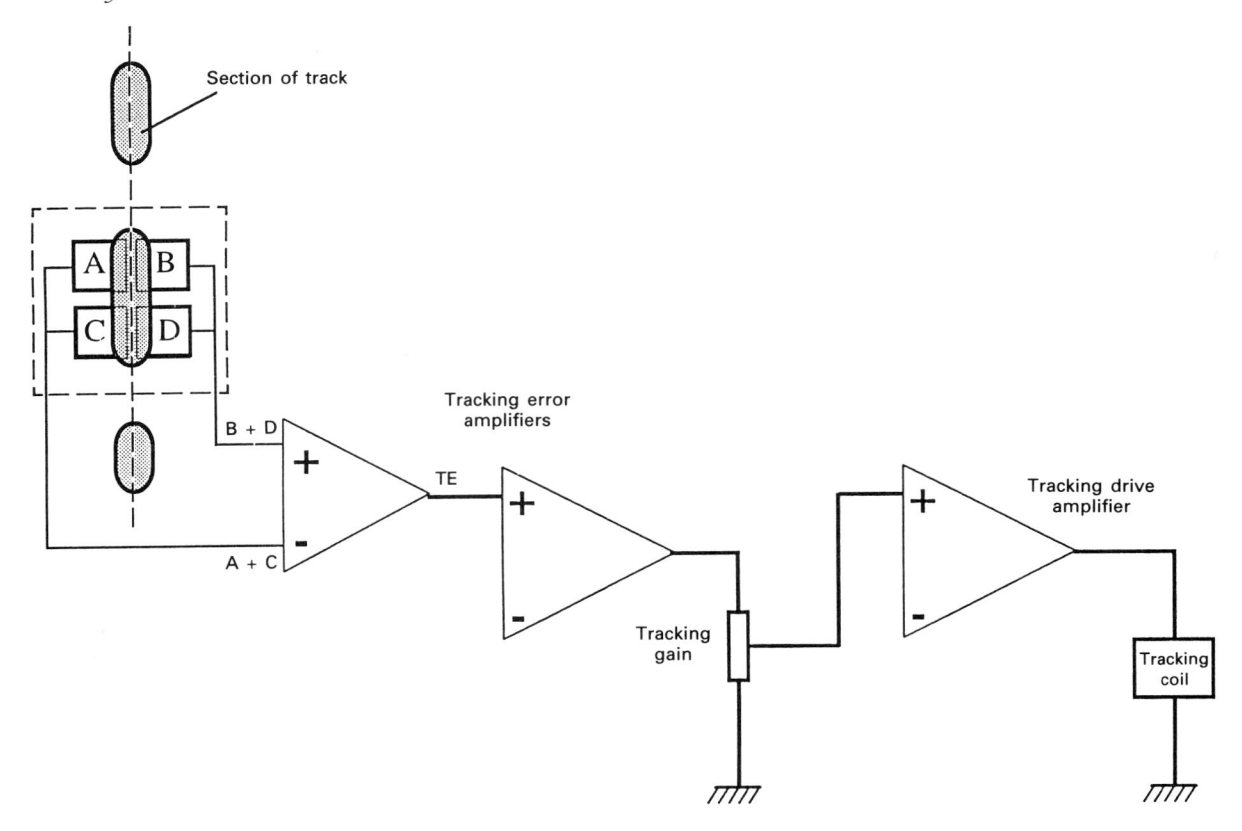

Figure 4.14 *Push–pull method of tracking servo*

that will relate to the amount and direction of tracking error, which in turn will enable the tracking coil to move in the correct direction to maintain its position on the track.

With the phase difference method of tracking servo (Fig. 4.15), again the photo-diodes are connected in pairs, but this time as (A + D) and (B + C). The outputs from the two pairs are fed to a differential amplifier and also a summing amplifier, with the differential amplifier producing at the output the difference between the two inputs, i.e. (B + C) − (A + D), and the summing amplifier producing an output of (A + D) + (B + C).

Referring to the graphical representation in Fig. 4.16, and considering the two waveforms drawn to represent the effect as a 'pit' passes through the photo-diode array, the output from the differential amplifier will indicate zero output for the 'on-track' condition, whilst a maximum amplitude signal will be produced from the summing amplifier.

For the 'off-track' conditions the differential amplifier will produce a varying amplitude signal, depending upon the 'pit' position, whilst the summing amplifier will produce a reduced amplitude signal. If a series of 'pits' were considered, then the resultant outputs would be similar, but the frequency intensity would increase in proportion to the periodicity of the 'pits'.

The output from the summing amplifier is passed to a signal processing stage to produce the RF eye pattern waveform, but an output is also passed to the zero cross detector, which will enable the sample and hold circuits to detect the the output from the differential amplifier as the RF signal passes through zero.

When the the laser beam is 'on track', with the output from the differential amplifier being zero, then the sample and hold circuits will detect zero levels, resulting in zero output from the tracking error amplifier.

During the 'off-track' situations, the phase of the differential amplifier outputs will be different

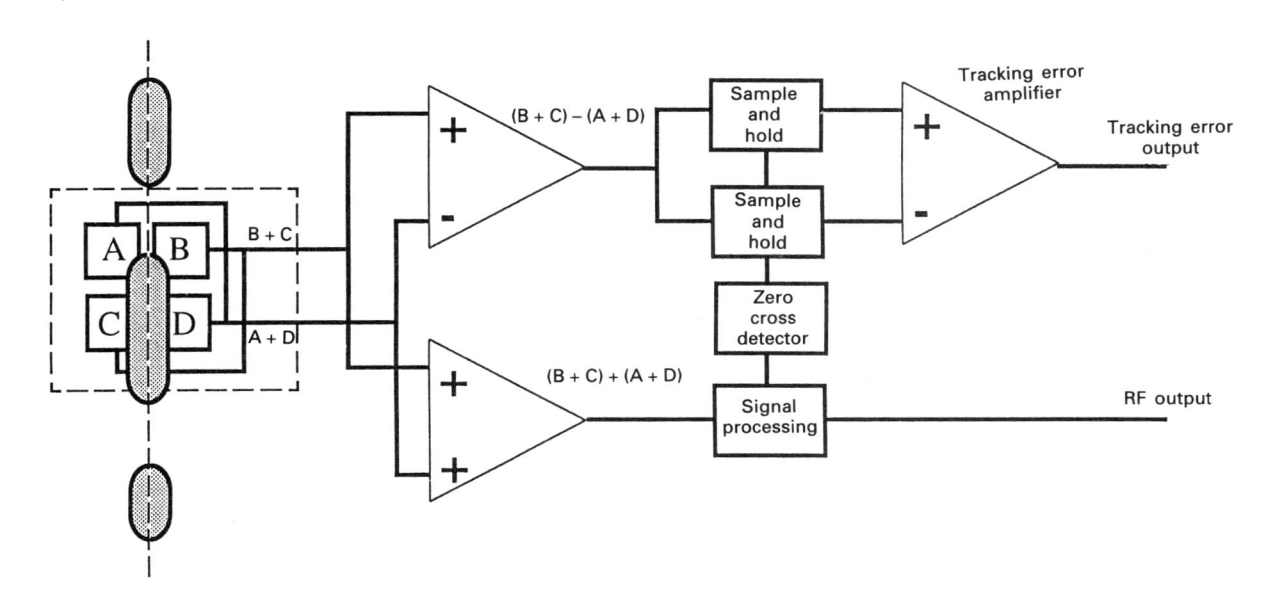

Figure 4.15 *Phase difference method of tracking servo*

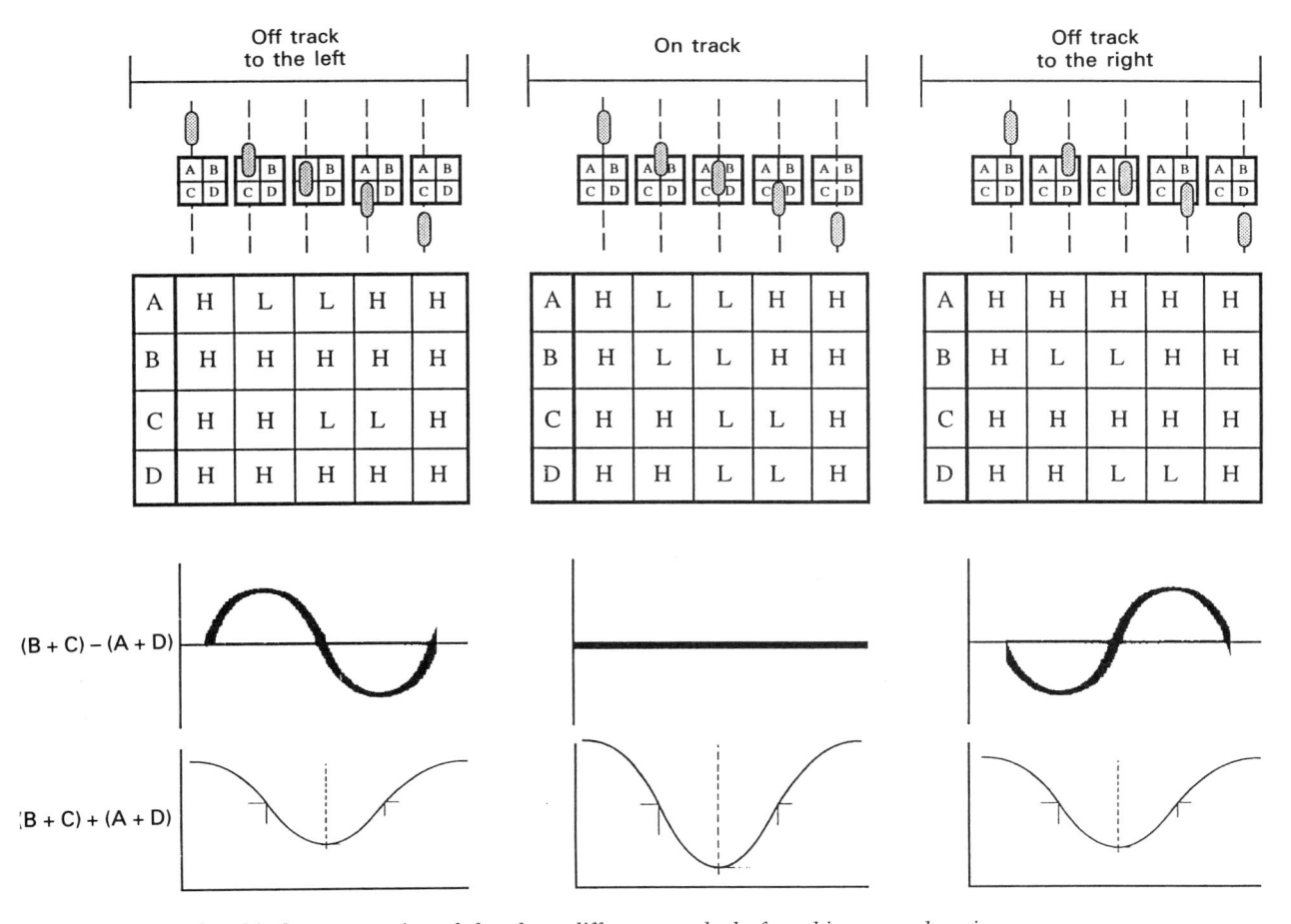

Figure 4.16 *Graphical representation of the phase difference method of tracking error detection*

67

Servo systems

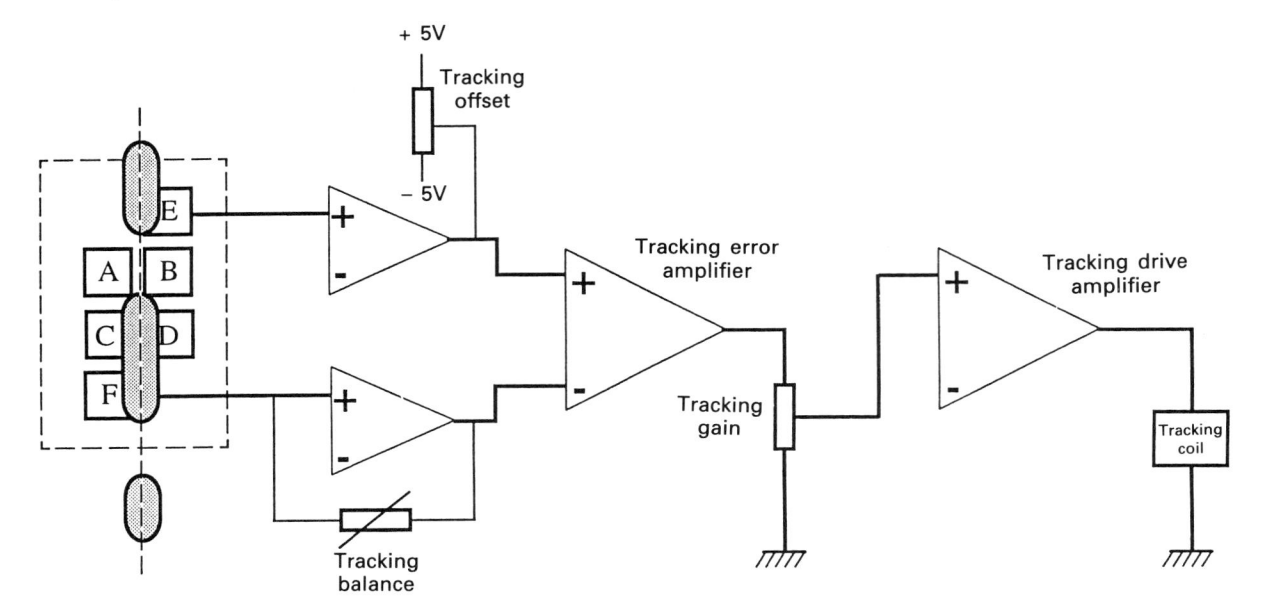

Figure 4.17 *Three-beam device tracking servo*

when 'off track' to the left is compared to 'off track' to the right. Therefore as the RF signal passes through zero, despite its reduced amplitude, it will be able to enable a positive or negative error signal to be developed in relation to the direction of tracking error.

This relatively complex method was used in earlier players, especially Pioneer in-car players, such as the CDX-1 and CDX-P1, and proved quite effective in maintaining track control in a somewhat hostile environment.

Three-beam device tracking servo

The three-beam device (Fig. 4.17) is probably one of the more common methods of tracking servo with the outputs from the (E) and (F) photo-diodes detecting any tracking errors as the track passes through the centre line of the photo-diode array. Each diode signal is amplified with the outputs being passed to the tracking differential amplifier to produce the tracking error output.

To ensure that the tracking error amplifier produces a signal that will be equal in value but opposite in phase when the same amount of error occurs either side of the track, it is necessary to ensure that the amplifier stages are balanced, this being achieved by the tracking balance control.

With the new Pioneer '92 optical block (Fig. 4.18) photo-diodes (D1) and (D5) fulfil the same function as the (E) and (F) diodes.

Carriage servo

Apart from the radial tracking assembly, all optical blocks allow the objective lens to move sideways

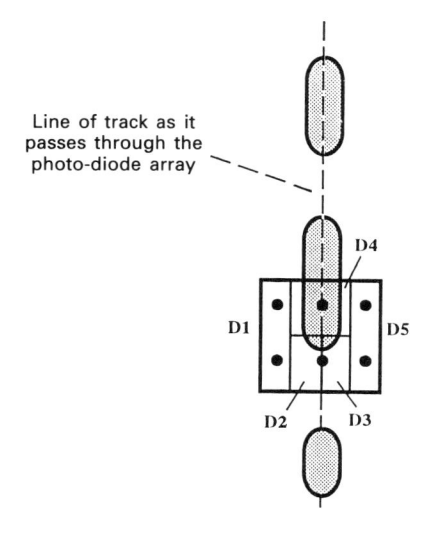

Figure 4.18 *Five-element photo-diode array*

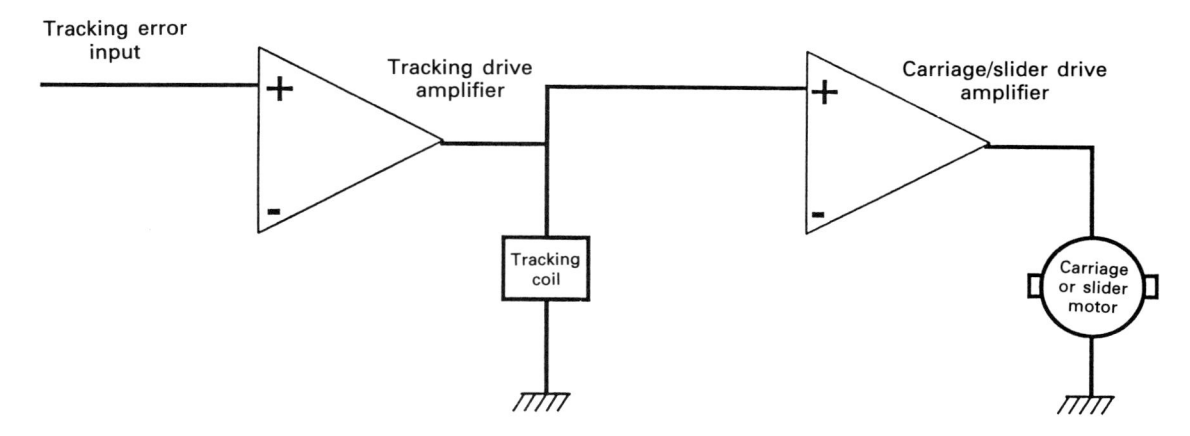

Figure 4.19 *Basic carriage servo*

to enable the tracking servo to ensure that the laser beam maintains its correct position on the track, but there is only a very small amount (~ 2 mm) of sideways or lateral movement.

In order that the optical block can gradually move across the surface of the disc, whilst tracking is being continually maintained, the tracking error signal applied to the tracking coil is continually monitored. As the lens moves laterally, the d.c. level of the tracking error signal will be increasing to cause the movement.

Applying this signal to the carriage drive motor amplifier (Fig. 4.19), a value will be achieved that will be sufficient to drive the carriage motor, which in turn will move the optical block an extremely small amount via reduction gearing, either in the form of a thread gear, or a gear and rack system.

Another method is the linear tracking system (Fig. 4.20(c)) where the conventional optical block moves across on a platform that is driven by a linear tracking motor. This type of motor provides an extremely efficient method of moving the optical block across the surface of the disc, as well as to specific areas of the disc when accessing different tracks.

Compared to the conventional gear and rack or thread-driven methods of moving the optical block, the linear tracking system provides almost instantaneous track access.

The varying d.c. potential that is derived from the tracking error signal to drive the carriage motor in the previous example is used to drive the linear motor (Fig. 4.21). This potential after amplification is applied to the drive coil of the linear motor, with the drive coil being set between two bar magnets which are in effect the length of the required travel that is necessary for the optical block to traverse across the surface of the disc.

A feedback coil provides an output that serves the two purposes of developing a speed signal to maintain velocity control, as well as a stop signal at the innermost rack position of the disc.

This method of carriage movement provides extremely fast access time to any required area on the disc, compared to the slower thread drive systems.

Spindle servo

The spindle servo (Fig. 4.22) is required to rotate the CD at the correct speed within the range of ~ 180–500 rpm. The data retrieved from the disc are compared to an internal reference within the CD player to produce a control voltage that will drive the disc at a speed compatible with the data rate from the disc.

The effective clock frequency from the disc is 4.3218 MHz, with 588 bits of the digital information comprising one frame of information. Dividing 4.3218 MHz by 588 provides the frame or sync. frequency of 7.35 kHz.

Most of the spindle servo circuitry is contained within the decoder, and despite its complexity its operation proves to be extremely reliable in the majority of CD players. When the CD is operating normally the disc will be running at a speed that will enable the surface speed passing the laser beam to be in the region of 1.2–1.4 m s^{-1}. As the optical assembly tracks towards the outer

Servo systems

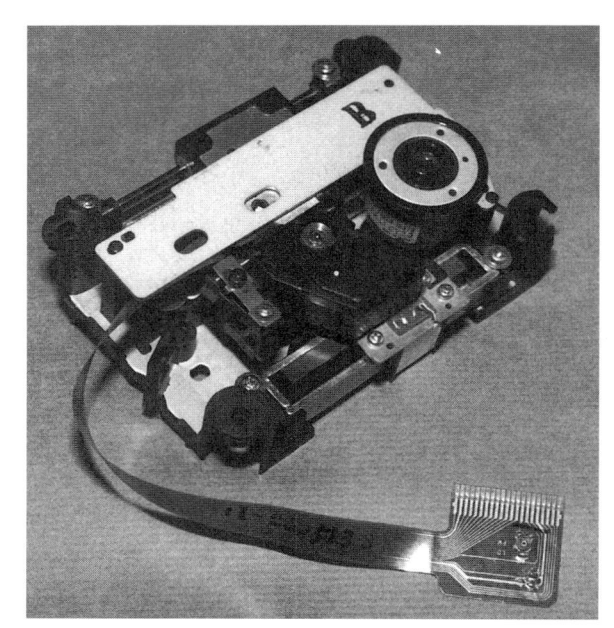

Figure 4.20 *Pioneer optical assemblies; (a) and (b) thread drive assemblies, (c) linear drive assembly*

Servo systems

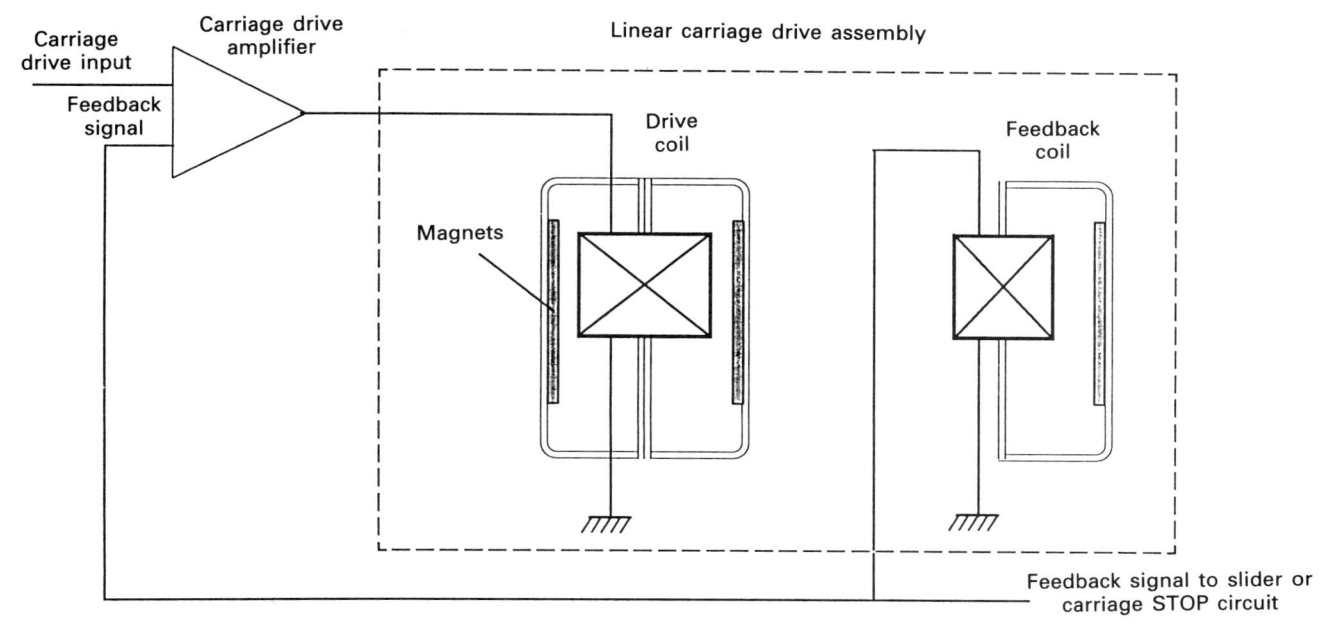

Figure 4.21 *Linear drive carriage systems*

edge of the disc, the disc will slow down as it maintains a constant linear velocity (CLV) of 1.2–1.4 m s^{-1}, which in effect is maintaining a constant data rate from the disc.

A crystal oscillator is required to provide a reference that can be compared with the data from the disc to obtain the required signal output which will enable the spindle motor circuit to maintain the correct speed. The spindle motor cannot maintain an absolute correct speed, due partly to disc and motor inertia as the disc naturally slows down during the playing process, but also because it is not essential to maintain an absolute exact speed.

In fact the disc speed will be slightly varying either side of the necessary speed to achieve the required linear velocity of 1.2–1.4 m s^{-1}, in order to maintain the correct level of digital data from the disc in the RAM. The absolute stability of the CD system of 'virtually immeasurable wow and flutter' is achieved by processing the data through the RAM, which is controlled by the reference crystal source, and any possible wow and flutter is now related in terms of crystal oscillator stability, which can be measured in parts per million.

Essentially the reference crystal oscillator is the 'master reference' with respect to the overall operation of the spindle servo. In most CD players the disc will commence spinning once the laser has been switched on and focus has been achieved. The system control circuit will 'kick start' the motor, via the MON (motor ON) signal, and as the motor gathers speed there will be a period during which the effective data rate coming from the disc as the EFM signal will approach 4.3218 MHz.

The EFM information is passed to various sections shown in Fig. 4.22. One particular path is via a phase comparator to enable a VCO to produce an output frequency of 8.6436 MHz, which is twice the effective frequency from the disc, to enable improved frequency lock and stability of the VCO.

The VCO frequency is divided by two to produce the CD clock frequency, which is passed to the VCO timing circuits to produce two frequencies of 612.5 Hz and 1.8375 MHz. At the same time the 'master reference' frequency of 8.4672 MHz is passed to the crystal oscillator timing circuits to produce the same two frequencies. The higher frequency of 1.8375 MHz is used to control the rough speed of the motor, whilst the lower frequency of 612.5 Hz becomes the phase or fine speed control.

The outputs from the coarse and fine servos

Servo systems

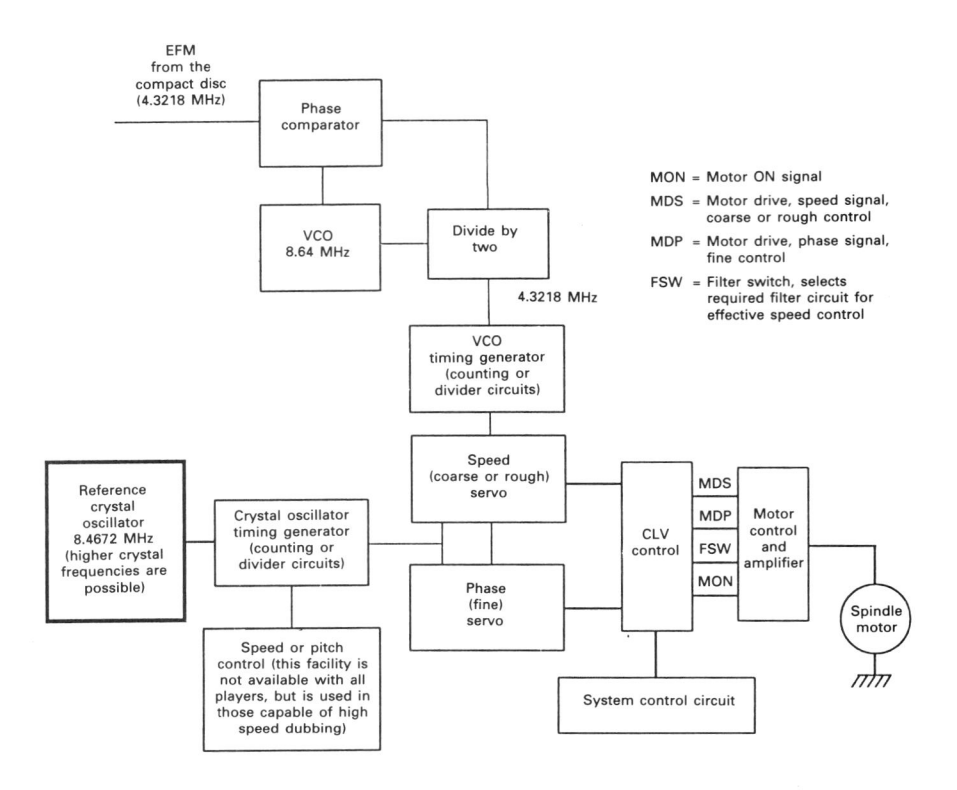

Figure 4.22 *Spindle servo system (similar to the early Pioneer in-car CD players CDX-1 and CDX-P1)*

are passed to the CLV control and then on to the motor control to provide the necessary signals to drive the spindle motor at the correct speed. The MDS and MDP provide the required motor speed and phase signals, with the MON providing the motor on/off control. The FSW signal controls the operation of a filter circuit within the motor drive circuit to enable its sensitivity to be varied, with a lower sensitivity whilst the motor is running up to speed, and a higher sensitivity once the motor has achieved the correct speed during normal playing of the CD to provide improved motor control.

As the data are extracted from the disc, and as the initial processing takes place of identifying the sync. signal and converting the 14 bit symbols back into the original 8 bit words, the restructured 16 bit words are fed to the RAM for the purpose of error correction and re-assembling the data words back into their correct order.

Within the RAM there are defined upper and lower limits of acceptable data storage, identifiable by the RAM address lines. Should the storage fall below the lower acceptable level the disc is momentarily increased in speed, to effectively top up the RAM to within the accepted limits. Likewise as the upper limit is reached then the disc speed is momentarily decreased to prevent overloading the memory with subsequent loss of information.

This fractional variation in speed is achieved by altering part of the dividing function within the crystal oscillator timing circuit to slightly alter the phase servo control and slightly increase or decrease the speed accordingly.

As previously mentioned, the spindle motor suffers excessive wow and flutter, but this is effectively corrected as the data are clocked through the RAM at a rate linked to the crystal reference oscillator, the stability of which may be measured in parts per million.

5

CIRCUIT DESCRIPTIONS

Chapters 2, 3 and 4 provided basic outlines of various aspects of the CD player aimed generally at block diagram level to enable a fundamental appreciation to be achieved.

This chapter is intended to provide an understanding of a wide range or circuits at circuit diagram level since many players of differing manufacture tend to utilise a common selection of Sony and Philips integrated circuits which are used for the RF processing, servo operation and decoding circuits, and this will prove useful as a basis of understanding for the majority of circuits used within most CD players.

Over the years the CD player has in many ways remained the same from one generation to another. The circuitry has evolved from a wide range of discrete components in the early players to just a few integrated circuits in the more recent models. However, the overall operation of the different players has remained relatively similar, and therefore a reasonable understanding of a range of selected circuits will generally enable a reasonable appreciation of many CD players to be achieved.

The following circuit descriptions are based on a group of integrated circuits (ICs) which provide the RF amplifier, servo control and signal decoding functions. There have been various generations of each group of these ICs, as indicated below, but experience has shown that there are similarities in their operation when comparing each generation, and that it is the new technologies as they have been developed which may have evolved the ICs into apparently much more complex packages.

Author's note:
Reference to specific component identifications, such as resistors, capacitors, transistors, and especially photo-diodes (i.e. A, B, C or D), may appear at variance to relevant circuit diagrams. Wherever possible it has been the author's intention to maintain consistency throughout this book to ensure cross referencing is as accurate as possible, and he apologises for any unintended errors that may 'slip through'.

Sony types (three generations)

RF Amplifier	Servo control	Decoder
CX20109	CX20108	CX23035
CXA1081	CXA1082	CXD1125 or
		CXD1135
CXA1471	CXA1372	CXD2500

Some later Pioneer and Sony players have a small sub-board fitted onto the optical assembly which has an IC that replaces the CXA1471, and may be identified as the M51593FP on Pioneer players, or the M52103FP on Sony players.

Philips types (three generations)

Photo-diode processor	Radial error processor	Decoder
TDA5708	TDA5709	SAA7010/
		SAA7020
TDA8808	TDA8809	SAA7210
TDA8808	TDA8809	SAA7310

Concerning the RF and servo processing circuits, there are many similarities in their general overall operation, and their various generations contain improvements in package density, circuit efficiencies in the areas of circuit processing techniques, and the voltage, current and temperature handling characteristics. It is the decoder circuits that have evolved more radically with more complex error correction, with the Sony CXD2500 having the RAM included internally, whilst many other decoders maintain the RAM externally. The latest Sony decoder to arrive on the scene (mid-1993) is the CXD2515, which combines not only all the decoder functions, including the RAM, but also the servo control, which operates digitally to provide automatic servo correction and gain control of the focus and tracking servos.

With this form of decoder, and similarly to the players using the Toshiba TC9220 and TC9221, servo and decoder ICs which provide automatic circuit control, many players will now have virtually no adjustments, with the possible exception

Circuit descriptions

of a couple of mechanical radial/tangential adjustments to ensure optimum operation of the optical block.

With respect to players that utilise the Sony-type circuits, the combinations of the CXA1081/CXA1082 and CXA1471/CXA1372 generally provide the same overall functions. Therefore their relevant circuit operations will be generally similar, and the CXA1081/CXA1082 combination will be used for the actual circuit descriptions, with any additional information included as necessary. But it is worth noting that the CXA1372 does contain some elements of the circuits used in the CXA1081, in particular the circuitry after the RF input (RFI) pin 1 as well as the servo control of the CXA1082.

The CXA1471 is generally a photo-diode processing RF amplifier, providing only the RF, focus error (FE) and tracking error (TE) signals. Some 1992 Pioneer and Sony players utilise a small sub-board attached to the optical assembly, with a small RF integrated circuit, either the M51593FP or the M52103FP producing these three signals direct to the CXA1372 as relevant.

With regard to circuits using the Philips ICs, the TDA8808/TDA8809 will be referred to as necessary, because of the widely differing circuit techniques that exist between players that use either the Sony or Philips type ICs.

Whilst players of other manufacture may appear to use totally different ICs, some are, in fact, similar under a different identification. It is the general circuit description that is intended to be applied to the majority of circuits that engineers will encounter.

Photo-diode array and RF amplifier stages

Philips circuit: photo-diode signal processor TDA8808

With reference to Fig. 5.1, the four photo-diodes, D1–D4, are connected to pins 22–25 of the photo-diode signal processor, TDA8808, and from these diodes will be formed the focus and radial tracking error signals. Also the sum of the four diodes, together with the guard ring photo-diode, are applied to pin 26, HF In, to eventually produce the RF eye pattern signal.

The signals derived from D1, D2, D3 and D4 are each amplified by gain controlled amplifiers, before being filtered and added together in pairs (D1 + D2) and (D3 + D4) to provide the two radial error signals RE1 and RE2. These two signals are also added together and fed to a gain control stage to provide an output for controlling the gain of the four diode input amplifiers.

The HF In signal at pin 26 is amplified and passed through an equaliser stage to provide the RF eye pattern signal output, HF Out, at pin 3. The inclusion of the guard ring photo-diode, together with the four main photo-diodes, enables this type of optical assembly, which is of the single-beam type, to compensate for any scratches or other possible problems on the playing surface of the disc.

Whilst the HF Out at pin 3 is passed to the decoder it is also fed back to pin 4 to be applied to the HF amplitude slicers and then in turn to the gain control circuit.

Four individual diode outputs, 1, 2, 3 and 4, appear after each filter, and are passed to the focus normalising circuit, to provide the focus error (FE) signal at pin 15.

At pin 6, communications from the system controller are fed to the starting circuit to switch on the laser diode, as well as commencing the focus search sequence in the focus normalising circuit.

The laser on (LO) output from the laser supply at pin 17 will go high to switch on transistor 6502 and in turn the laser diode. Stabilisation of the laser diode light intensity is maintained from the monitor diode by being applied via pin 18 to the laser supply circuit.

Laser power is controlled by the preset 3520 which varies the working point at pin 18.

Sony CXA1081S RF amplifier: photo-diode input

The integrated circuit shown in Fig. 5.2 appears in many players, and the general principles of operation can be applied to earlier, and even later, circuits that fulfilled these functions, as well as very similar versions occasionally identified under different type numbers.

Laser power control

When the LDON signal from the system control applied to pin 29 is LOW, the output at pin 5

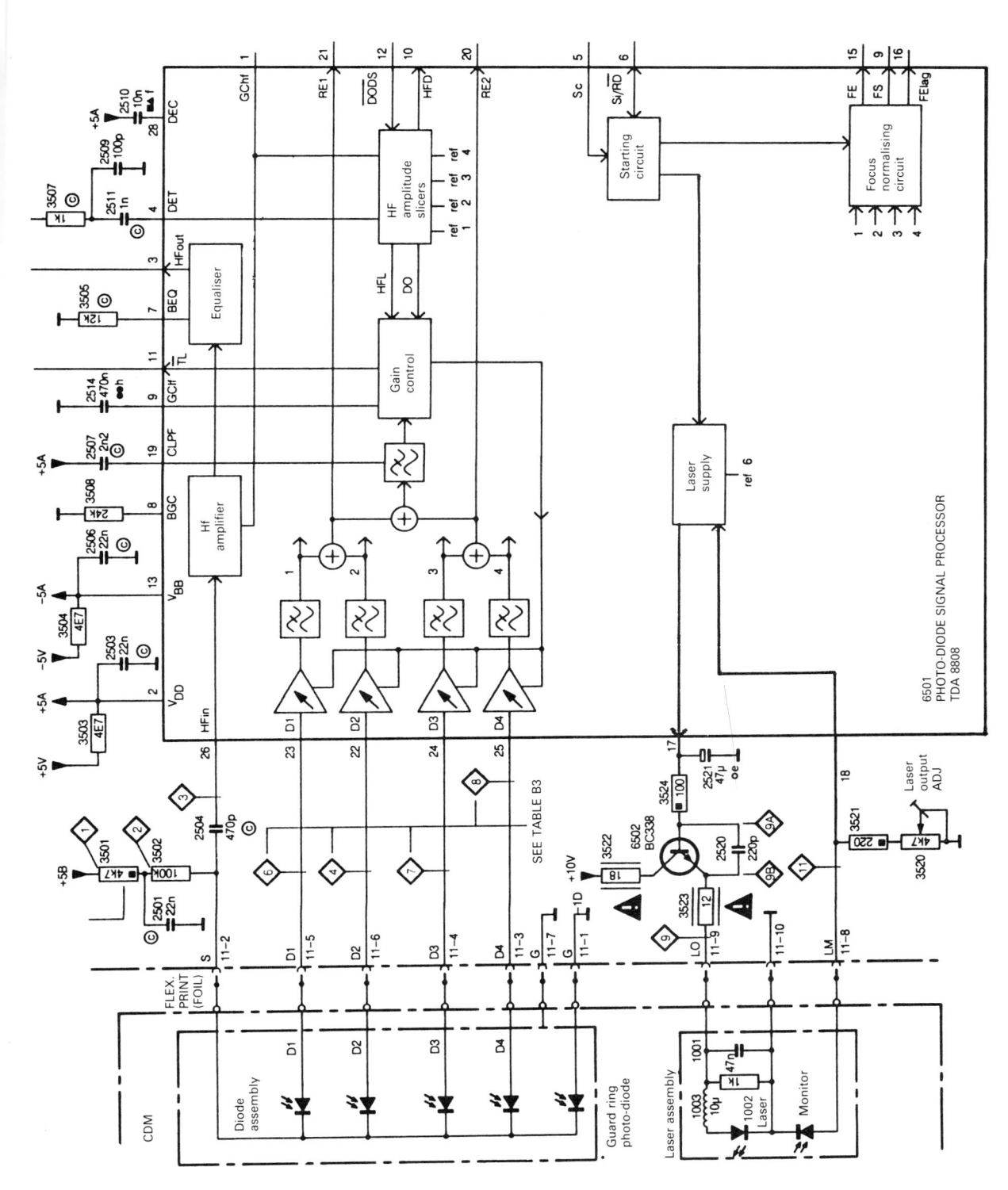

Figure 5.1 *Philips circuit – photo-diode signal processor TDA8808*

Circuit descriptions

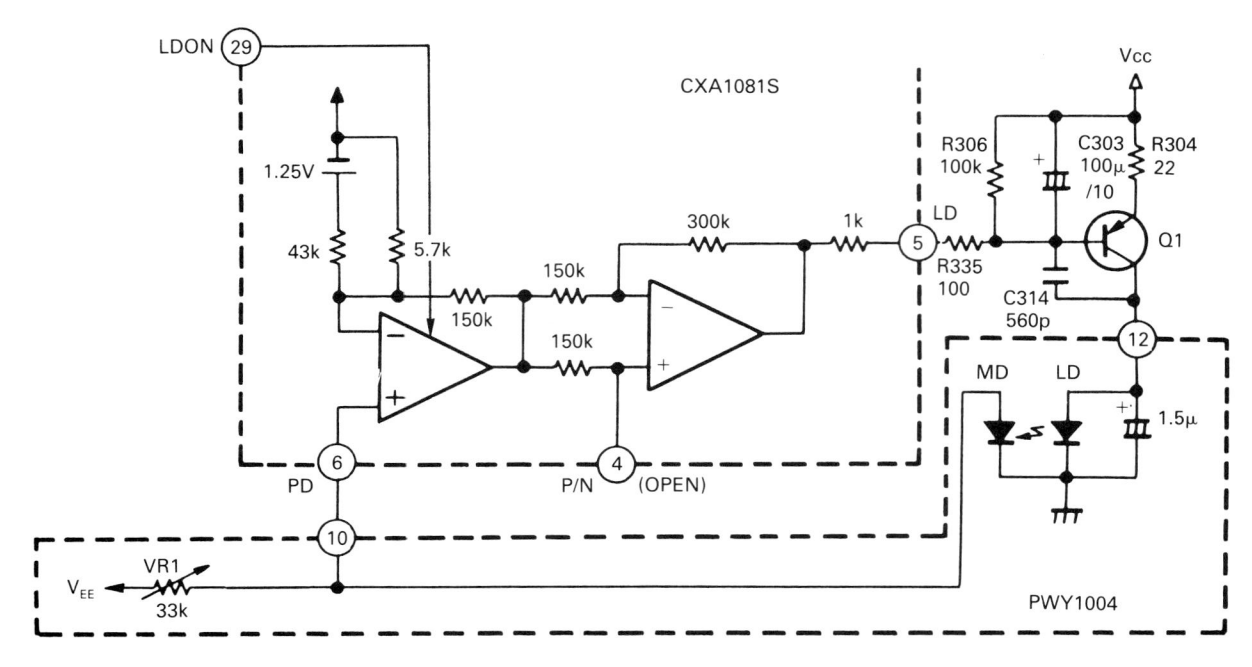

Figure 5.2 *Laser automatic power control (APC) circuit*

will also go LOW allowing transistor Q1 to conduct, switching on the laser diode. The monitor diode MD passes to pin 6 a measurement of the light intensity from the laser diode which enables the automatic power control circuit (APC) formed by the two internal op amps to maintain a stabilised laser diode output.

VR1 provides the operating point for the circuit via pin 6 to determine the necessary laser intensity from the laser diode.

RF amplifier circuit (Fig. 5.3)

The output from the photo-diodes is initially amplified before being applied as two pairs of signals (A + D) and (B + C) to pins 7 and 8 as PD1 and PD2, where they are further amplified before being summed together in the RF summing amplifier to provide (A + B + C + D), which will become the RF eye pattern signal (RFO) at pin 2.

A certain variation occurred in some Pioneer players with the introduction of a delay line in the preceding photo-diode circuits, as illustrated in Fig. 5.4.

Accufocus system

The purpose of the delay lines is to minimise any signal distortion that may occur because in effect the 'pits' from the disc will reach the preceding photo-diodes A and B first, and provide an output which will be fractionally ahead of the outputs from C and D, which may appear to develop an inconsistency in the quality of the total output signals. Though this method may have had advantages in reducing distortion and noise, together with a more accurate type of signal, it was however a system that was not continuously maintained throughout all the models that use this particular type of IC.

Focus error amplifier

Within the IC shown in Fig. 5.5 the two internal signals VA and VB, which are the result of the opposite connected diode pairs (A + D) = VA and (B + C) = VB, are applied to the focus error amplifier to provide the focus error signal (FE) output at pin 19.

The FE bias control connected to pin 18 is usually identified as the focus offset control in many players, and is a potential which is variable usually between Vcc (+ 5 V) and Vee (− 5 V) to

Circuit descriptions

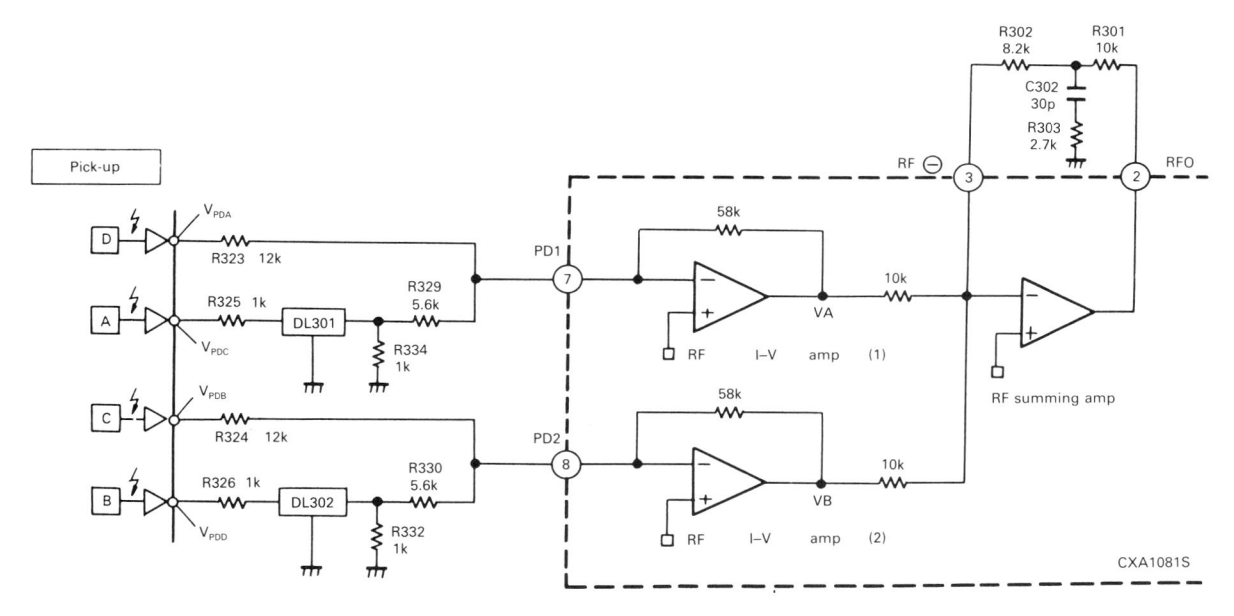

Figure 5.3 Sony CXA1081S RF amplifier – photo-diode input

neutralise any undesirable d.c. offset potentials with the focus servo system.

Focus OK circuit

When the CD player is operated after a disc has been inserted, it is necessary for a particular sequence of logical events to take place before complete playing of the disc can commence. These are usually as follows:

1 laser on: difficult to focus in the dark;
2 focus search: to detect that a disc is present, and enable focus to be achieved;
3 rotate the disc: enable the disc to reach roughly the correct speed; and
4 switch on the tracking servo: sequence complete and disc now playing.

Some variation in this sequence may occur in relation to when the disc commences rotation, with some manufacturers enabling the disc to rotate before the focus operation.

After switching on the laser and the focus search sequence taking place, the RFO signal from the RF summing amplifier is fed to one side of the focus OK amplifier, and then from pin 2 via a capacitor to pin 1 to the other side of the amplifier. At the correct focal point the RF signal will be at a maximum level, which will enable a high level to be obtained at the FOK output at pin 28. This FOK signal is passed to the system controller to inform it that a disc is present and that focus has been achieved, enabling the next relevant sequence to take place (Fig. 5.6).

Tracking error amplifier

With the common three-beam optical assembly, two photo-diodes (E and F in Fig. 5.7) are used for the purpose of developing the tracking error signal. The signals from these two diodes, after initial amplification, are applied to pins 10 and 11, for further amplification. The outputs from this stage of amplification are now passed to a differential amplifier, to provide the tracking error (TE) signal at pin 20 (Fig. 5.7).

The signal output at this pin will be a potential, the value and polarity of which will be related to the amount and direction of tracking error. It is essential when an error occurs to one side of the track, which will produce a potential of a specific level and polarity, that an equal error on the other side of the track enables a level to be produced that is equal in amplitude, but opposite in polarity. For this to be achieved, the gain of one of the input amplifiers can be adjusted. Connected between pins 12 and 13 is the tracking balance control identified in Fig. 5.7 as VR5.

Circuit descriptions

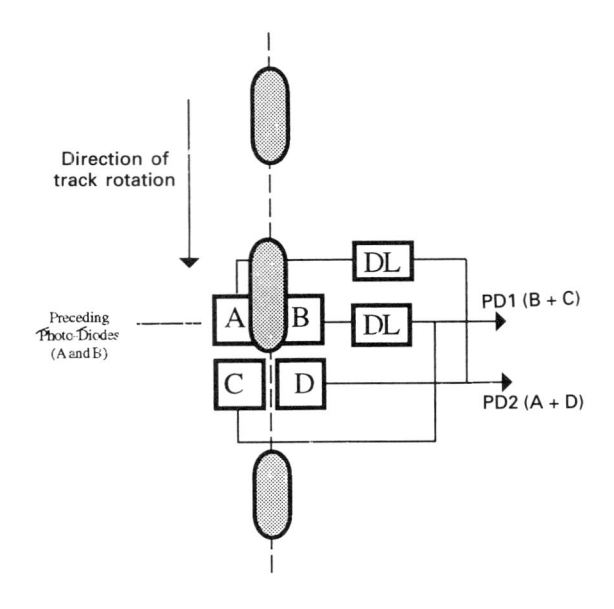

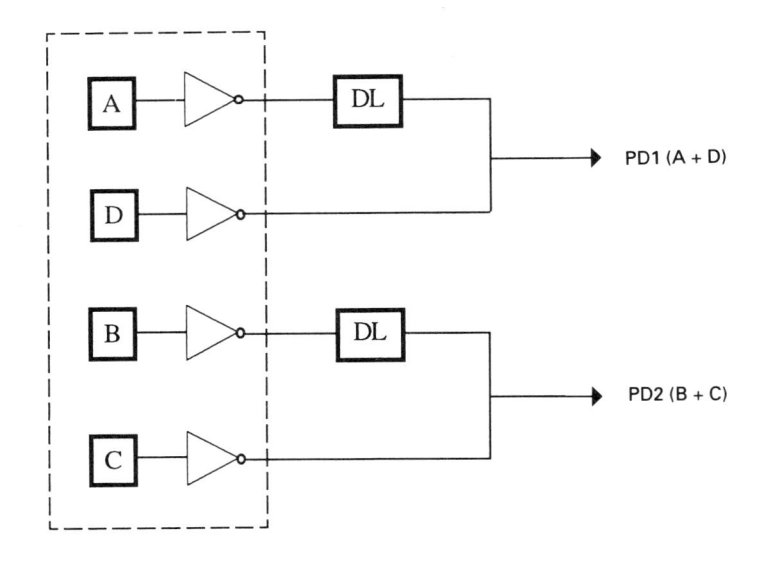

Figure 5.4 *Accufocus system*

Circuit descriptions

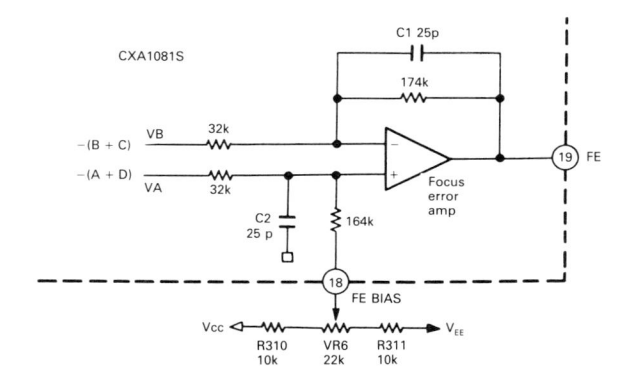

Figure 5.5 *Focus error amplifier*

signal building up to a maximum level as the laser locks onto the track correctly.

The signal is amplified to provide the inverted signal at (G), with the upper peaks of this waveform relating to 'on-track' conditions, and the troughs being related to almost 'off-track' conditions.

Further processing of the signal through the peak and bottom hold circuits will now develop the signals (H) and (I), that relate to the average of each of the signals above and below the 0 V level, and provide the signal (J) from the comparator section.

This signal is now applied to the mirror hold amplifier, where the level (K) is developed in relation to the hold capacitor at pin 23, and the mirror comparator where the combined effect of the signals (J) and (K) produces a series of high and lows that relate to the off-track (high) and on-track (low) conditions.

Mirror circuit (Fig. 5.8)

It is essential that the laser beam effectively follows the track on the disc. However, it is possible for the beam to follow the space between the tracks, and to ensure this does not occur continual monitoring of the RF signal is maintained. If the beam was tracking between tracks, the reflected beam would be maintained at a consistently high level, and this level would be used to 'inform' the system control and tracking servo to 'kick' the beam onto the track.

The RF input (RFI) to pin 1 is in fact the eye pattern waveform (RFO) monitored with the oscilloscope timebase operating at a relatively low speed to compress the waveform as shown, and will fluctuate either side of 0 V due to the capacitor C1. The waveform illustrated shows the RF

EFM comparator

The RF eye pattern waveform (RFI) is derived from pin 2 and applied to pin 1 via a capacitor. The effective aim is to turn the signals which consist of a series of amplitude variations (depending upon the lengths of the pits and spaces) that are being retrieved from the disc, and turn them into a series of square waves. However, it is necessary to consider the possibility of asymmetry caused during the manufacture of the disc, where the quality of the pits or bumps on the disc may not be consistent from one disc to another. The capacitor connected between pin 2 and pin

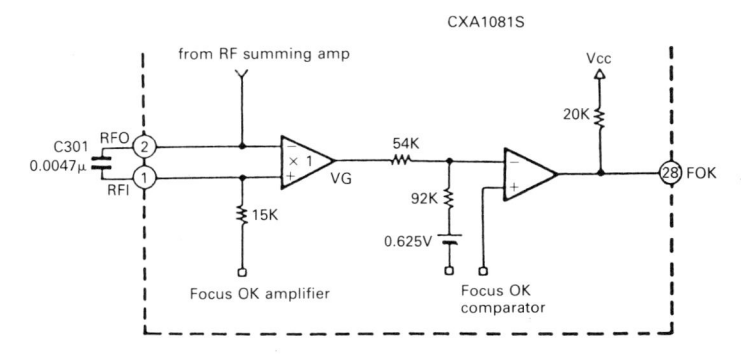

Figure 5.6 *Focus OK circuit*

Circuit descriptions

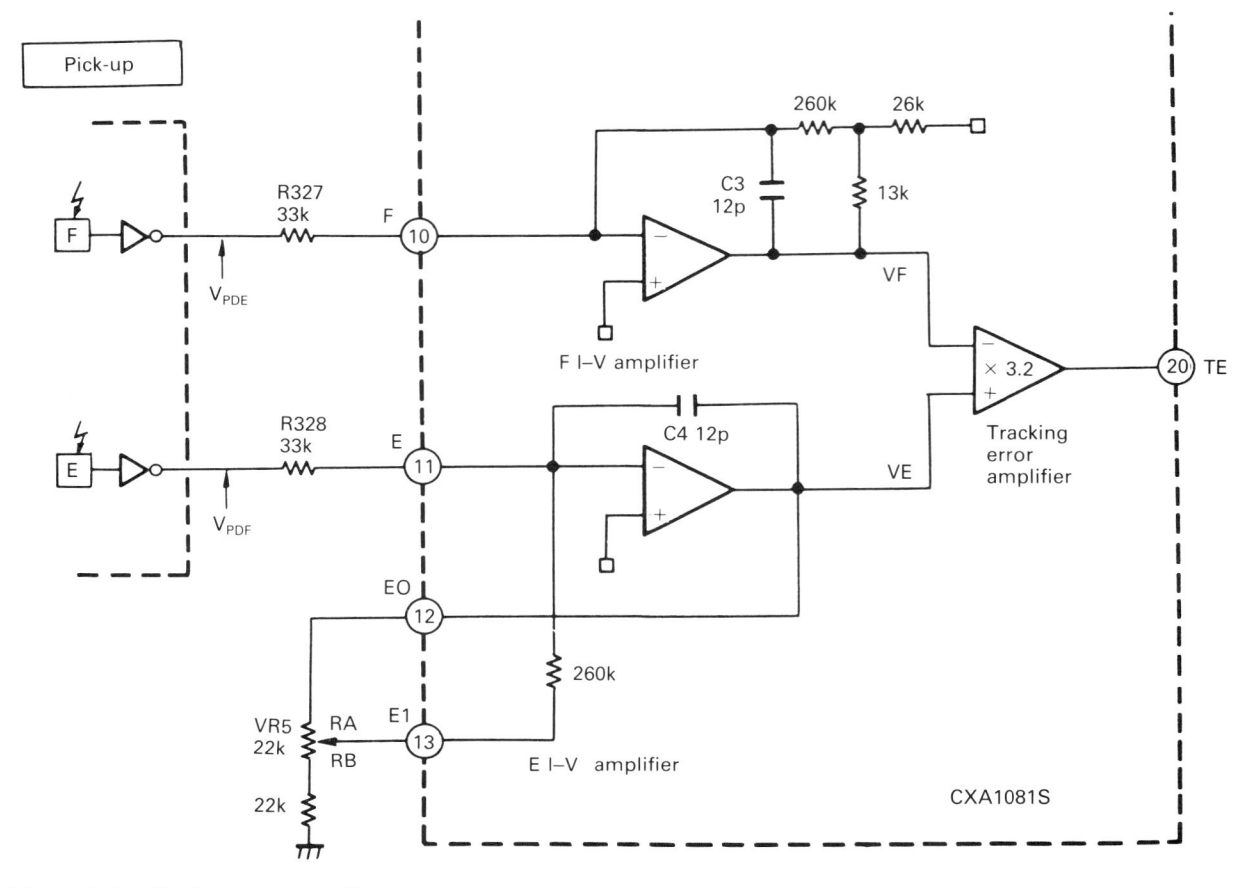

Figure 5.7 *Tracking error amplifier*

1 can help to overcome asymmetry problems, but further processing is usually necessary.

The output from the EFM comparator (Fig. 5.9) is fed from pin 27, through a buffer stage which is usually within the decoder, and then through a filter back to pin 26 (ASY), where via an internal buffer and amplifier stage it becomes the reference or working point from which the comparator can operate.

Usually 2.5 V will be present at pin 26 which is derived from the decoder (CXD1135) pin 6. The EFM comparator comprises a d.c. loop from pin 27, through the buffer in the decoder to pin 26 and via the internal ASY buffer and amplifier stages back to the EFM comparator. The RF eye pattern input to pin 1 appears at pin 27 as square waves, but the leading and lagging edges of the waveform (which are in fact the 1's information of the digital data) can be affected by the quality of the 'pits' or 'bumps' on the disc, Fig. 5.10.

The filter between pin 6 of the decoder and pin 26 will remove the square wave variations, to develop a slight variation of the nominal 2.5 V at pin 26 due to the charge developed by C316. This resultant potential will be related to the disc quality, which in turn is passed to the EFM comparator to ensure that a high quality of square waves is maintained for further processing in the decoder.

The filter will cause C316 to charge up, or integrate the signal, to the mid-point level of the maximum and minimum levels, to provide the previously mentioned reference or working point, and become an automatically variable level, auto asymmetry, which is related to the pit quality of each disc as it is played.

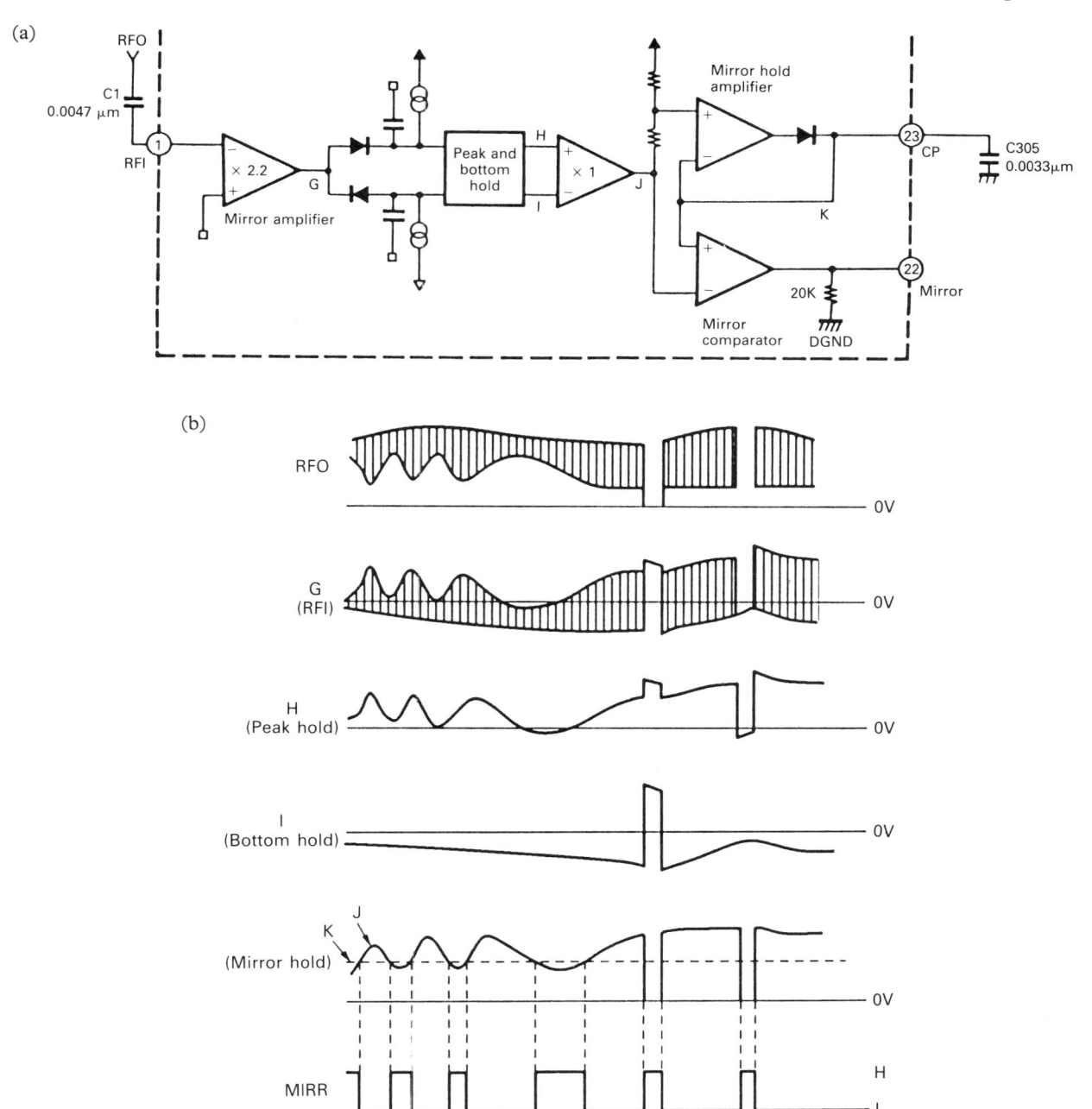

Figure 5.8 *Mirror circuit; (a) circuit (b) waveforms*

Circuit descriptions

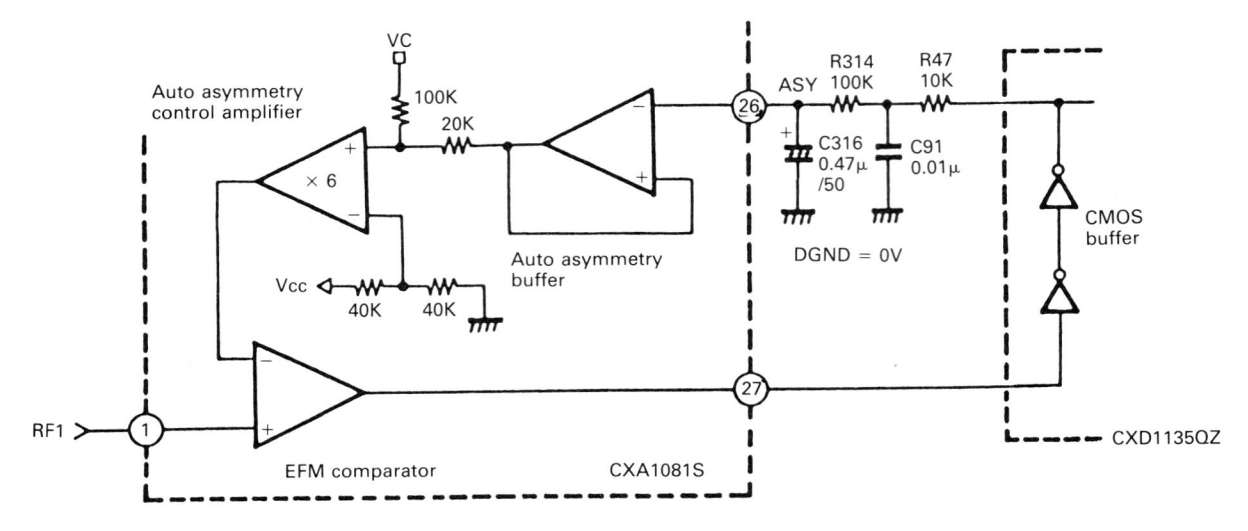

Figure 5.9 *EFM comparator*

Defect circuit (Fig. 5.11)

It is essential that when a defective disc (i.e. with scratches) is played some form of damping occurs within the tracking servo circuit; otherwise it is possible for track jumping to occur when scratches pass through the laser beam.

The RF eye pattern waveform (RFO) at pin 2, (a), is inverted at the output of the the defect amplifier, waveform (b), which becomes the bottom hold signal. As the effect of the scratch now appears positive going both of the diodes, D1 and D2, will be prevented from conducting for the period of the positive going scratch pulse.

During periods when no major scratches are present, both D1 and D2 will slightly conduct as a result of point (b) being low, and a current flowing from both the positive current references and capacitor CB will be charged slightly positively. When scratches occur D1 and D2 will stop conducting.

As a result of D1 not conducting, the positive current reference will now appear at pin 16 (CC1) and is coupled via the capacitor, CC, to pin 15 (CC2), causing the pulse to appear at (c) which is applied to the positive input of the defect comparator.

When D2 is prevented from conducting, CB will commence to charge up more positively, but due to the short duration of the first or small scratch pulse this will not cause any increase of the level occurring at pin 24. As a result, a short duration square pulse (e) will appear at pin 21 (defect).

For a more excessive scratch, CB will be able to charge up more positively, as shown on the dotted line for the signal (d), and the longer square pulse at (e) will terminate when the level at (d) becomes greater than that at (c), causing the comparator output to change over. This is to ensure that there is a maximum allowable defect pulse length of 1.4 ms to occur for scratches that are of a longer duration; 1.4 ms is equivalent to a scratch width of approximately 1 mm.

Philips circuit: radial error processor, TDA8809

The radial error signals from the TDA8808 (Fig. 5.12) are applied via pins 27 (RE1) and 28 (RE2) to a summing and dividing network to provide a range of outputs for further processing.

Signals (re1) and (re2) from the summing and dividing network each pass to a multiplier, where an input from the offset integrating amplifier is also applied. This input is related to the level of tracking error determined from the detected phase of the wobble frequency.

The second pair of multipliers have an input from the gain integrating amplifier which is the result of comparing the phase of the wobble frequency at 45° and the phase of the returning radial

Circuit descriptions

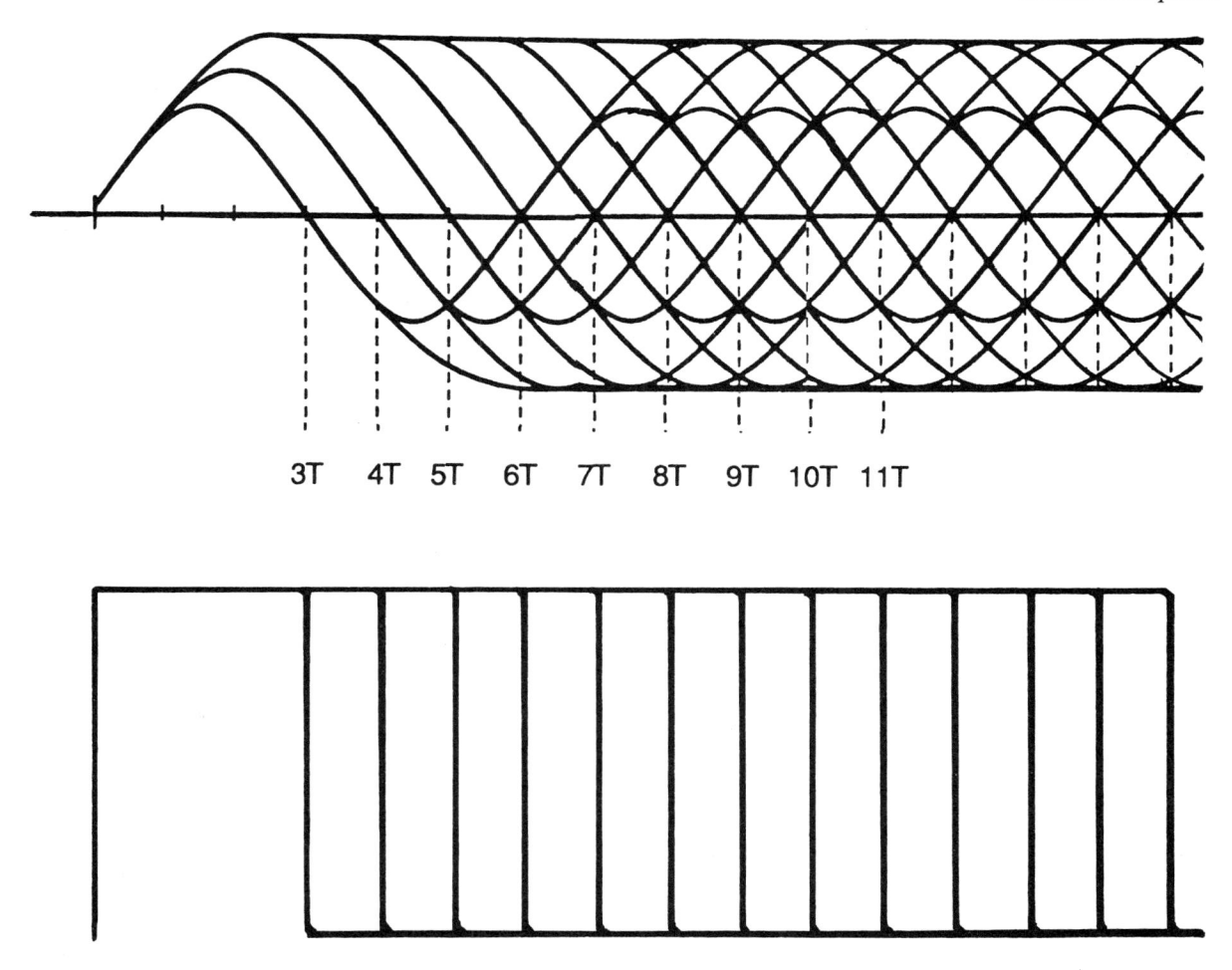

Figure 5.10 *Eye pattern and square wave illustrations*

error signal which will also contain 650 Hz. The phase relationship will be related to the quality of the disc reflectivity, laser power and the radial position on the disc, factors which can affect the overall stability, and therefore it is possible to achieve a level which can control the gain of the radial error servo to maintain the necessary stability.

After the multipliers, the two radial error signals (re1) and (re2) pass through a subtractor or comparator to provide the radial error signal (RE), which then passes through the adder where the 650 Hz wobble frequency is injected.

The resultant radial error signal (RE) appears at pin 19 and passes via an anti-skating filter to pin 16 for further amplification to finally appear at pin 15 for application to the radial assembly after final amplification in the radial drive amplifier.

The function decoder receives information from the system controller through pins 8, 9, 10 and 11, which controls the operation of various switches within the IC as well as the 3.5 bit DAC which controls track jumping via the final adder preceding pin 15.

When track jumping occurs, due to selecting another music track or when selecting 'fast forward' or 'reverse', it is necessary to reduce the overall gain of the circuit during this process, and to ensure that the laser beam is directly on track before restoring the circuit back to normal operation. This is achieved by monitoring the (RE dig) signal available at pin 7 (which is related to the

Circuit descriptions

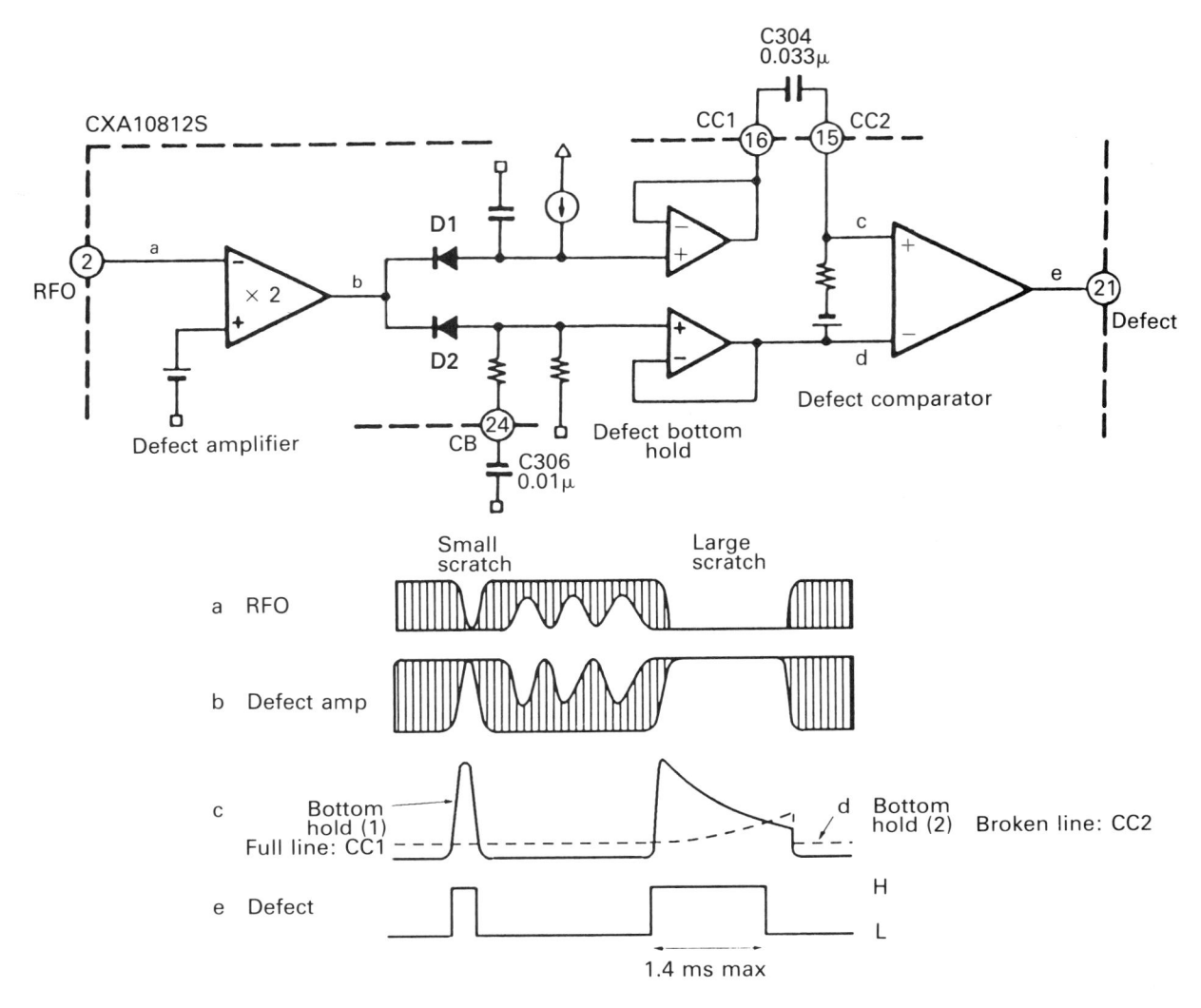

Figure 5.11 *Defect circuit*

position of the laser beam compared to the track), and feeding this signal to the system controller.

Of the various internal switches which are controlled at various stages of enabling the radial error servo to operate correctly, switch 7 when in the up position feeds the basic radial error signal into the offset circuit at the commencement of playing the disc when 'tracking lock' is being achieved, and then for the fine control the switch will change state to the low position to feed in the detected 650 Hz phase detected level.

The internal radial switch between pins 16 and 15 will open and close the radial error servo loop to ensure that tracking commences at the correct point. It will also be operated in the test mode to control the servo as necessary.

Sony CXA1082

Focus servo system (Fig. 5.13)

The focus error signal (FE) from pin 19 of the RF amplifier IC is applied via the focus gain control to pin 6 and passed through to the focus phase compensator, which will apply correction to the overall circuit in relation to the specific frequency range within which the focus servo

Circuit descriptions

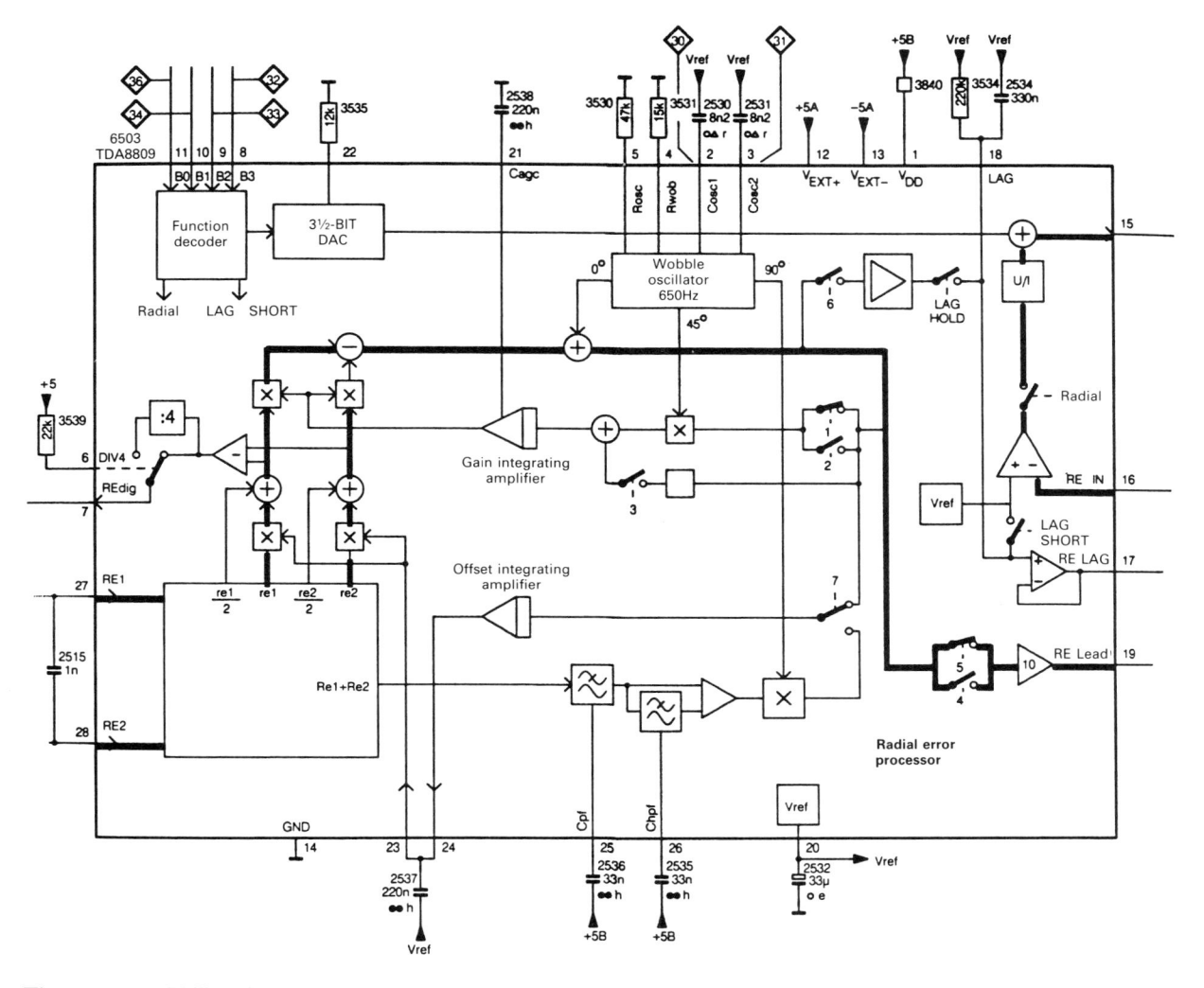

Figure 5.12 *Philips circuit: radial error processor TDA8809*

operates, to finally appear as the focus error output (FEO) at pin 11.

A range of switches within the servo IC are controlled by output from the internal logic circuits, which in turn are controlled by the system controller via a CPU interface circuit which is usually incorporated within the decoder. These switches will control all the required operations of the focus, tracking and carriage servos.

The focus zero cross (FZC) is passed to the internal logic control circuits which, together with the focus OK signal (FOK), will determine the correct time to operate the focus servo by opening (FS4), and enabling the complete loop to be closed.

The (DEFECT) and (FS3) switches reduce the servo gain either when a defect occurs, or in the case of (FS3) when focus is operating normally and a lower gain is normally required (compared to a higher gain when focus is initially acquired).

The focus phase compensator output after further amplification is fed from pin 11 to the focus drive circuit and the focus coil to complete the overall circuit. Occasionally a complementary push–pull pair will drive the focus coil instead of the op amp circuit illustrated.

Circuit descriptions

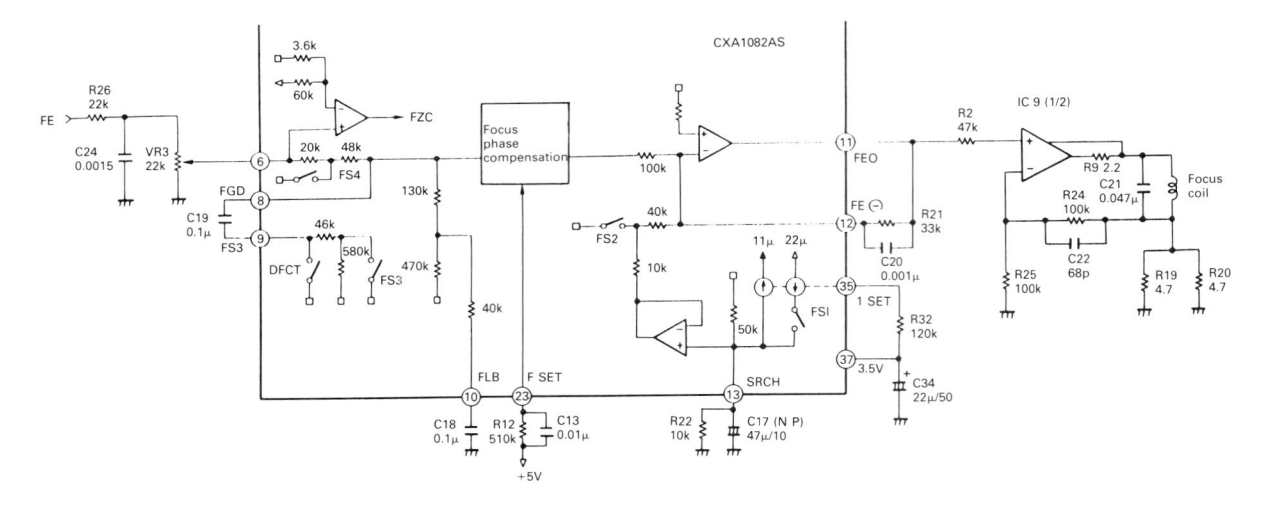

Figure 5.13 *Sony CXA1082 focus servo system*

A focus return signal developed across the two 4.7 Ω resistors is fed back to the inverting side of the op amp as a focus error rate signal to stabilise the focus servo operation and prevent the focus coil operating erratically as focus errors occur. With circuits that contain the complementary push–pull pair, the focus return signal is usually applied to pin 12.

When a disc is inserted into the player it is necessary for a focus search sequence to take place, by raising and lowering the lens, to enable the correct focal point to be achieved. This is performed by the focus search (FS2) and focus up/down (FS1) switches, which, together with the focus open/close (FS4) switch, are controlled by the internal logic circuits.

During focus search the sequence of operation is as follows:

FS4 will be CLOSED
FS2 will OPEN
FS1 will be CLOSED

At this point the lens will move to one end of its travel.

When FS1 is CLOSED, the current reference of −11 μA and +22 μA will provide +11 μA to flow through the internal 50 K resistor, providing 0.55 V (22 μA + (−11 μA) × 50 K = 0.55 V) to be applied to the input of the internal op amp and also C17, the search (SRCH) capacitor, which is connected to pin 13. The output from the op amp will appear, via another internal amplifier, at pin 11, and after feeding through the focus drive amplifier will be applied to the focus coil thus causing the lens to move to one extreme end of its travel.

After a brief period of time (~1 s), FS1 will OPEN removing the +22 μA internal reference, and now only −11 μA will be applied to the internal op amp and C17. This will enable the lens to move to the other extreme, at a rate controlled by the charge/discharge of C17 as it adjusts to the change in potential. This sequence usually occurs twice if no disc is present or if the laser circuit is defective, and is controlled by data interchange between the system controller and the servo IC.

If a disc is present, and assuming that the laser circuit is operating correctly, only one focus search sequence is necessary as the lens moves through the correct focal point, a maximum reflected signal will be obtained from the surface of the disc, which will result in the focus OK (FOK) signal going HIGH. At the focus zero cross (FZC) point FS4 will be opened to switch on or close the focus servo loop, and FS2 will be closed to remove any focus search potentials.

An interesting addition to the input of the focus drive circuit can occasionally be observed in some earlier Pioneer multi-disc players, where, typical of these players, the optical block was installed upside down, i.e. with the lens facing downwards, and was therefore subject to gravitational effects. With the process of selecting each disc from a six-disc magazine, each disc would 'sweep' past the lens, which could have a possible detrimental 'de-capping' effect on the lens as well as causing physical contact between lens and disc surface during certain stages of the operation of selecting and de-selecting the discs.

Focus correction circuit

The effect of the correction circuit (Fig. 5.14) is to feed a residual bias current through the focus coil to raise it into its normally neutral position.

Later models refrained from using this type of circuit, and achieved the raising of the lens to its neutral position by adjusting the focus offset adjustment to a pre-determined level, usually a negative potential, which provided the same compensation.

Tracking and carriage servo circuit (Fig. 5.15)

The tracking error (TE) from pin 20 of the RF amplifier IC, CXA1081, is applied via the tracking gain control to pin 3, which then passes through the circuit to emerge at pin 17 as the tracking amplifier output (TAO).

The tracking error (TE) is also applied via C27 to pin 4 as the TZC signal, to provide the correct time to apply the brake operation to the tracking servo during track jump sequences.

As with the focus servo circuit, various internal switches, which are controlled by outputs from the internal logic circuits, provide the various required control functions for the tracking and carriage circuits.

The two TG1 switches, and TG2, provide the required gain control functions that are needed during the initial track search process. Usually a higher gain is required as tracking is achieved, and a reduced gain when tracking is functioning normally. Also, different gain levels are required during the various track jump sequences.

TM1 switches the tracking servo on (TM1 open) and off (TM1 closed).

TM2 controls the carriage servo in a similar manner to TM1.

TM3 and TM4 provide positive and negative current references for track jump requirements.

TM5 and TM6 operate in a similar manner to provide forward or reverse carriage movement; an extremely useful function in the test mode function to ensure that the carriage servo can be operated when either a high or low can be monitored at pin 20, which should cause the carriage to function.

TM7 provides a smooth brake function to stop further sideways movement of the objective during track search sequences. TM7 will be momentarily closed when the laser beam is precisely on track, a fact that is determined by comparing the mirror (MIRR) input at pin 48, and the TZC signal within the internal logic circuits to provide the specific moment to close TM7 (as shown in Fig. 5.16).

The tracking error (TE) signal at pin 3 passes through the various gain control circuits and the tracking phase compensation circuit (which will provide the necessary frequency correction), to appear at pin 17 to be applied to the tracking drive circuit, which will usually be either a complementary push–pull transistor pair, as shown by Q6 and Q7, or as in many players, an op amp to drive the tracking coil.

The signal applied to the tracking coil is also fed to pin 19 via a filter circuit comprising R50, R51, C39 and C40, which will remove the high frequency tracking error variations, and provide a relatively low frequency d.c. variation, to provide an output at pin 20 (SLO), for operating the carriage or sled motor to move the carriage assembly. As the tracking coil moves the lens across the surface of the disc, it is necessary to gradually move the optical block to compensate for the sideways movement of the objective lens, and ensure that the carriage motor moves the optical assembly before the lens reaches the limit of its travel, which could result in skipping or jumping of the sound.

Disc, spindle or turntable motor servo circuits

Generally speaking the decoder will provide the actual motor speed information, whilst the system control will provide the relevant motor control information. The motor speed is a function of the data information from the disc, after processing within the decoder, compared with a

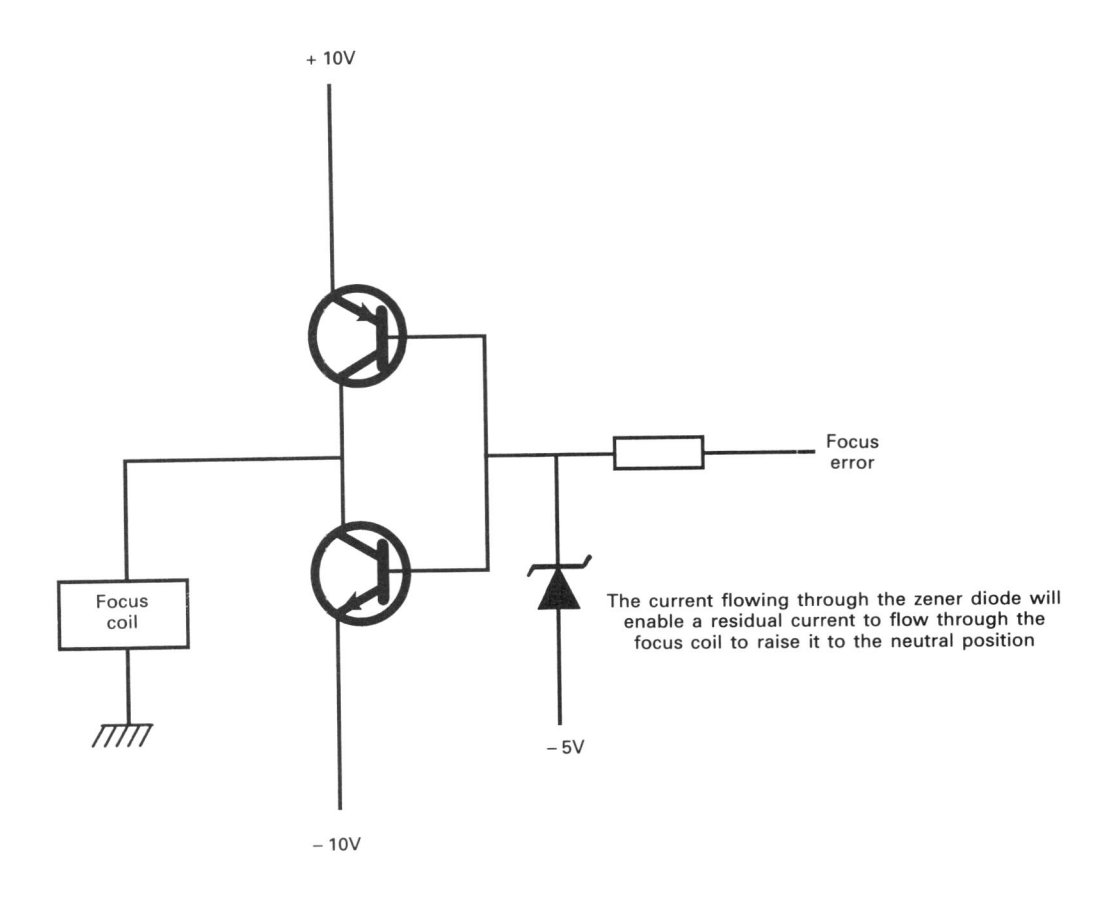

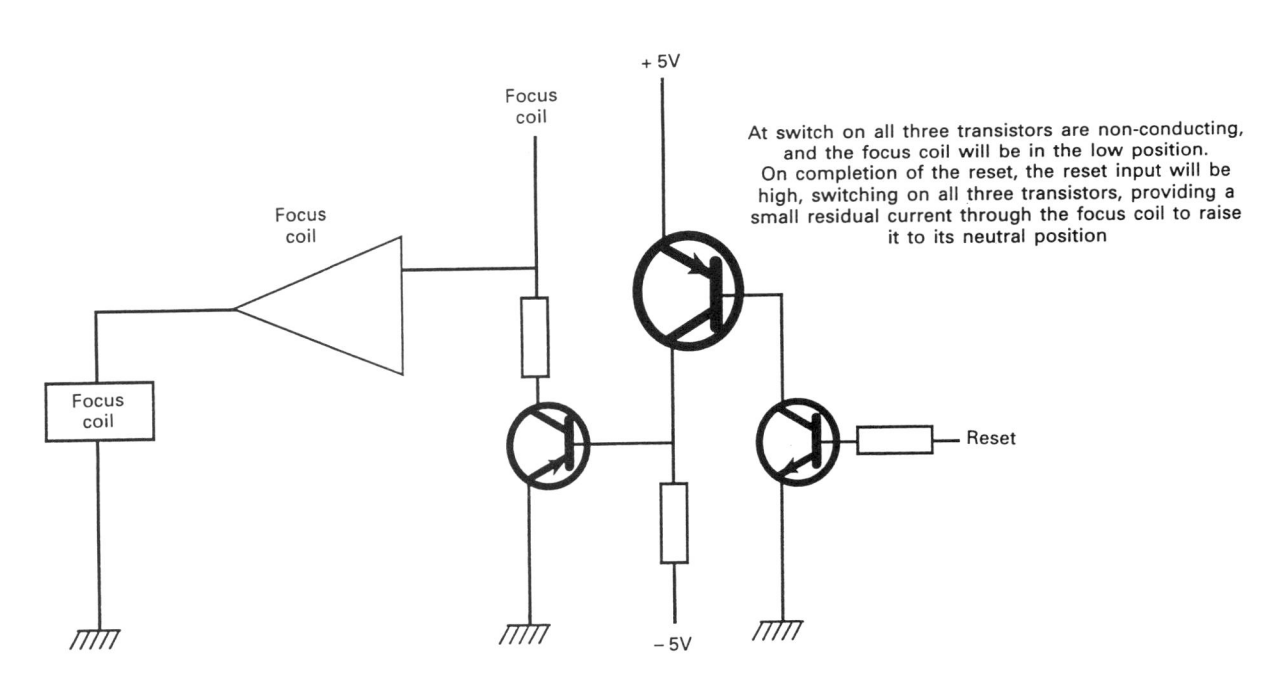

Figure 5.14 *Focus correction circuit*

Circuit descriptions

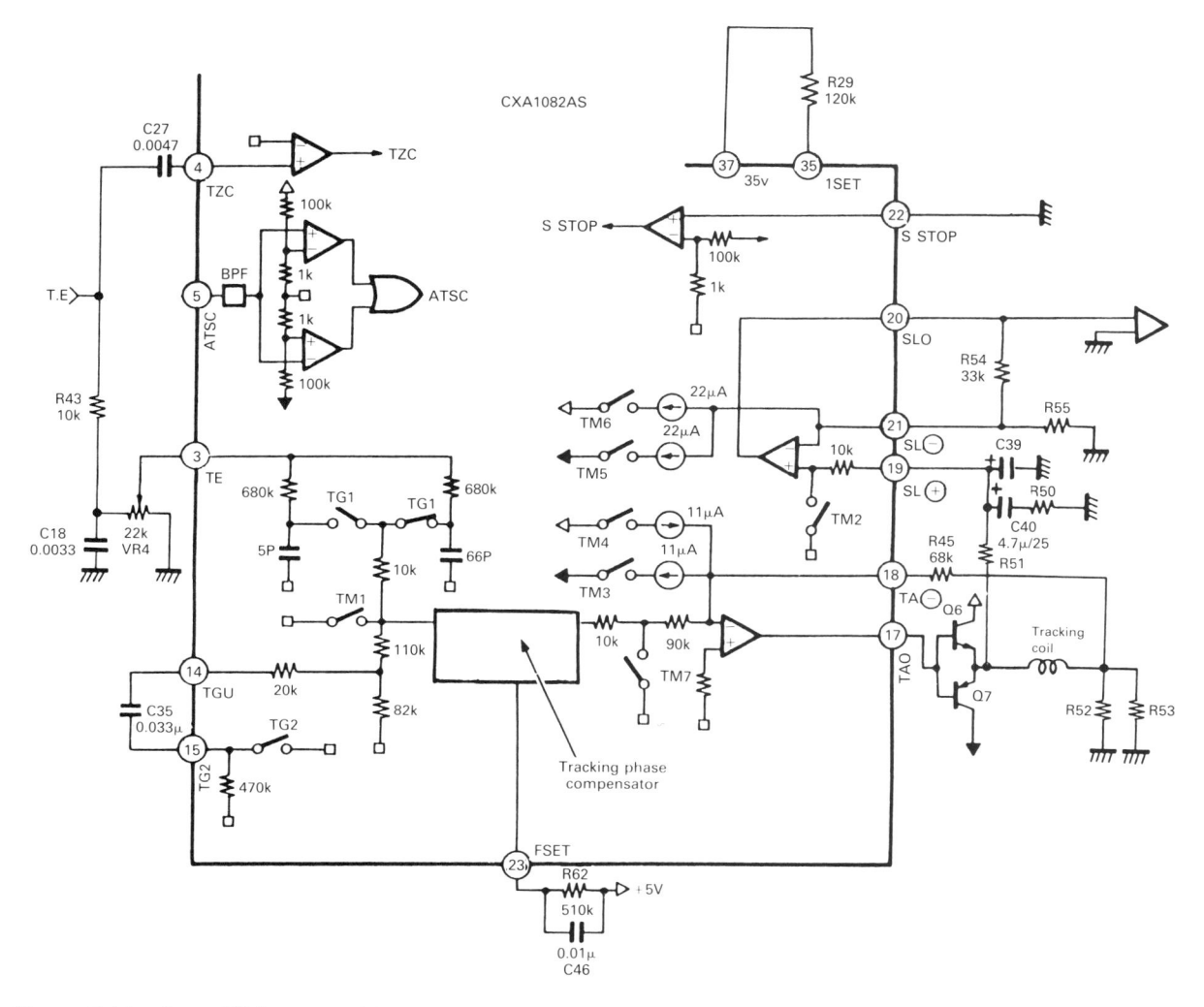

Figure 5.15 *Sony CXA1082 tracking and carriage servo circuit*

crystal source which is also applied to the decoder. It is necessary to switch on the motor at a pre-determined point, which is a function of the system control circuits. This point is usually dependent upon the manufacturer. The usual sequence is where the disc motor commences operation after the laser and focus have achieved correct operation by ensuring the laser beam is correctly focused onto the playing surface of the disc. An alternative method is to commence rotating the disc just before the laser and focus commence operation to ensure the laser beam can correctly focus onto an already rotating disc, which in some cases may be warped.

Philips disc/turntable motor servo circuit

The circuit shown in Fig. 5.17 provides a useful guide to the operation of a typical circuit used within the Philips based players. Two inputs are provided:

1 MCO – the 'motor control on' signal
2 MCES – the motor speed control signal

The motor control on (MCO) signal will be high to switch off transistor 6106 when the laser beam has achieved focus onto the disc. This is achieved when a ready (RD) signal from pin 21 of the

Circuit descriptions

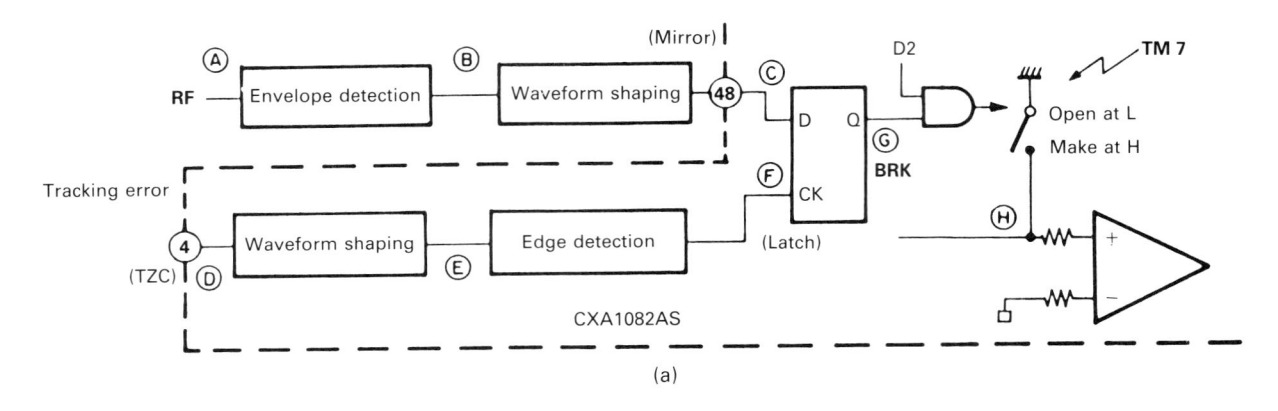

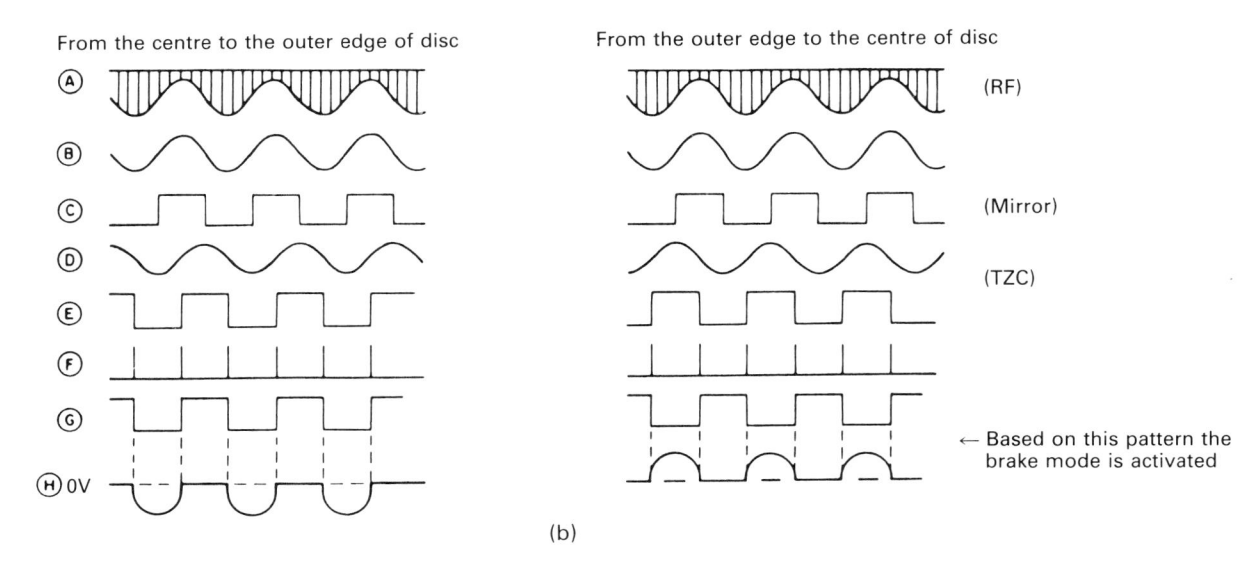

Figure 5.16 *(a) Switch TM7 brake circuit and (b) related signals*

photo-diode signal processor (TDA5708) starting circuit, which will be high when focus has been achieved, is passed to the error correction IC. (Please refer to the relevant circuit diagram.)

Switching off transistor 6106 will in turn switch off transistors 6117 and 6107, enabling the turntable motor control circuit to commence operation. The MCES signal will comprise a 50% duty cycle square wave at a frequency of 88.2 kHz when the player is first switched on or when in standby condition (as shown in Fig. 5.18(a)).

When the turntable is instructed to commence operation, the MCES signal square wave radio will alter as shown in Fig. 5.18(b), and will now have an approximate frequency of 44.7 kHz, and is in fact the disc run-up signal to achieve the correct nominal running speed, which when achieved will change over to the waveform shown in Fig. 5.18(c) during normal play operation. These square wave signals are integrated by resistor 3128 and capacitor 2125 to provide a signal which is processed by the IC6103 A and B. The output from pin 7 of 6103B is fed, via resistor 3116, as Vc, to the turntable brushless motor circuit.

An alternative more simplified circuit (Fig. 5.19) is used to drive the common and conventional small spindle motor, so common in the majority of players.

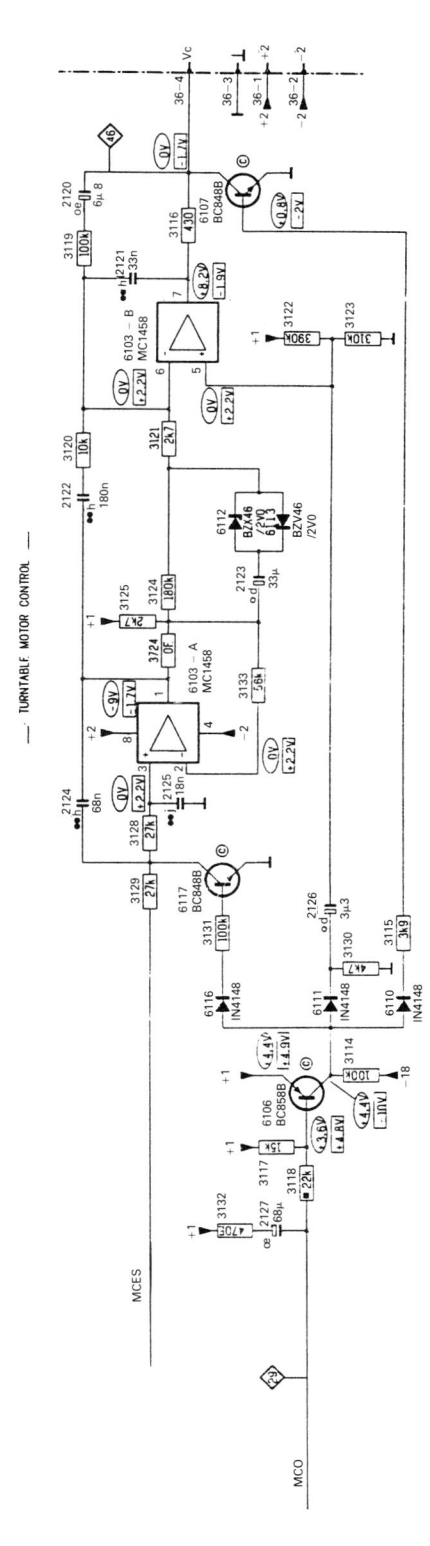

Figure 5.17 *Philips disc/turntable motor servo circuit*

Circuit descriptions

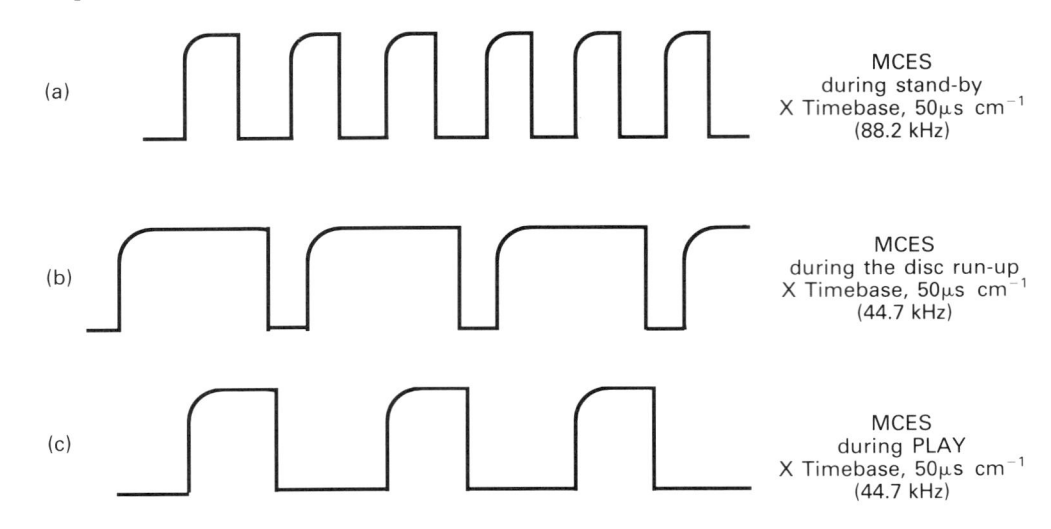

(a) MCES during stand-by
X Timebase, 50μs cm^{-1}
(88.2 kHz)

(b) MCES during the disc run-up
X Timebase, 50μs cm^{-1}
(44.7 kHz)

(c) MCES during PLAY
X Timebase, 50μs cm^{-1}
(44.7 kHz)

Figure 5.18 *Turntable motor control waveforms*

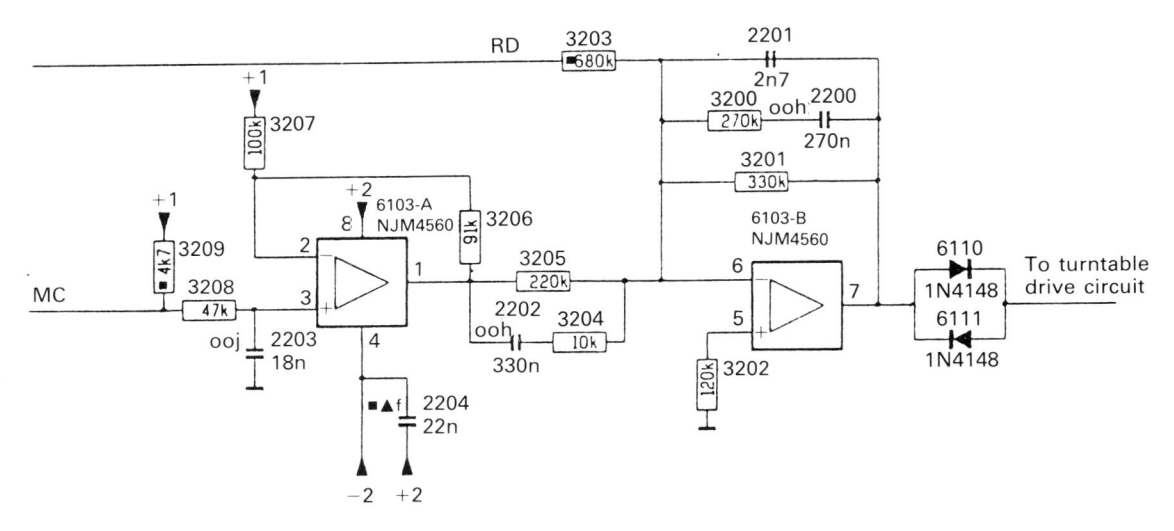

Figure 5.19 *Alternative Philips turntable motor circuit*

Alternative Philips turntable motor circuit

With this circuit again two inputs are provided, which comprise the ready signal (RD) and the motor control signal (MC), which provide an identical function to the MCES signal previously described.

The ready signal, as previously described, will be high when the laser has achieved focus on the playing surface of the disc, and this will switch on the turntable motor control circuit of IC6103B.

The motor control signal (MC) is applied from the decoder to pin 3 of 6103A, via the integrating circuit 3208 and 2303, with the final motor control output, Vc, from pin 7 of 6103B, via diodes 6110 and 6111.

Pioneer spindle motor circuit (Fig. 5.20)

Four outputs are applied from the decoder, CXD1130 or CXD1135, to the spindle motor

92

Circuit descriptions

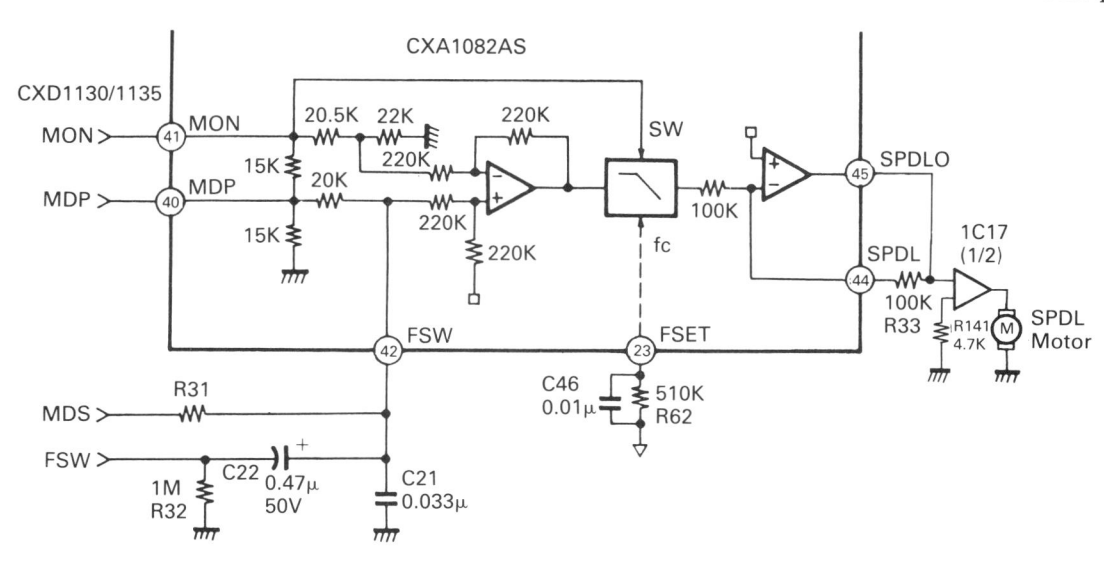

Figure 5.20 *Pioneer spindle motor circuit*

servo circuit contained within the servo IC, CXA1082. These are as follows:

1 MON: the motor on or off signal
2 MDS: the motor drive speed or coarse control signal
3 MDP: the motor drive phase or fine control signal
4 FSW: the filter switch signal, to control frequency servo response during coarse and fine operation

A basic description of the spindle motor servo is included in Chapter 2, with a signal being derived as a result of a comparison between data from the disc and a crystal source; additional relevant information is also included in the decoder circuit description, later in this chapter.

The basic conditions and waveforms that can be monitored at the MON, MDS, MDP and FSW points are illustrated in Table 5.1 and Figs 5.21 and 5.22.

During the initial run-up of the disc, after the laser and focus circuits have enabled the laser beam to become focused onto the playing surface of the disc, the disc or spindle motor is given a kick start with the application of the MON signal, which will be high, and together with the MDS signal the motor will commence to increase in speed. At this point FSW will be low to introduce C32 at pin 42 to ensure that the circuit will not respond to any high frequency components that may be present.

As the disc runs up to speed the decoder will be 'looking' for the sync. signal and comparing it with the crystal frequency of 8.4672 MHz.

Although the frame sync. signal is referred to as having 24 bits, which comprise a 1 followed by ten 0's, another 1, followed by 10 more 0's and another 1, with a 0 inserted at the ends to ensure that the sync. signal 'slots' into an frame sync. window within the decoder (usually a 23 bit shift register), it is looking for the first 23 bits of the sync. word, which will develop a waveform to equal 22 bits when the disc is running at approximately the correct (rough) speed.

When these 22 bits are compared to 22 cycles

Table 5.1 *Spindle motor servo control signals and sequences*

	Stop	Start	Play
MON Pin 41	Low	High	High
MDS R31	Low	Refer to waveform 1	Refer to waveform 2
MDP Pin 40	Low	Similar to waveform 1	Refer to waveform 3
FSW Pin 42	Low	High reduces to 3V	3V level with ripple
FSW R32/C22	Low	Low with ripple	Low with ripple

93

Circuit descriptions

Waveform 1
MDS and MDP during start-up
as disc runs up to the correct speed

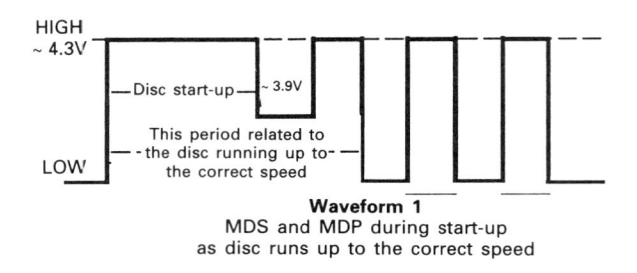

Figure 5.21 *Spindle motor servo MDS and MDP signals during disc run-up*

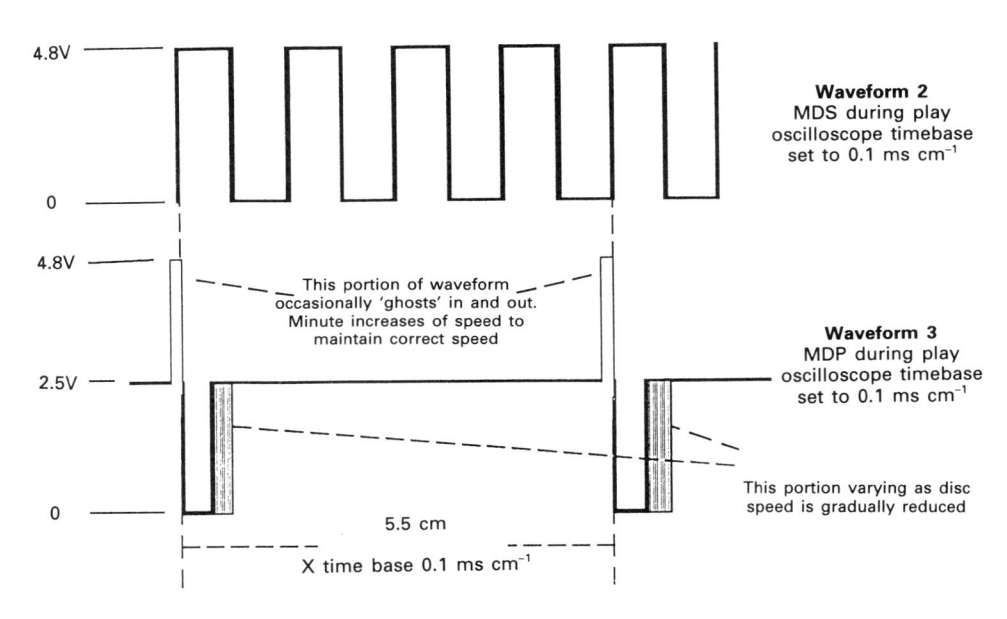

Waveform 2
MDS during play
oscilloscope timebase
set to 0.1 ms cm^{-1}

Waveform 3
MDP during play
oscilloscope timebase
set to 0.1 ms cm^{-1}

Figure 5.22 *Spindle motor servo MDS and MDP signals during play*

of the 8.4672 MHz crystal frequency which is in effect driving the decoder, the disc is considered to be running at approximately the rough or coarse speed.

At this point the tracking servo will be closed, and the MDP signal will now be the result of comparing the data from the disc, which will lock the VCO (see below) to provide an output of 4.3218 MHz, which after further frequency division, can be compared to an identical frequency which is derived from the 8.4672 MHz crystal frequency reference.

Comparison of these two frequencies in both frequency (speed/coarse control) and phase (fine control) will develop a combination of MDS and MDP signals to provide the final speed control of the spindle motor.

The FSW signal will be high when the coarse or rough speed has been achieved and the tracking servo is closed, increasing the frequency response of the filter connected to pin 42.

The resultant output at pin 45 is applied to a current driven amplifier stage to finally drive the spindle motor.

A slight variation can occur in this area where the signal from pin 45 is applied to a voltage-

94

Circuit descriptions

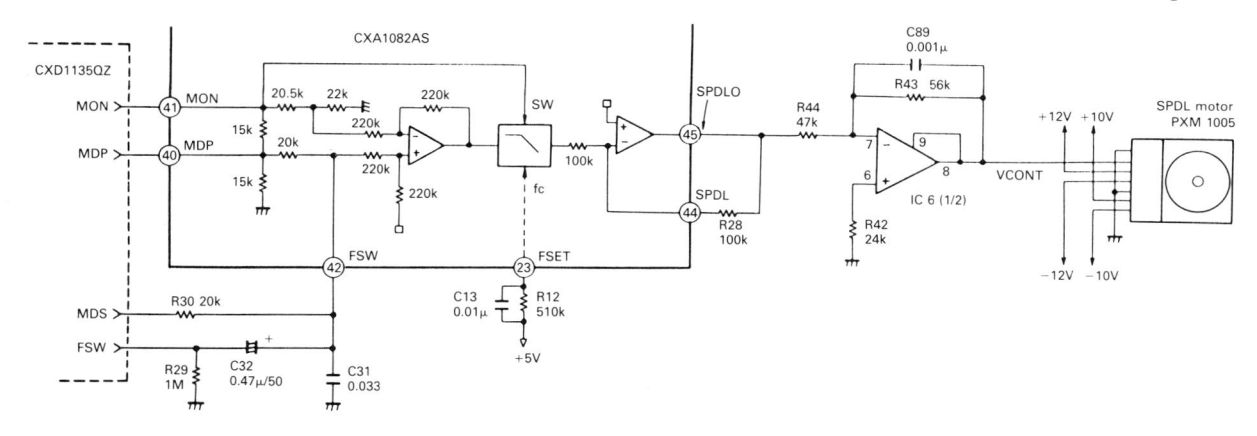

Figure 5.23 *Sony CXA1082 focus servo circuit*

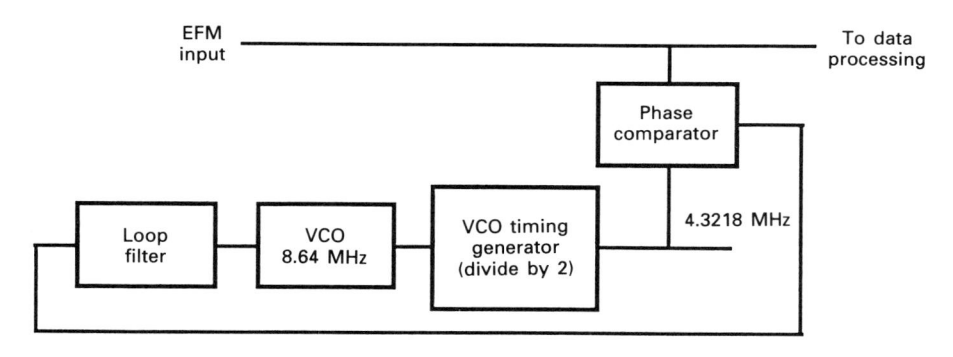

Figure 5.24 *Basic block diagram of VCO/PLL*

driven amplifier circuit to drive a brushless spindle motor (Fig. 5.23).

Voltage-controlled oscillator – phase locked loop circuit

A voltage-controlled oscillator (VCO) is necessary within the CD player to provide a disc data frequency of 4.3218 MHz for the main purpose of deriving signals for the operation of the disc or spindle motor circuits.

The EFM signal from the disc contains a fundamental data clock frequency of 4.3218 MHz, which is applied to the phase comparator (Fig. 5.24). Also applied to the phase comparator is 4.3218 MHz which is derived from the 8.64 MHz VCO after being applied to the VCO timing generator which is in effect a divide by two circuit.

The resultant output from the phase comparator is applied, via the loop filter which will remove unrequired high frequency variations and provide a basic d.c. signal, which in turn will control the VCO to provide the required resultant oscillator frequency.

VCO circuit and loop filter (Fig. 5.25)

The phase detector (PD) signal from the decoder is the phase comparator output which is applied to pin 34 as phase detector input (PDI) which passes to the internal low pass filter circuit to produce the required potential to control the VCO. The VCO frequency can be set by the

Circuit descriptions

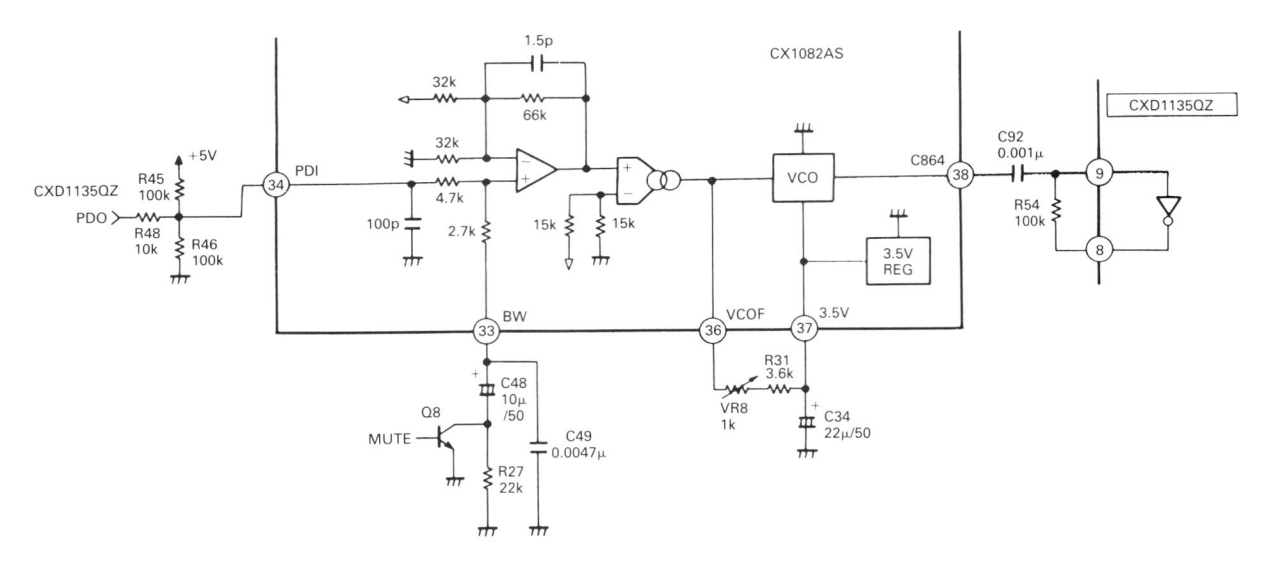

Figure 5.25 *VCO circuit and loop filter*

control VR8. The transistor Q8 is fed by a MUTE signal, which causes it to be switched on during the process of running the disc up to speed, during which time R27 will be shorted to ground, introducing C48 from pin 34 to ground, and effectively grounding the phase detector input to prevent any unnecessary variations of the VCO during the disc run-up process. When the rough speed has been achieved, Q8 will switch off, enabling the VCO to adjust almost to the correct frequency, which will be finally achieved when the disc is running at the correct speed.

An alternative circuit arrangement for the VCO in later players is included in the description for decoders CXD2500 and CXD2515 which derive the VCO internally by a frequency counting process from a crystal oscillator source.

The decoder

Sony CX23035, CXD1125 and CXD1135 decoders

These decoders, which are generally very similar, have been used in a wide range of players, and provide excellent examples of which a basic understanding can be achieved. They have a wide range of functions as follows:

1 demodulates the EFM data back into 16 bit data words
2 phase locked loop (PLL) for the VCO
3 produces the necessary bit clock from the EFM–PLL circuit
4 detect the frame sync., and sync. correction as necessary
5 de-interleaving, in conjunction with the RAM
6 error detection and correction
7 sub-code (control word) detection, and related error correction
8 spindle (disc) motor CLV servo
9 timing information for display purposes
10 determination of the table of contents
11 system control CPU interface
12 digital audio interface output

The EFM input to pin 5 (see Fig. 5.26) is applied to the ASY buffer to provide an output at pin 6 which will enable the necessary correction in relation to disc quality to be derived. Also the EFM signal is applied to the EFM–PLL stage to enable a phase detected signal (PDO) to be available at pin 11 for phase correction of the VCO, which is contained within the servo IC, CXA1082, and applied to pin 9.

The frequency of the VCO at pin 9 will be in the region of 8.64 MHz, and after passing through the buffer it is applied to the VCO circuit timing generator, to provide an output of 4.3218 MHz,

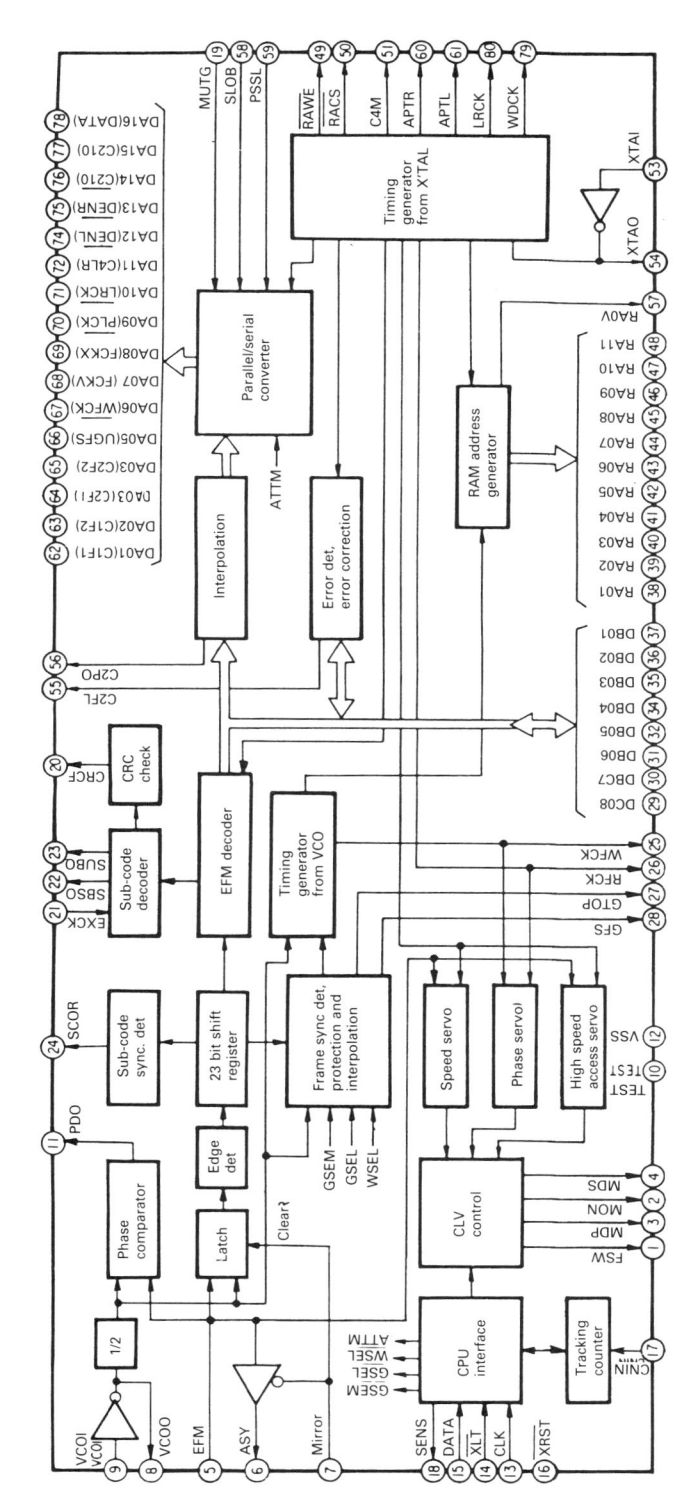

Figure 5.26 Sony CX23035 decoder

Circuit descriptions

0 1 0 0 0 0 0 0 0 0 0 0 0 1 0 0 0 0 0 0 0 0 0 1 0

Additional 0 to slot the 23 bits into the register

23 bits of the sync. signal

22 bits

Sync. signal developed from the 23 bit shift register

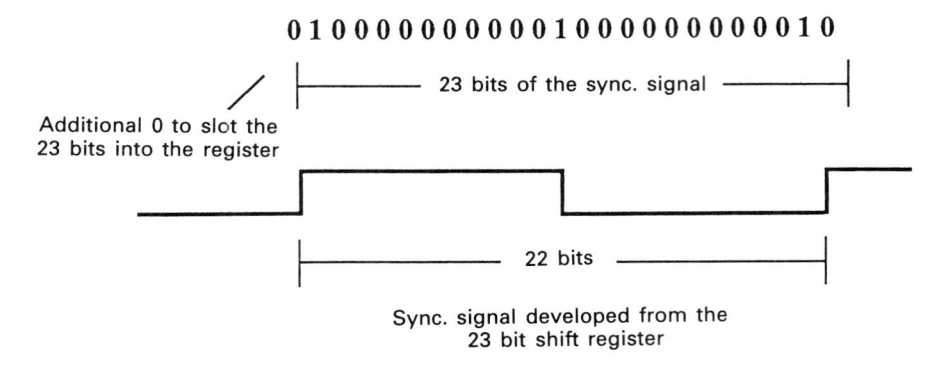

Figure 5.27 *Sync. signal*

the effective bit or data rate on the CD, which is applied to the EFM–PLL stage, as well as the CLV servo control, EFM demodulator, and frame sync. detector stages.

A crystal of 8.4672 MHz is connected to pins 53 and 54 to provide the reference oscillator for the crystal circuit timing generator, which is in effect a series of counting circuits to provide a range of outputs for the data processing circuits, which contain error correction and interpolation, RAM address control, and digital filter and data output control. Another output from the timing generator is applied to the CLV servo control to enable the necessary signals to be developed to provide the correct speed of the disc.

As the disc commences to run up to speed the EFM data from the disc will approach a bit rate of 4.3218 MHz, which will enable the VCO timing generator to also provide the same frequency as a result of the phase-locked loop action, an output of which is in turn passed to the CLV servo control.

The 4.3218 MHz input frequency to the CLV servo control is counted down to a value that is compared to a similar frequency derived from the crystal timing circuit, to produce outputs at pins 3 (MDP) and 4 (MDS). These are used to determine the coarse (speed–MDS) and fine (phase–MDP) control signals to control the disc (spindle) motor speed.

A 'kick start' for the motor is derived from the CLV servo control via the CPU interface, to enable MON, MDS and MDP to go high and commence the motor operation, as described above for the Pioneer spindle motor circuit. When the motor is running at the correct speed the EFM data from the EFM–PLL stage are applied to the 23 bit shift register which is in effect looking for the sync. signal from the disc (Fig. 5.27).

The sync. word or signal actually comprises 24 bits; 23 bits are required for the shift register, and the extra 0 is required to actually slot the sync. signal into the register, to provide a square wave output, which is passed to the frame sync. detector and then to the phase servo via the VCO timing generator. The 8.4672 MHz crystal frequency is fed to the speed (coarse) and phase (fine) servo via the crystal timing generator.

When this square wave has reached a length equivalent to 22 pulses of the 8.4672 MHz crystal frequency the disc will be running at almost the correct speed, i.e. the coarse or rough speed, and when a certain number of frame sync. square waves have been recognised by the frame sync. detector, a guarded frame sync. (GFS) output is passed via pin 28 to the system control, which feeds data back via the CPU interface to switch the CLV control from the speed servo to the phase servo, which will now control the speed at the correct rate. With the disc rotating at the correct speed, 14 bit data words, or symbols from the disc, are now passed to the EFM decoder, to restore them back to their original 8 bit form.

The first 14 bit word after the sync. signal is the sub-code information which is routed up to the sub-code detector, and the following 8 bit data words from the EFM decoder are now formed in pairs to make 16 bit data words which are now processed via the error correction, external RAM and interpolation circuits to provide the serial 16 bit data word output from pin 78 to be passed to the D to A conversion circuits.

Circuit descriptions

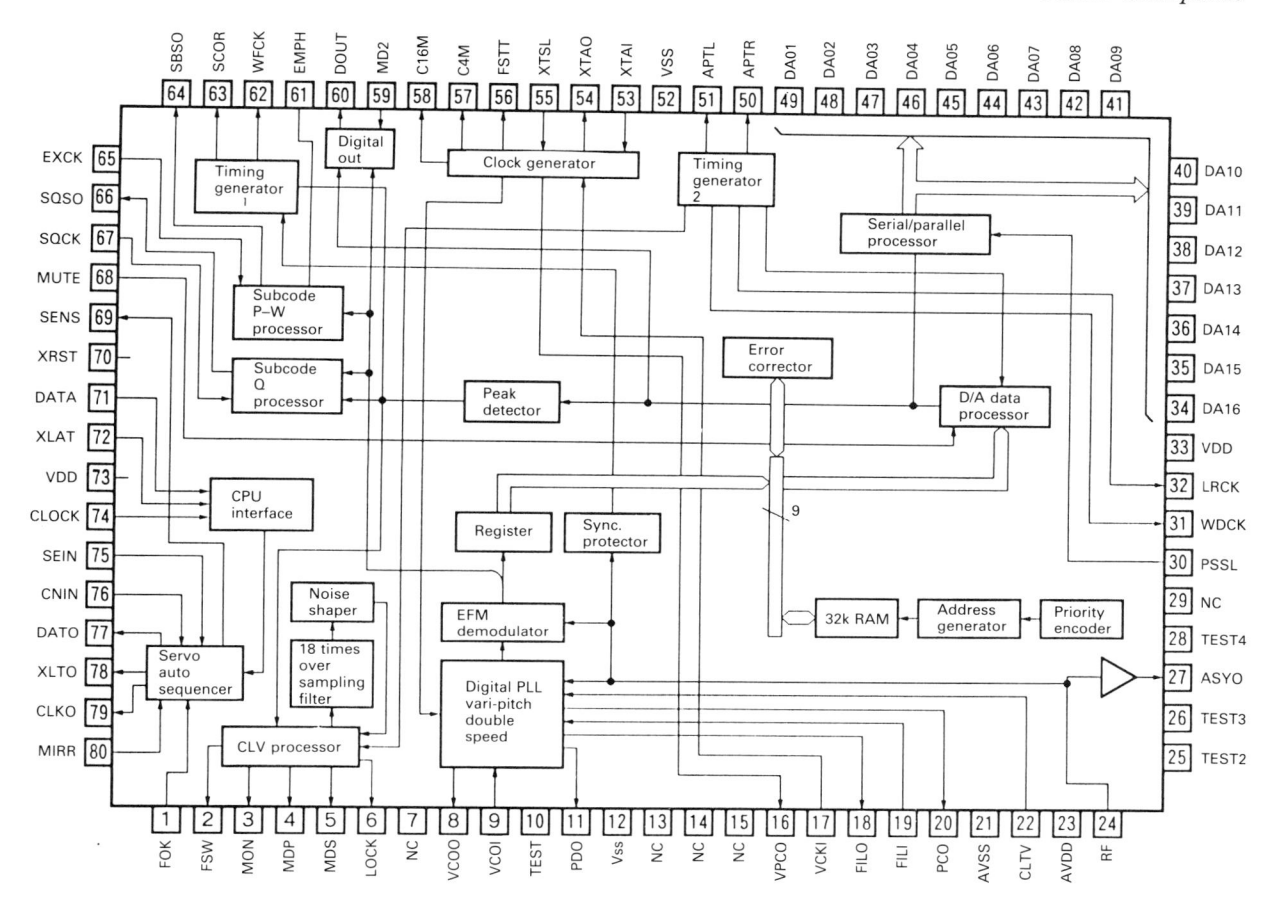

Figure 5.28 *Sony CXD2500Q decoder*

Sony CXD2500Q decoder

The general operation of the Sony CXD2500Q decoder (Fig. 5.28) is very similar to that of the types previously described, with three main differences: the RAM is now built internally into the decoder; this decoder is capable of variable disc speed operation; and the VCO frequency is derived from the main crystal, instead of a conventional PLL oscillator.

Variable speed or pitch operation

Some manufacturers provide a high speed dubbing of disc to tape facility (Fig. 5.29), whilst some also provide a variable pitch facility which can be used for Karaoke purposes.

Hi-speed dubbing

CD players that have a hi-speed dub facility (Fig. 5.29) are normally part of a 'stack' or 'midi' system, together with a tape deck facility. When this feature is selected, the disc and tape will operate at twice the normal speed. Providing all CD player processing circuits are linked to the main reference crystal, usually of 16.9334 MHz, then the crystal timing circuits are altered to provide twice the normal operating frequencies, which will enable the disc to rotate at twice the normal speed, as well as all the relevant digital processing circuits.

The two HCCRS signals from the system controller (one is high, the other is low) will operate the relevant circuits.

It is, however, necessary to alter the gain of both the RF amplifier and CLV stages to provide

99

Circuit descriptions

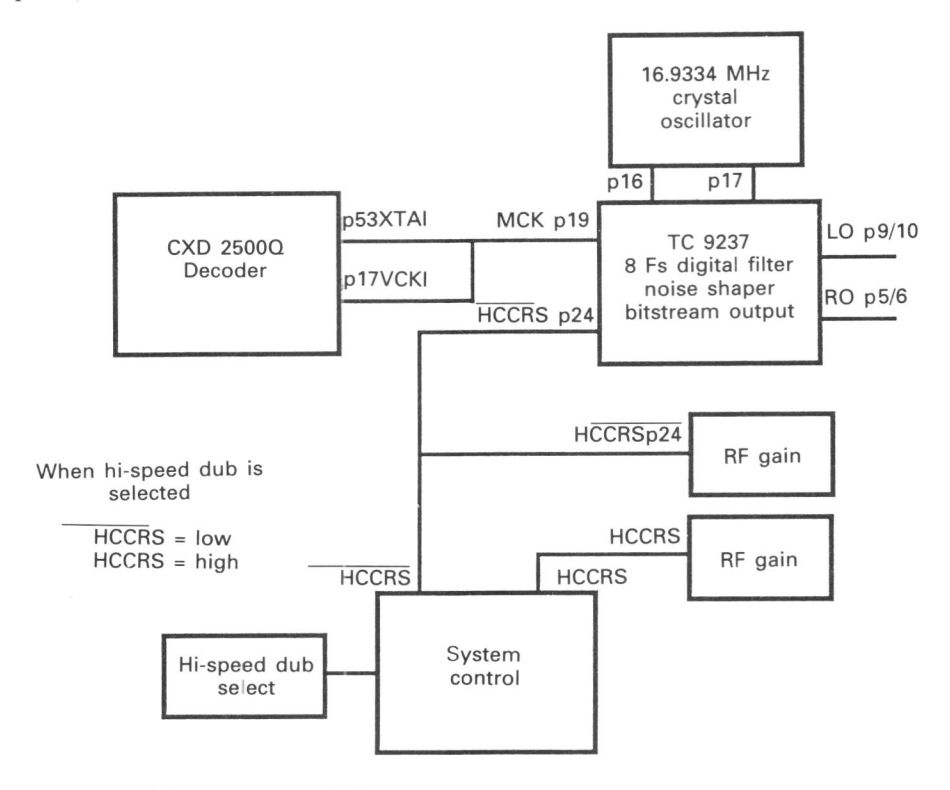

Figure 5.29 *High speed dubbing basic block diagram*

the required operating characteristics when the disc is functioning at twice the normal speed.

Variable pitch operation

As shown in Fig. 5.30, an output from pin 16 of the decoder, VCPO, is part of a PLL circuit, and is in fact the phase detector signal which is the result of the VCO signal, and is centred on 16.9344 MHz, but variable between 13.26 and 19.78 MHz, being compared with the crystal frequency of 16.9344 MHz. Within the decoder the resultant frequency modifies the crystal timing circuits to provide slightly different frequencies at which the disc speed and data processing circuits function, thereby providing a variation in the pitch of the output signal.

Digital VCO operation

As with previous players employing the earlier decoders, with the CXD2500Q decoder there is no facility for adjusting the VCO. The circuit illustrated in Fig. 5.31 shows an internal VCO which is operating at 34.5744 MHz (8 × 4.3218 MHz). A crystal oscillator input of 16.9344 MHz is divided by 48 to provide a frequency of 352.8 kHz, and the same frequency is obtained by dividing 34.5744 MHz by 98. The resultant output from the phase comparator at pin 20 is passed via the filter circuit to pin 22 to provide accurate control of the VCO. The main reason for using this method is that the VCO can also be used in the variable pitch function of some of the later players.

Sony CXD2515Q decoder

The Sony CXD2515Q decoder (Fig. 5.32) is one of the latest decoder ICs to appear in the more recent CD players, and is an indication of component density, with this IC being a 100 pin quad

Circuit descriptions

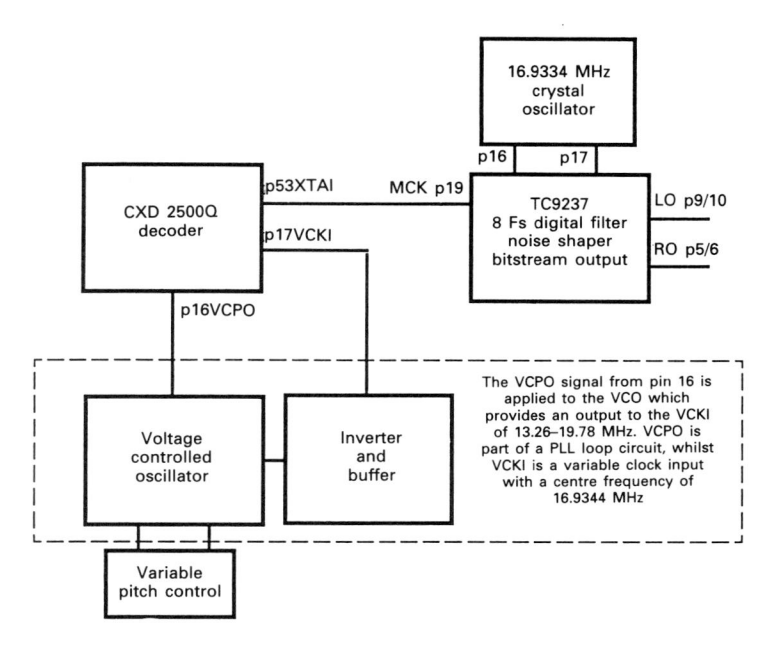

Figure 5.30 *Variable pitch basic block diagram*

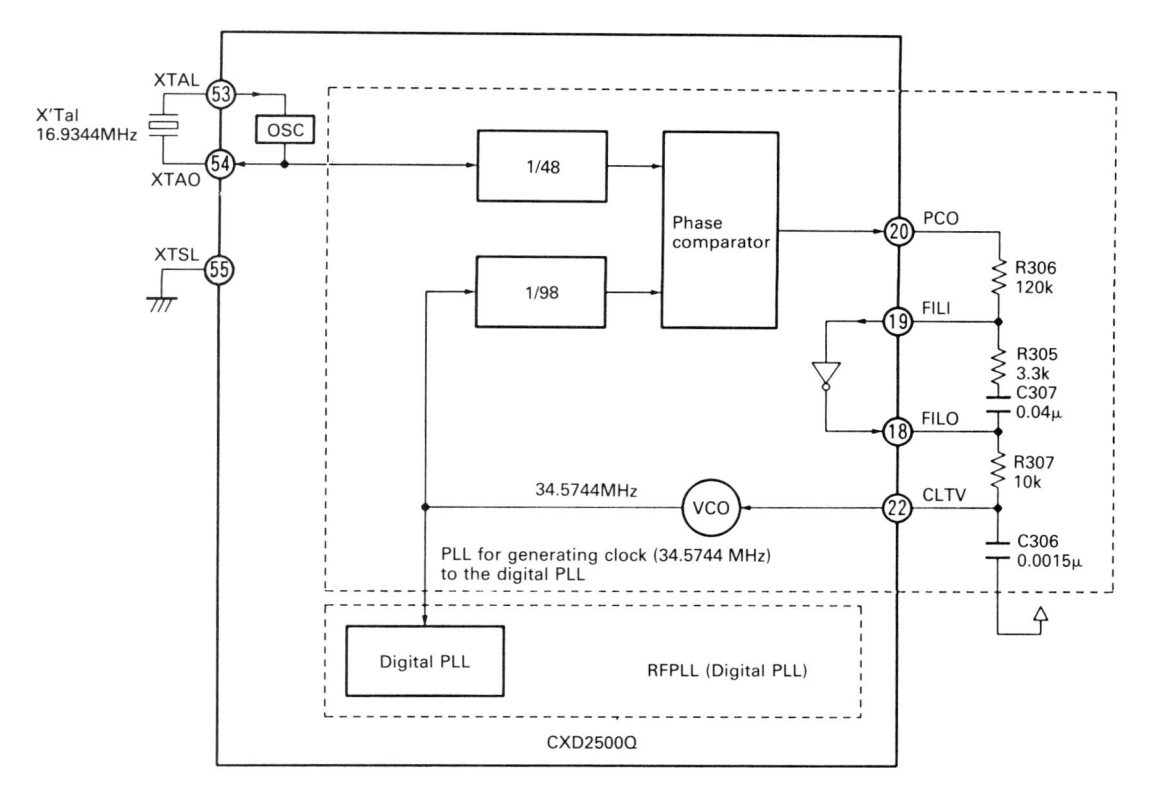

Figure 5.31 *Digital VCO operation*

101

Circuit descriptions

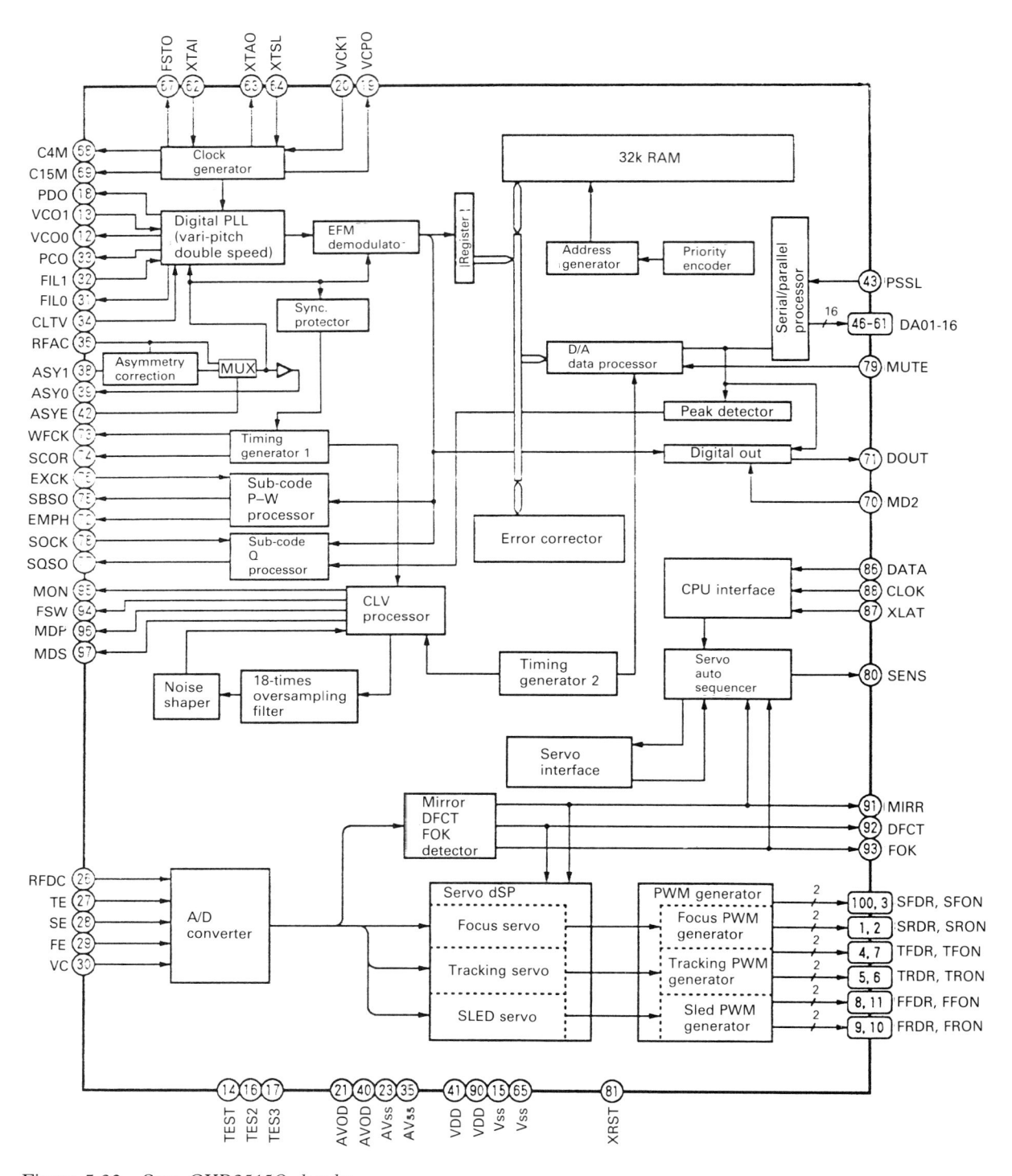

Figure 5.32 Sony CXD2515Q decoder

Circuit descriptions

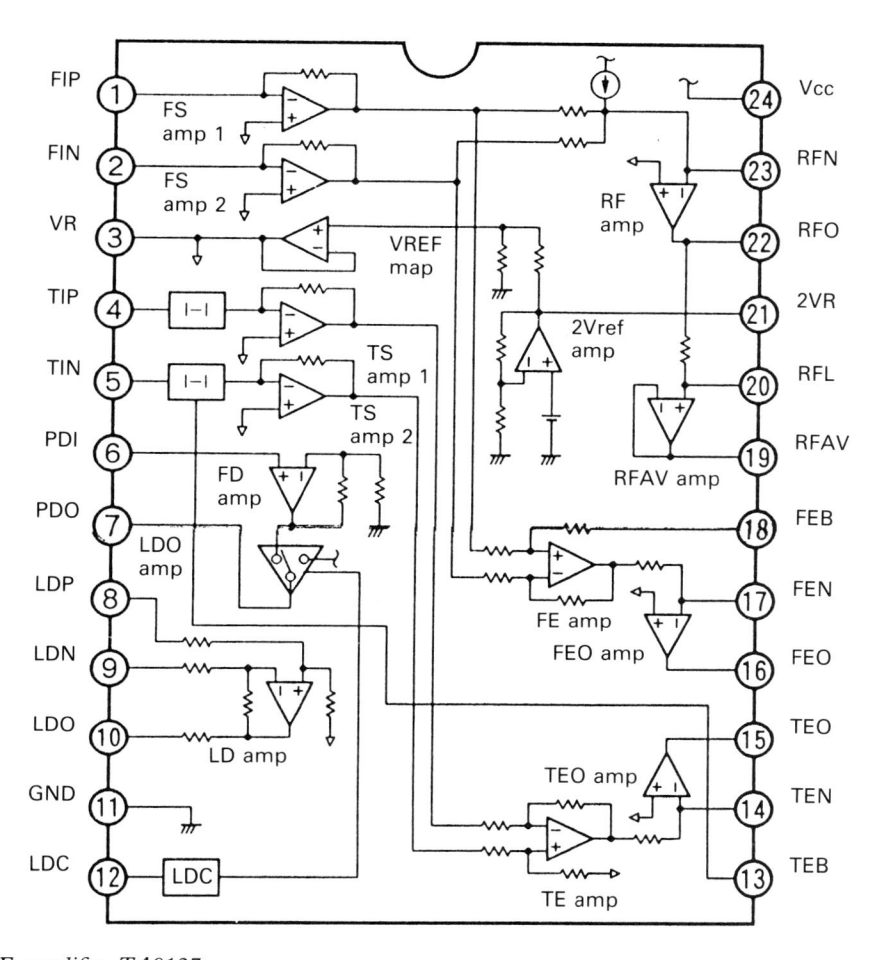

Figure 5.33 *RF amplifier TA8137*

package. The actual size is virtually the same as that of previous types that appear on the printed circuit board. Included in this package are all the previous functions, plus the digital filter circuits, as well as the servo processing section.

The servo section contains automatic adjustment of all relevant servo correction requirements. The players that include this IC will have virtually no adjustments, with the possible exception of lateral/radial adjustments as and when a new optical block is fitted.

The various circuit sections that have been described in this chapter are intended to help engineers gather an appreciation of other types of ICs that perform similar functions, a selection of which have been included in Figs 5.33–5.39.

Circuit descriptions

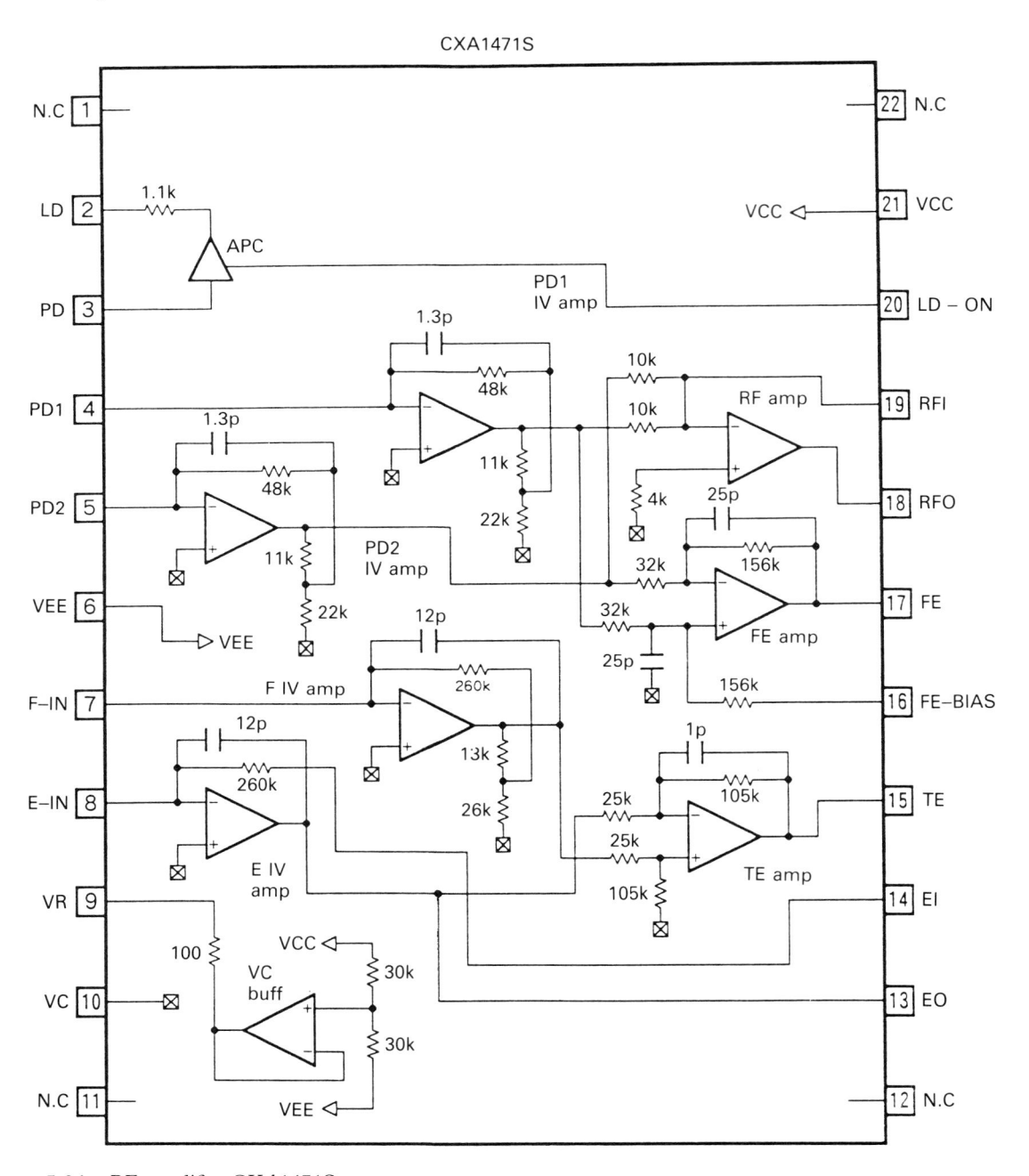

Figure 5.34 *RF amplifier CXA1471S*

Circuit descriptions

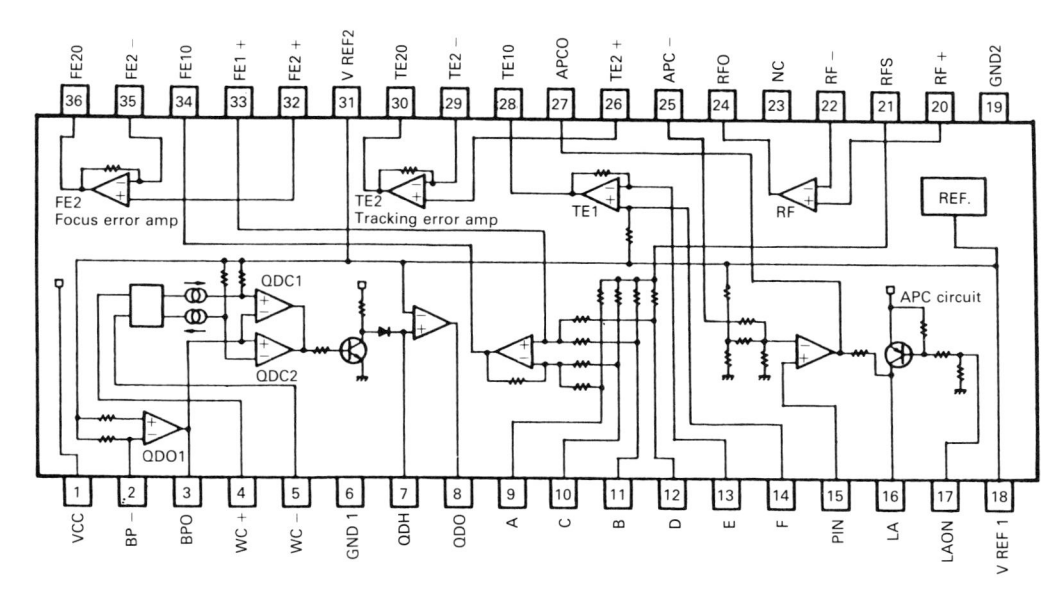

Figure 5.35 *RF amplifier UPC1347GS*

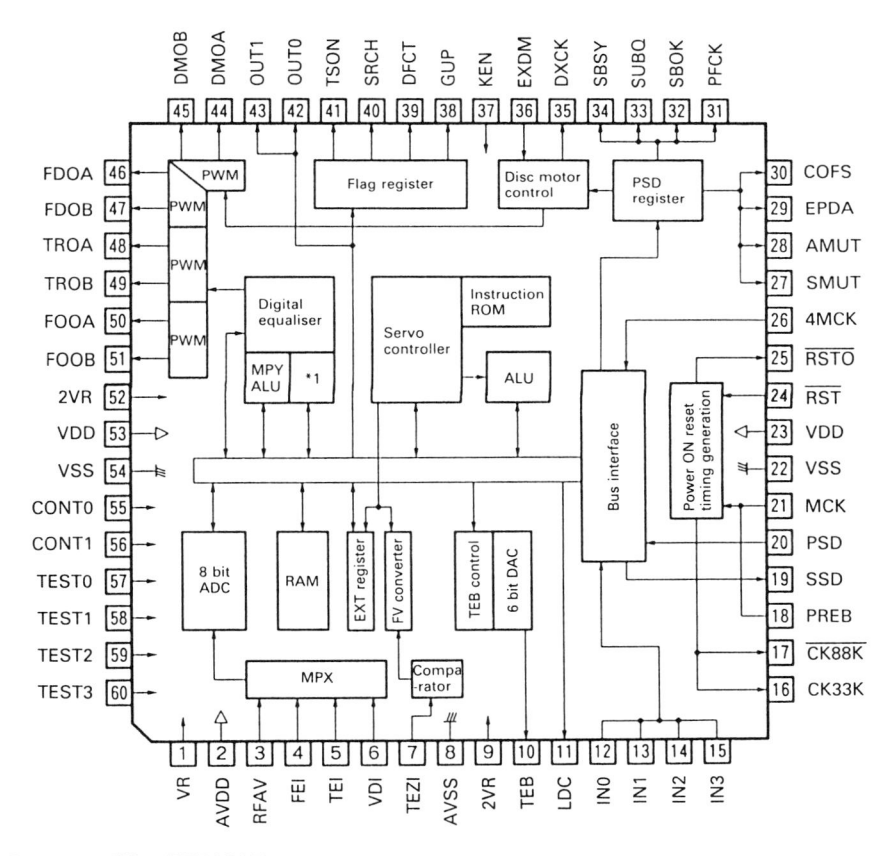

Figure 5.36 *Servo amplifier TC9220F*

Circuit descriptions

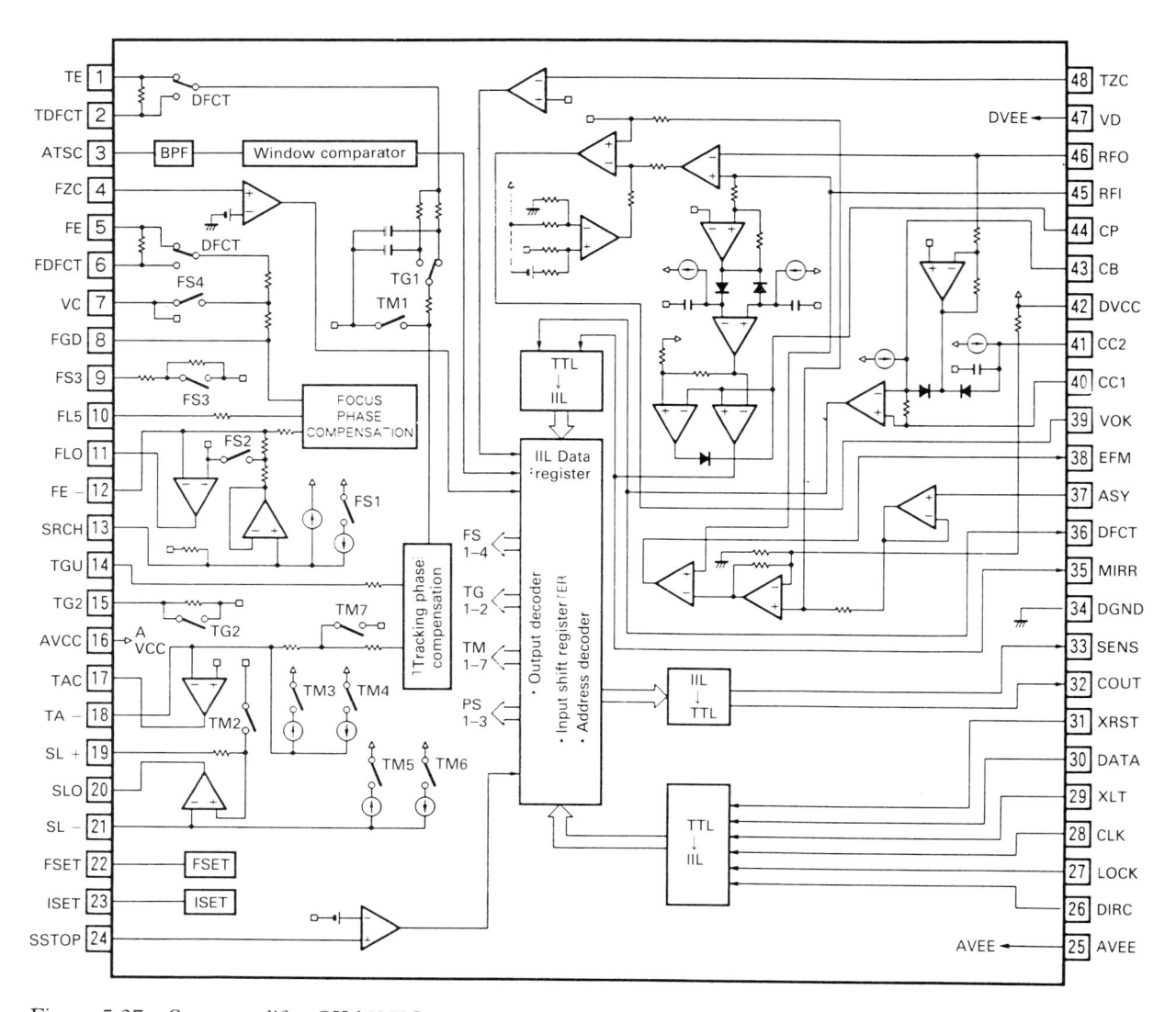

Figure 5.37 *Servo amplifier CXA1372S*

Circuit descriptions

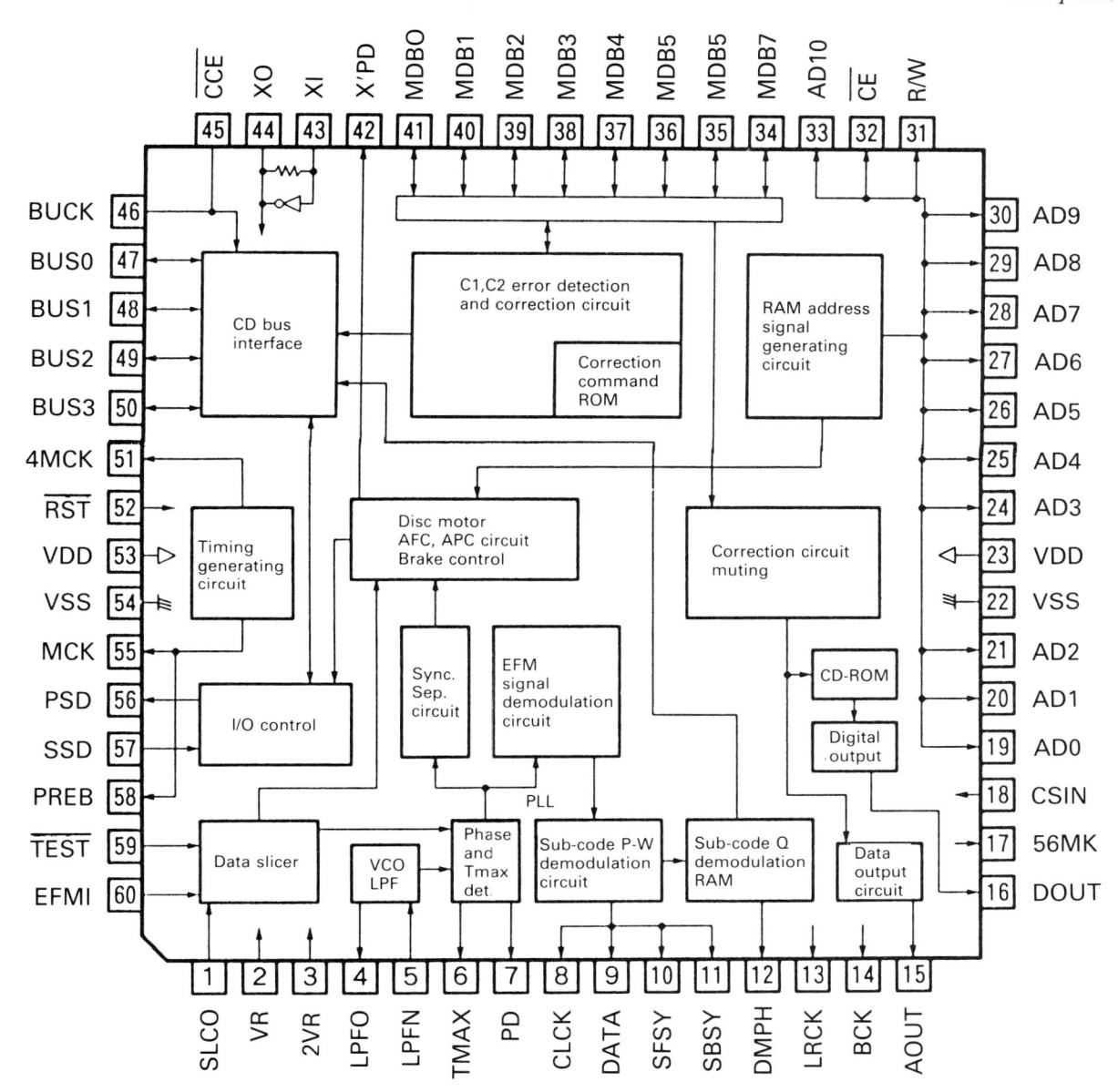

Figure 5.38 *Decoder TC9221F*

Circuit descriptions

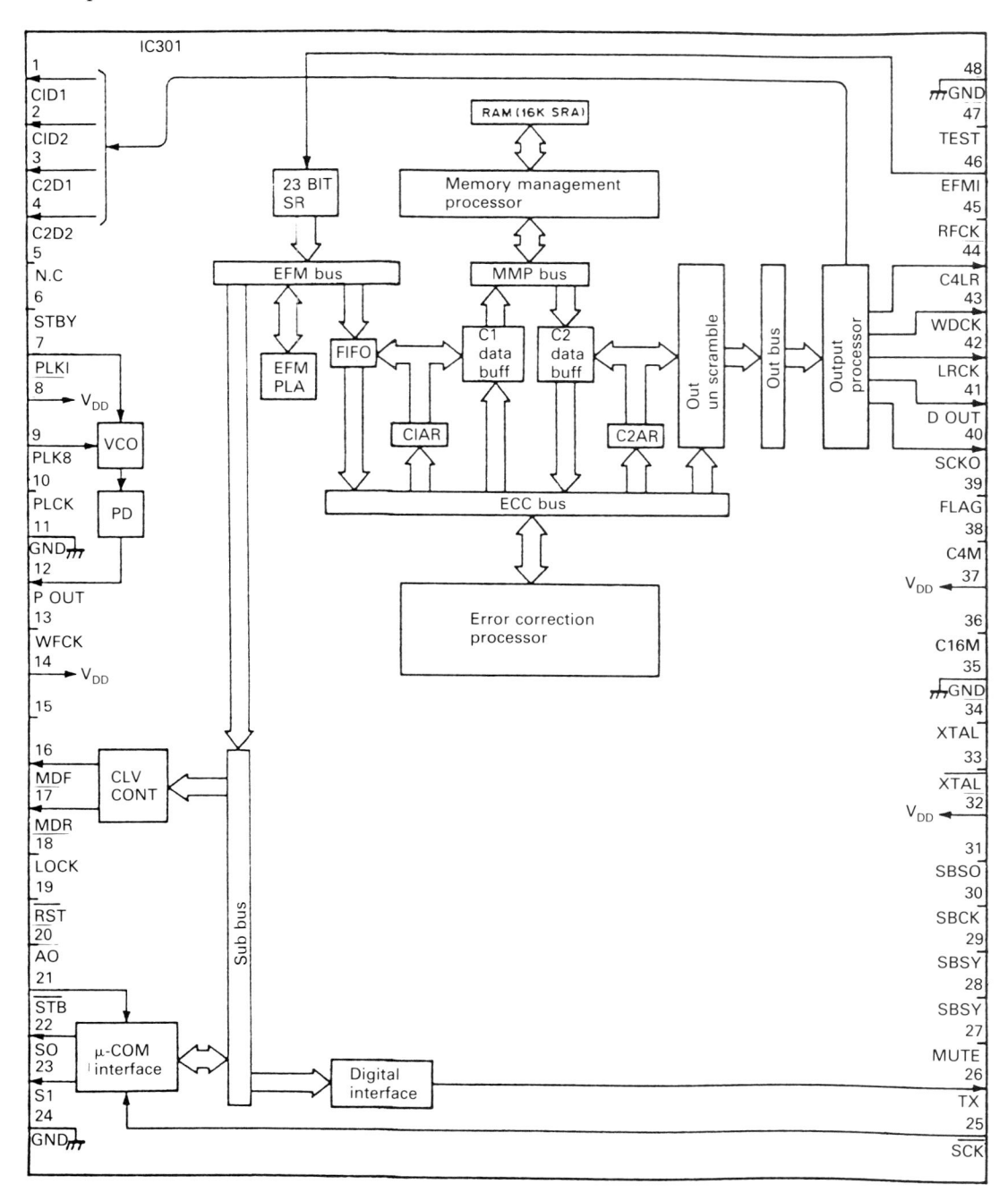

Figure 5.39   *Decoder UPD6375CU*

6

TEST MODE AND ADJUSTMENT PROCEDURES

The majority of CD players require a series of mechanical and electrical adjustments to ensure optimum operation of the player.

Mechanical adjustments are usually related to the optical assembly whilst the electrical adjustments will be either for the laser power, the servo systems or the VCO.

Any mechanical adjustments will normally only be required to be undertaken when an optical assembly is replaced. With some players these adjustments are either none or minimal, whilst with others there can be quite an involved procedure.

The electrical adjustments will again vary in how many there are and, like the mechanical adjustments, their complexity will usually relate to the type of CD player as well as the manufacturer.

To be able to achieve the correct completion of any adjustment procedures a practical range of tools and certain essential items of test equipment are necessary.

Tools

Whilst it is not the intention to teach granny to suck eggs, and the majority of engineers will be in possession of a comprehensive range of tools, it is worth mentioning selected items that have been found necessary, based upon experience:

1   A good selection of flat bladed, Philips, Posidriv and Torx drivers, especially the small to middle range of sizes. In particular, a fine flat blade type with a 2.5 mm blade width and an insulated handle, preferably not an instrument screwdriver, for the grating adjustment of many Pioneer optical assemblies.
2   Allen keys or drivers, especially the smaller sizes. Again with many Pioneer players a 1.5 mm ball type of Allen driver with an insulated handle is extremely useful when adjusting the tangential setting on certain models, but more about that later. This particular item is available from RS Components, Part No. 662–670.
3   The usual range of fine pointed pliers, snippers and tweezers, of course, always have their place amongst any engineer's collection.
4   A selection of specialist tools, such as grating drivers and height gauges as recommended by manufacturers in their relevant service manuals, will be necessary in order to achieve success with some specific mechanical adjustments.
5   Test CDs, of which many types are available. Reference to the various service manuals will indicate some of the types to be recommended. These will include the Sony YEDS7, YEDS18 or YEDS40, as well as Philips' and Technics' own test discs. A favourite of the author is the Pierre Verany Digital Test (Ref. No. 788031) – a twin disc set of French origin.

Recommended test discs at reasonable prices do appear from time to time in various hi-fi magazines, and will no doubt be useful as test discs, but there is an important factor to consider when obtaining a disc that is to be used when checking or setting up certain areas of a CD player; this is related to laser power and the amplitude of the RF eye pattern waveform. Whilst conventional music discs fulfil the majority of requirements for the engineer to effectively service a CD player, the quality of the reflectivity of the disc, though not necessarily apparent to the naked eye, can vary from disc to disc, and whilst many engineers' favourite 'test discs' may well suffice for the majority of servicing requirements, it is essential to have a known acceptable standard from which laser power and RF eye pattern can be determined. More on this subject will be given under the areas on laser power and RF eye pattern amplitude.

Test mode and adjustment procedures

Other than recommended test discs, and the engineer's favourite discs, whether scratched or otherwise, another type of disc is an 8 mm disc with a maximum playing time of 20 min, which may take some searching for as they do not seem to be readily available, but they do prove extremely useful with respect to certain types of player.

Test equipment

A minimum range of test equipment is absolutely necessary to effectively service CD players, and most of it will be available in the majority of service workshops.

Oscilloscope

The oscilloscope is the most important (and the author's favourite) item of test equipment when it comes to CD player servicing. A twin beam with a minimum bandwidth of 20 MHz is recommended, though some service manuals suggest a minimum of 50 MHz. A pair of ×10 probes will also prove necessary to minimise unnecessary 'loading' of frequency conscious circuits.

Multimeter

A general purpose digital or analogue type of multimeter will be adequately effective, though the author, now becoming somewhat aged in years, has a preference for the ancient analogue type.

Laser power meter

A laser power meter can prove to be an unnecessary luxury as laser power can usually be determined by the use of a recommended test disc and obtaining a specific RF eye pattern amplitude. With some players measurement of the laser diode current can indicate whether the laser power is within specified limits.

On occasions when the service manual recommends that the laser power be measured with a laser power meter, final adjustment is usually achieved with the RF eye pattern amplitude.

With some players it is not possible or even recommended to adjust the laser power.

Test jigs and filters

Some service manuals recommend specialist jigs and filters of which the jigs are generally used for the purpose of servo gain adjustment, and the filters to minimise noise when monitoring certain waveforms.

As the jigs, wherever recommended, will generally differ from one CD player manufacturer to another, many engineers have frequently determined methods of successfully completing adjustments without jigs, due in the main to their cost as well as their limited availability.

Concerning filters, the author has found that monitoring relevant waveforms without filters has not impeded the required result from the related adjustment.

Further comments will be made concerning these items as suggested procedures are outlined for relevant adjustments.

Test or service modes

Before proceeding with any adjustments, it may well prove necessary as well as useful to determine whether a specific CD player has some form of test mode, which will no doubt assist engineers in servicing these units. Of course, reference should be made to the relevant service manuals.

The test or service mode can also be extremely useful when it is necessary to carry out fault diagnosis on CD players.

To highlight the facilities available with the test or service modes, the methods adopted by Pioneer and Philips will be outlined, but there are other manufacturers that have either a limited facility or have introduced more extensive facilities in recent years for which further information can be obtained in relevant service manuals.

It is, however, possible to achieve reasonable success in many of the adjustments with CD players that do not incorporate any form of test mode facility.

Pioneer test mode procedures (Fig. 6.1)

With Pioneer CD players the test mode is selected by either depressing a special non-latching test button, or shorting two links marked 'test,' either of which will be found on the main PCB, and applying the mains supply.

As with many products the mains supply will either be controlled via a mains on/off switch, or be applied directly to the mains transformer, on/off control being determined by a standby switch in conjunction with the system control. Therefore the two options to achieve the test mode will be as follows.

1 Players with mains on/off switch

a. depress test switch or short test links
b. switch on mains supply switch
c. count to three to ensure test mode is achieved
d. release test switch or short to the test links
e. the front panel display should have minimal illumination

If the illumination appears normal, the player is probably not in the test mode: repeat the above sequence.

2 Players with permanent supply to the mains transformer

a. depress test switch or short test links
b. apply the mains supply
c. count to three, to ensure test mode is achieved
d. release test switch or short to the test links
e. switch on the standby or on switch
f. the front panel display should have minimal illumination

If the illumination appears normal, the player is probably not in the test mode: repeat the above sequence.

Philips service mode procedures

To achieve the service mode, the methods for three particular models will be outlined:

— CD150. Simultaneously depress the following keys: PREVIOUS, NEXT and TIME/TRACK.
— CD582. Simultaneously depress the following keys: SEARCH, PAUSE and REPEAT.
— CD840. Simultaneously depress the following keys: NEXT and PLAY.

Now switch on the mains supply, and count to three before releasing the three keys, to ensure that service mode is achieved. This is the standby mode, and '0' will appear on the display. In this condition it should be possible to move the radial arm from one extreme to the other by depressing the 'Search FWD' and 'Search REV' keys.

Further stages in the service mode can be achieved by subsequent depressions of the NEXT key.

Service position 1

With no disc inserted

Depress the NEXT key. The laser will switch on, and the focus lens will move vertically 16 times before stopping; the player will revert to the standby mode.

With a disc inserted

Depress the NEXT key. The laser will switch on, and the focus lens will achieve focus. When this has been achieved, '1' will appear on the display.

Service position 2

Depress the NEXT key. The turntable will commence rotating, and '2' will appear on the display.

Service position 3

Depress the NEXT key. The radial control servo is switched on, and '3' will appear on the display. Providing the radial arm is within the music area of the disc, sound output should be achieved. The service mode is cancelled when the player is switched off.

Adjustment procedures

The range of adjustments, electrical and mechanical, that are necessary can vary between ranges of models as well between different manufacturers. Whilst each method of adjustment will be described, it should be emphasised that not all the listed adjustments will appear at the same time in one player – much to the relief of many engineers, no doubt.

Unfortunately with CD players a range of preset controls for adjustment purposes often provide a 'tweaker's paradise', and random adjustment can often provide further problems, some of which may appear quite catastrophic.

Test mode and adjustment procedures

Electrical adjustments

There are a wide range of adjustments, and when they are carried out, those that are relevant to a particular player should be implemented in the order listed below, unless otherwise recommended, in order that optimum results may be achieved:

1. focus offset
2. tracking offset
3. RF offset
4. laser power
5. tracking balance
6. focus balance
7. focus bias
8. photo-diode balance
9. RF level
10. focus gain
11. tracking gain
12. VCO frequency

Mechanical adjustments

The following range of adjustments will not necessarily appear in all players, but again there is a recommended sequence with respect to particular players.

1. turntable height
2. tangential adjustment
3. lateral adjustment
4. diffraction grating adjustment

Methods of electrical adjustment

If a CD player has been incorrectly adjusted at some stage, frequently during fault diagnosis, it is often worthwhile to look at the settings of the electrical adjustments. Generally the majority of these adjustments will be near to the centre of the range of the adjustment setting when the player is operating normally; if any of the presets are virtually at one extreme then the possibility exists that there is an additional 'man-made' problem in addition to the original fault. With the exception of the laser power and also the VCO presets, all other presets can usually be set to their mid-point before the adjustment procedure is commenced.

Essential oscilloscope checks

Before commencing any electrical adjustment or measurement with an oscilloscope, it is essential that the calibration of the oscilloscope (especially the 10:1 probe) is checked and adjusted as necessary.

Always allow for the 10:1 probe when measuring levels such as the RF eye pattern, and check that if the Y amplifier is set to 50 mV in reality, 500 mV will be the value of the actual measurement at the end of the 10:1 probe.

Also, and again never wishing to 'teach granny to suck eggs' when measuring offset potentials, always ensure that the input selector switch is set to d.c. Though this may seem obvious, it has always proved interesting on the many technical training courses attended by the author, how many oscilloscopes are set to a.c., where of course if the trace is set to zero, any adjustment to the offset control *always* returns to zero after any movement of the preset!!!

One final point is worth mentioning: if measurements of waveforms are being monitored with the input selector set to a.c., ensure that the internal input capacitor of the oscilloscope is discharged, by just grounding the tip of the probe. Otherwise any voltage charge in the capacitor from one measurement can be carried across to a voltage sensitive area, and a few laser beams have been known to 'disappear' through a nasty short sharp shock from the end of a 'scope probe.

Focus offset adjustment

As with all servo systems, it is necessary to minimise undesirable d.c. potentials within the system to ensure that it operates with optimum efficiency (Fig. 6.2). If this adjustment is not correctly observed then problems such as the inability to focus, skipping or jumping can occur.

The oscilloscope is connected via a 10:1 probe to the focus error (FO ER) test point, or equivalent, and the focus offset adjusted to neutralise any voltage or set to the manufacturer's specifications.

The player does not necessarily have to be in the test or service mode and, unless specified, a disc is not inserted.

Examples of some different manufacturers' procedures are detailed below.

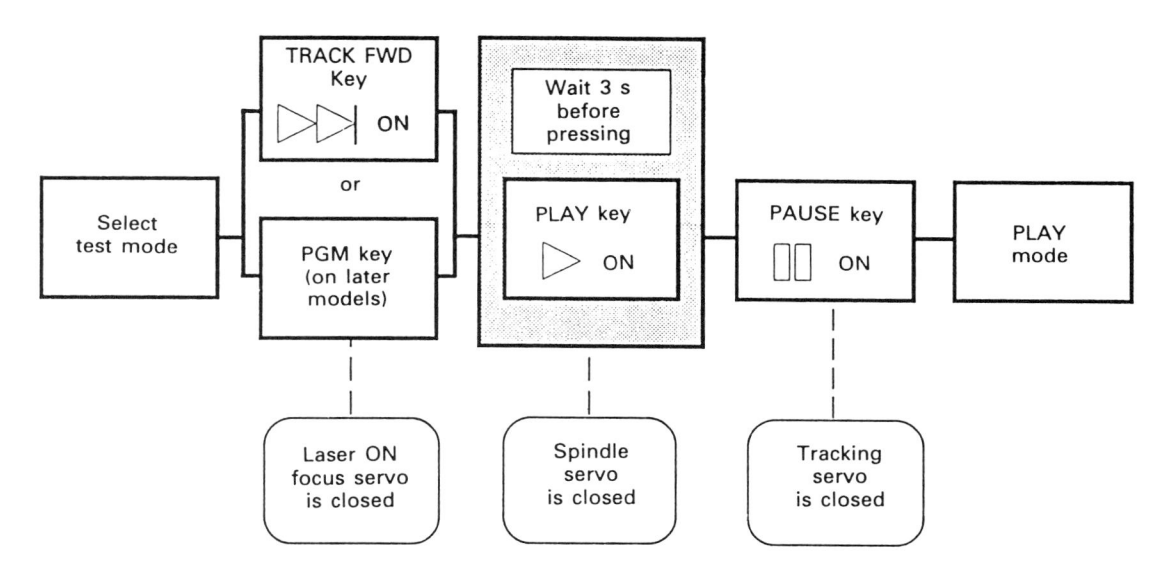

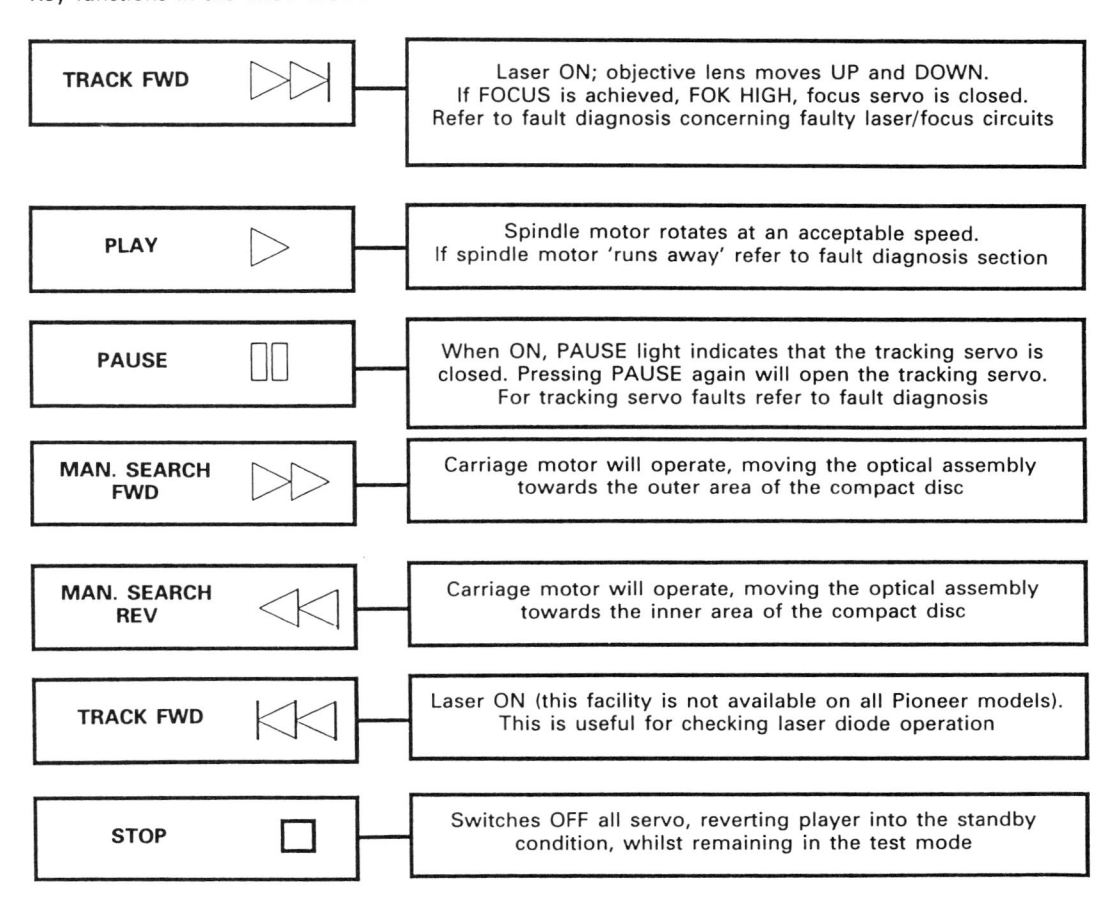

Figure 6.1 *Test mode procedures (Pioneer)*

Test mode and adjustment procedures

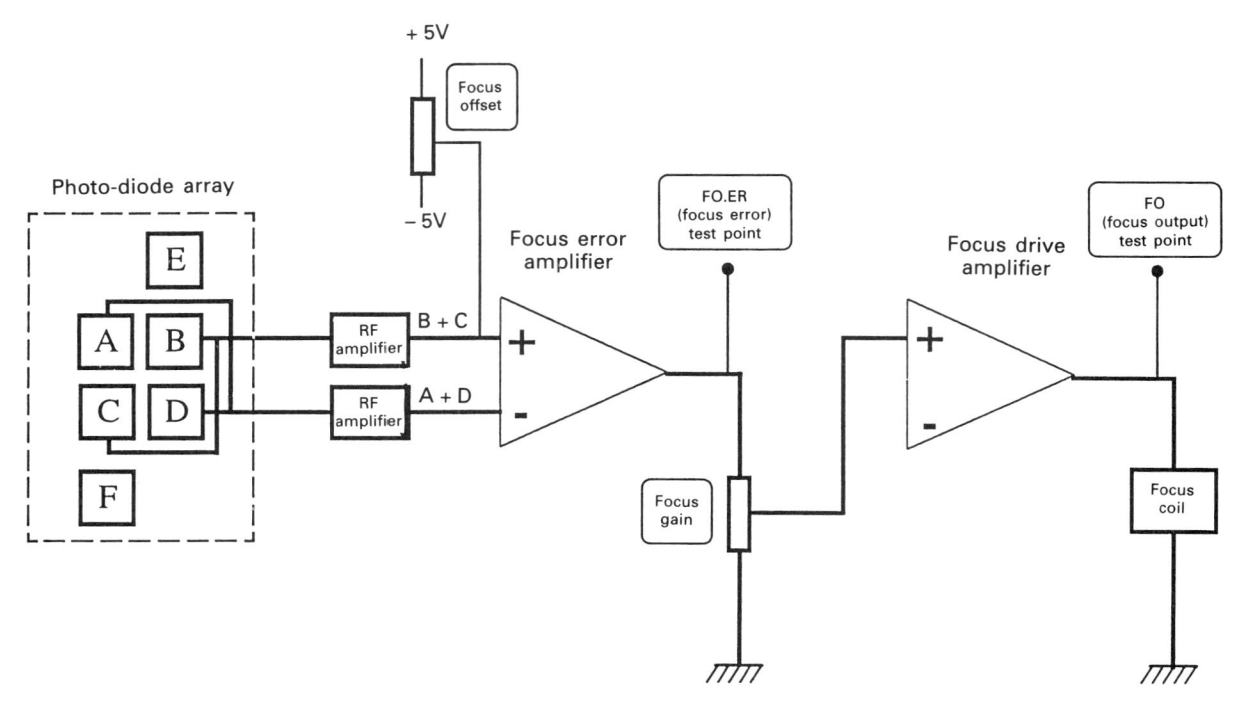

Figure 6.2 *Basic focus servo*

Philips players

Not all Philips players have a focus offset, the CD150 being one of these. The section of circuit shown in Fig. 6.3 relates to the CD840 and is similar to that in other models containing this adjustment.

1 The oscilloscope is connected via a 10:1 probe to the FE LAG connection, usually referred to as Test Point 27.
2 The player is switched on with no disc inserted.
3 The focus offset is adjusted for the optical mid-position of the objective lens.
4 Now the test disc is played and the focus offset is adjusted until 400 mV is obtained at Test Point 27.

Pioneer players

With the Pioneer players, variations in the value of the offset potential can be experienced depending upon which type of optical block is fitted, and in some instances whether the player is a single disc type with the objective lens facing upwards, or a multiple disc player, or even the stable platter single disc type where the objective lens faces downwards (see Fig. 6.2).

1 The player is switched on. For this adjustment the test mode is not essential, and no disc is inserted.
2 Connect the oscilloscope, via a 10:1 probe, to the FO.ER or FCS ERR. test point, depending upon the model.
3 Adjust the focus offset until the following level is achieved in relation to the optical block that is fitted:

PWY1003 to PWY1011	0 V
PEA1030 (single disc and twin tray players)	− 50 mV
PEA1030 (multi-disc players)	−150 mV
PEA1030 (single disc stable platter players)	−150 mV
PEA1179 (see below)	0 V

Note: The PEA 1179 is the new Pioneer optical block fitted to some of the most recent players, and is also referred to as the '92 Optical Block. This particular optical block is supplied with a small PCB attached with some small preset controls which are factory set, and therefore are not recommended for any adjustment. With this optical block the focus offset is verified to be at the zero level.

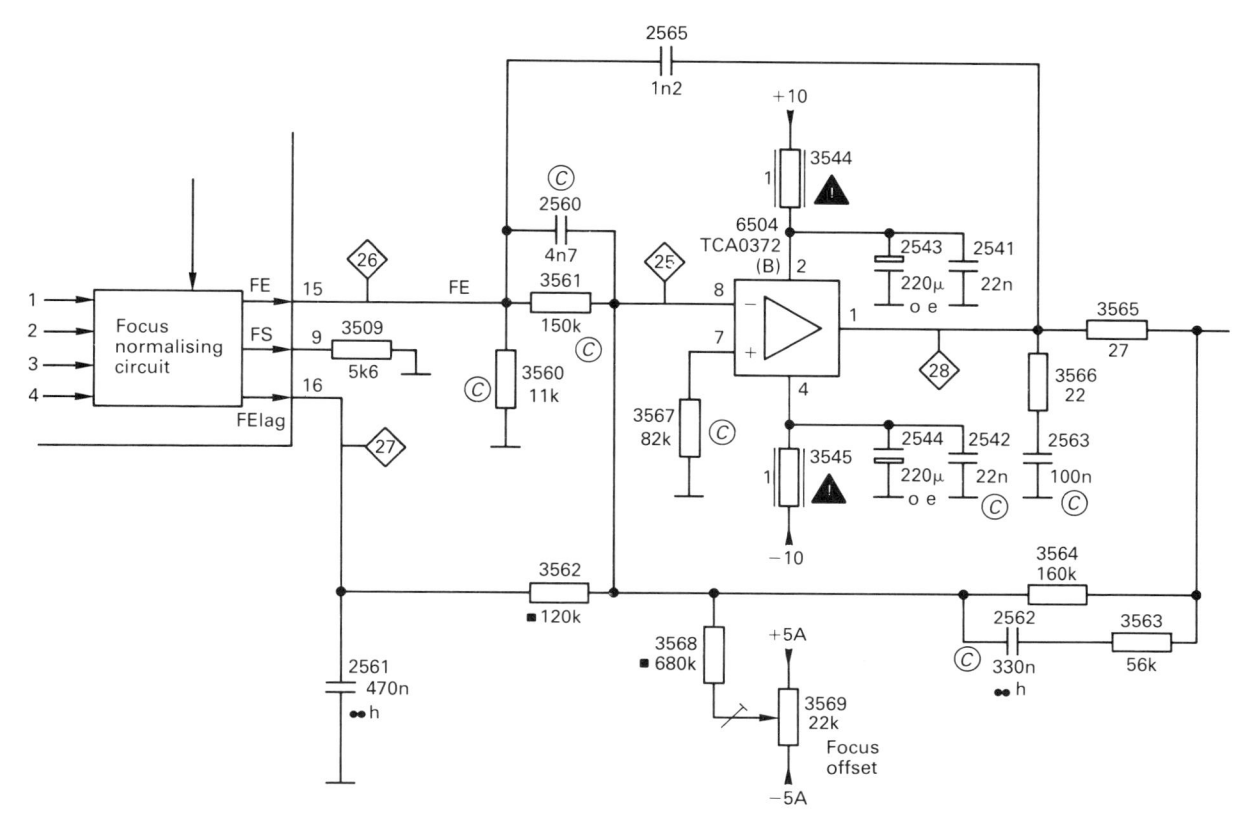

Figure 6.3 *Philips focus offset*

Other players

Not all players have the focus offset adjustment, but many of those that do have this adjustment fall into one of three categories:

1 The player has a radial optical assembly, and the procedure outlined for the Philips players will usually provide an acceptable method (Fig. 6.3). The Sharp DX-361 and Technics SL-P111/7 fall into this category, and are identical in achieving the correct focus offset.
2 Unless otherwise specified in the relevant service manual, the offset control is adjusted to 0 V as outlined for the basic servo above.
3 The offset preset is set to the mid-position, the disc is played, and whilst monitoring the RF eye pattern with an oscilloscope the focus offset is adjusted for the clearest waveform to ensure that optimum adjustment is achieved.

With reference to 3 above, it can frequently prove useful to monitor either the RF eye pattern waveform, or the focus error waveform, which is in effect a 'noise' waveform, and to carefully adjust the preset to optimise either waveform. This in effect is 'fine tuning' the focus coil in the optical block to the servo system and can occasionally assist in endeavouring to overcome some 'skipping' and 'jumping' problems that have a tendency to occur.

Further reference to the fault diagnosis section will provide further information regarding these type of problems.

Tracking offset adjustment

Similar to the focus offset adjustment, the objective is to remove unnecessary d.c. potentials that would otherwise affect the operation of the tracking servo (Fig. 6.4).

Generally, for those players that have the

Test mode and adjustment procedures

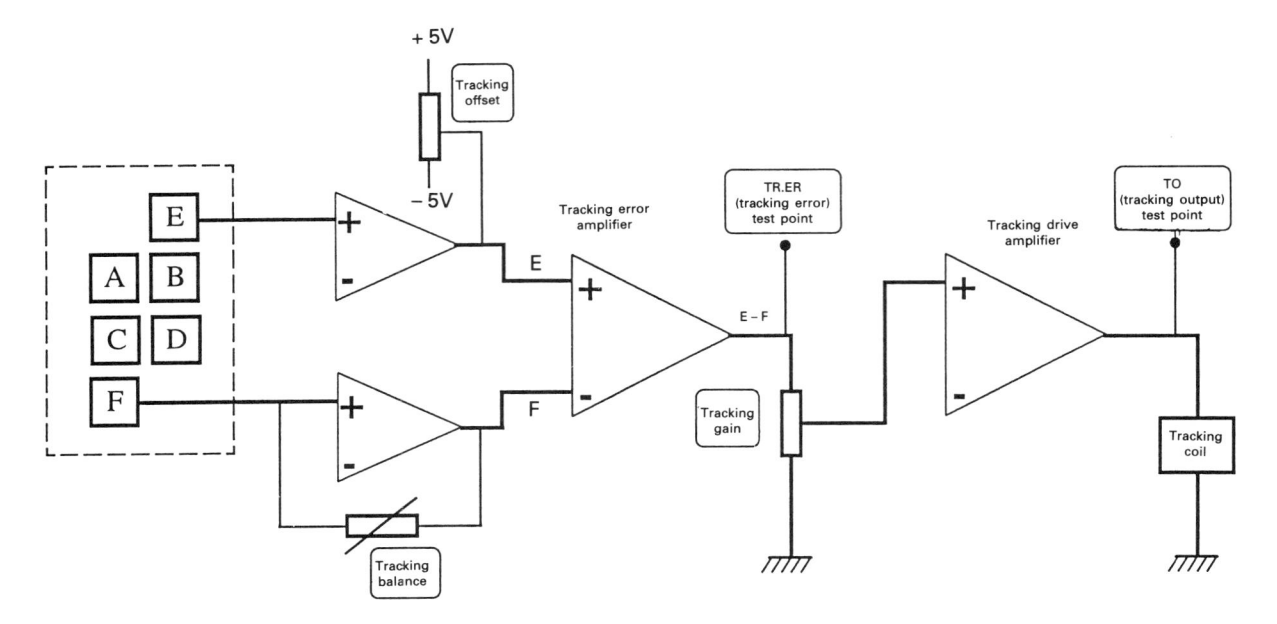

Figure 6.4 *Basic tracking servo*

tracking offset adjustment the procedure is quite straightforward:

1 Connect the oscilloscope to the tracking error test point (TE or Tr Err) or equivalent.
2 With the player switched on, and no disc inserted, adjust the preset for 0 V as accurately as possible.

It is possible to obtain a fine adjustment by playing a disc and monitoring the RF eye pattern waveform, and carefully adjusting the preset for the clearest waveform, which will be the point of virtual maximum amplitude.

RF offset adjustment

This adjustment is not present in many players but was necessary in a range of Pioneer models (Fig. 6.5), and also in the early Sony models such as the CDP-101.

The adjustment is carried out by switching on the player, with no disc inserted.

1 Monitor the RF test point:
    Pioneer players: TP1 pin 1
    Sony CDP-101:   TP3
2 Adjust the RF offset for:
    Pioneer players: 100 mV
    Sony CDP-101:   1.35 V

Laser power adjustment

Not all players have the facility of laser power adjustment, and even if there appears to be a method of altering the laser power, care should be taken as experience has shown that when a laser diode has aged increasing the power of an apparently weakened laser beam can more often as not hasten its final demise.

Usually the laser power is only adjusted when an optical assembly is replaced, and the level will normally be maintained by the automatic power control circuit.

Antistatic precautions

Though all replacement optical blocks have some form of static protection, it is recommended that normal precautions are taken to avoid static electricity damage when handling optical blocks as well as many of the integrated circuits used in many of today's sophisticated electronic products. These precautions can involve body and workbench grounding facilities, as well as the minimal use of clothing subject to static.

Precautions concerning laser beams

With respect to laser beams themselves, care should be taken if deciding to view the laser beam

Test mode and adjustment procedures

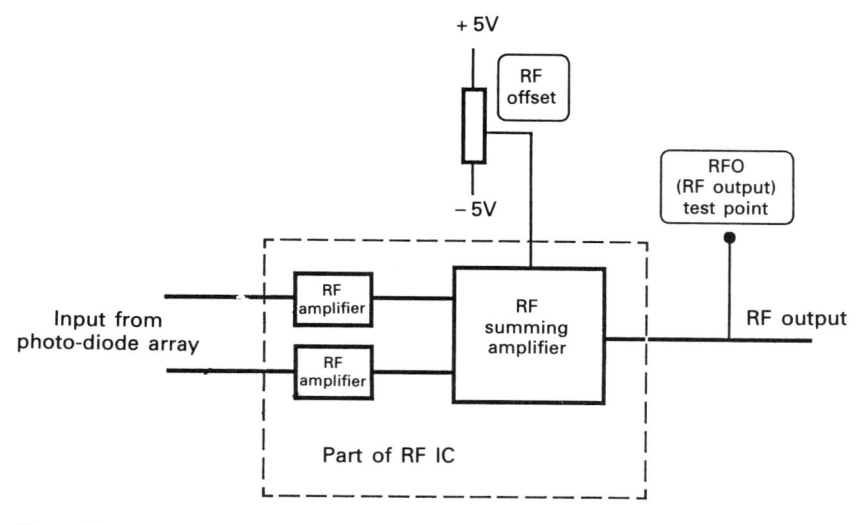

Figure 6.5 *RF offset adjustment*

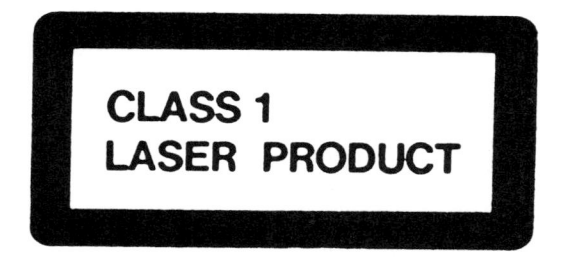

Figure 6.6 *Laser warning diagram*

when endeavouring to determine whether the laser diode is functioning (see Fig. 6.6). Ideally laser beams should not be viewed, as the majority of the power is invisible and the frequency at which the laser operates is within the infra-red spectrum, so all that can normally be seen is usually a faint red spot of light.

For many engineers the quickest method of determining whether a laser is functioning is in fact to actually view its emission from the objective lens. If this is the chosen method then applying a few basic rules will minimise any problems.

1 Ensure that the laser beam is low power, and in the case of CD players this power is usually no more than 250 µW with many players operating at around 120 µW.
2 Maintain a minimum distance between the lens and eye of 30 cm.
3 Do not view the lens directly, but slightly to one side.
4 Do not view the laser for more than 10 s. (After all is said and done, how long does one have to view an ordinary lamp before deciding that it is switched on?)

Generally the laser power only requires adjustment when an optical block is changed, and then only if it is possible and recommended. Many replacement optical blocks come with the laser power factory-set, and therefore do not require any further adjustment.

As previously mentioned, the initial setting of the laser power with a laser power meter is usually finalised by monitoring the RF eye pattern waveform for the correct amplitude. Many service departments do not possess a laser power meter – or if they do, it often becomes a redundant

Test mode and adjustment procedures

piece of test equipment. Furthermore, access to monitor the laser can prove difficult with some CD players.

The author's preferred method is to monitor the RF eye pattern waveform for the correct amplitude, but there is merit in measuring the laser current in order to determine the laser power. Examples of some different methods recommended by different manufacturers will be highlighted as follows, but further information is available in the relevant service manuals and other supporting technical information frequently supplied by manufacturers.

Philips players

The recommended procedure for the CD150 (which is similar to other Philips models) is to measure the voltage across a resistor which is the result of the power of the laser beam reflected from the CD onto the photo-diode array (Fig. 6.7).

1 Insert a recommended test disc.
2 Set the laser power control to the mid-point setting (if the optical block has been replaced).
3 Put the player into service mode 1 (laser will be on and in focus).
4 Monitor the voltage across test points 1 and 2. Adjust the laser power control to obtain 40 mV (this is a preliminary adjustment).
5 Play the disc.
6 Adjust the laser power control to obtain 50 mV across test points 1 and 2.

Pioneer players

The player will need to be put into the test mode, and a test disc inserted (refer to Figs 6.1 and 6.8).

1 Select TRK FWD or PGM (programme) depending upon the model of the player, followed by PLAY after ~3 s; the disc should now rotate at a reasonable speed.
2 Monitor the RF test point to determine the peak-to-peak value of the 'blurred' eye pattern waveform. (The reason for observing a 'blurred' eye pattern is that the disc cannot be played correctly if the grating adjustment is incorrect when replacing an optical block, or if someone else got there first and adjusted it incorrectly) (Fig. 6.9).
3 Adjust the laser power control (usually VR1) until the following peak-to-peak value is obtained depending upon the type of optical block fitted:

PWY1003–PWY1011	1.5 V
PEA1030	1.2 V
PEA1179	1.2 V (see Note below)

Note: The PEA 1179 is the new Pioneer optical block fitted to some of the later range of players, and is also referred to as the '92 Optical Block. This particular optical block is supplied with a small PCB attached with some small preset controls which are factory set, and therefore are not recommended for any adjustment. As will be highlighted elsewhere, this optical block minimises the number of adjustments in the relevant players to which it is fitted.

A final check of the laser power is made on completion of all other adjustments to ensure that the recommended level is still maintained when a disc is being played, and some final adjustment may prove necessary.

Sony players

With Sony optical blocks details are usually supplied with each optical block concerning specified laser power current.

The information is given on a label fixed to the block, and examples of these are given in Fig. 6.10.

Specification limits are: the current value printed on the label +11 mA to −5 mA at 25°C.

As a laser ages the current will increase, and if the current exceeds the specified limit, the optical block should be replaced. Any attempt to adjust the laser power will usually hasten its demise.

JVC players

Adjustment is similar to the Sony method described above. An example detailed in a JVC service manual may prove of interest to engineers.

When the life of the laser has expired the following symptoms will appear:

1 The level of the RF eye pattern waveform will be low.
2 The drive current required by the laser diode will be increased.

See Fig. 6.11.

Measurement of laser diode current

With reference to Fig. 6.12, open the relevant link and insert a 1 Ω resistor. With the laser diode operating, monitor the voltage across the

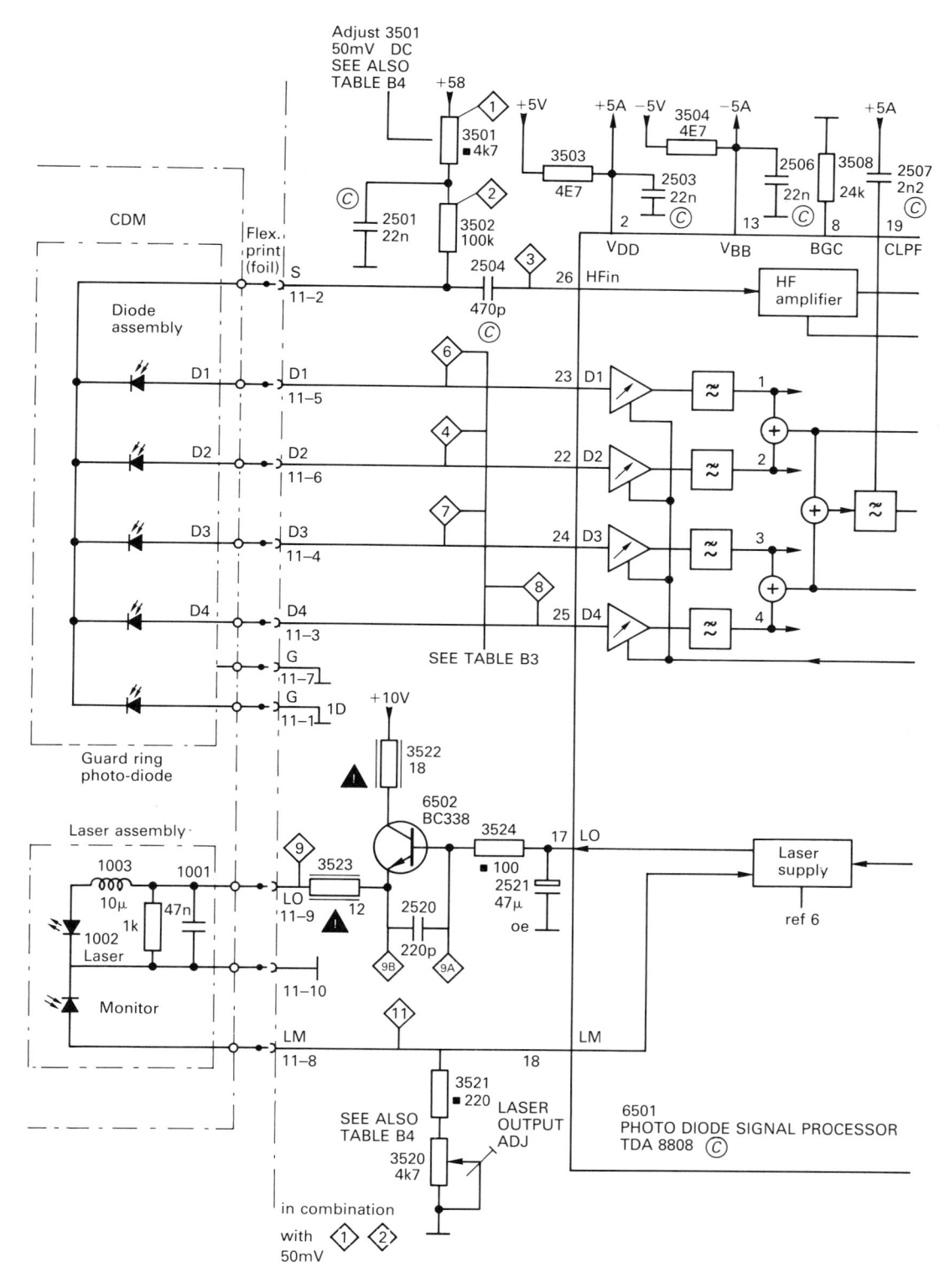

Figure 6.7 *Philips laser power adjustment*

Test mode and adjustment procedures

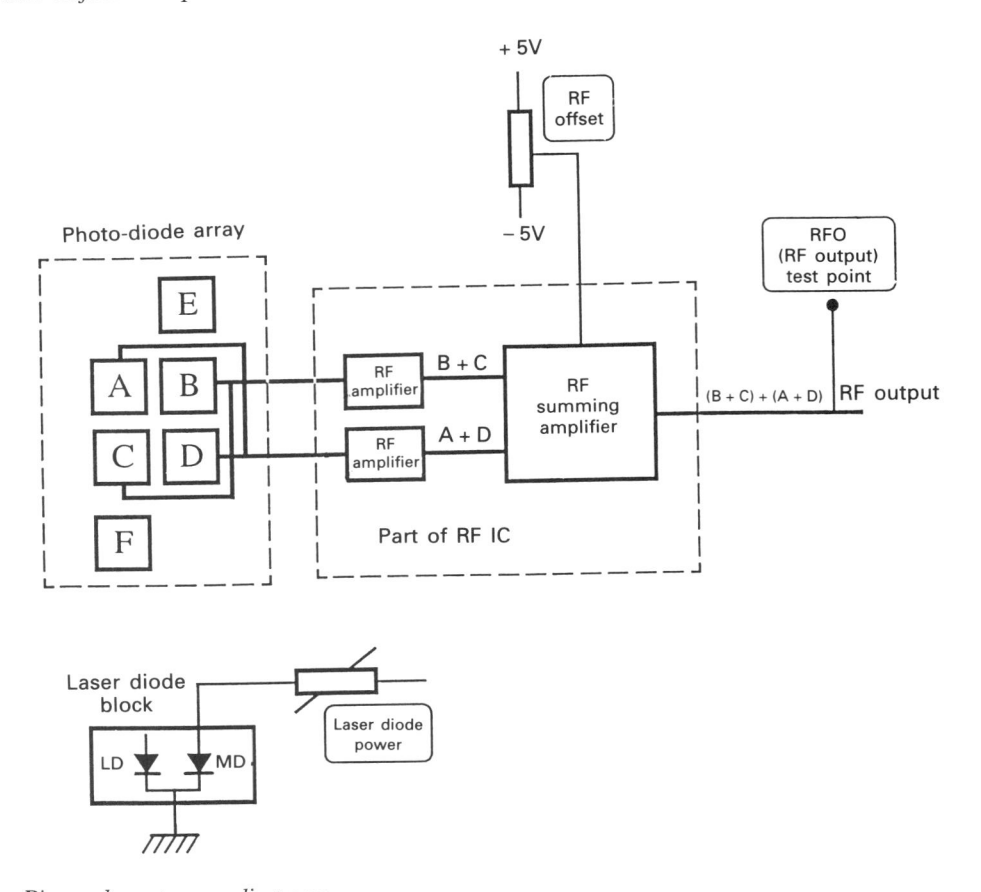

Figure 6.8 *Pioneer laser power adjustment*

resistor. If the voltage is excess of 120 mV (which is equivalent to a current of 120 mA), the life of the laser diode has expired, and it should be replaced.

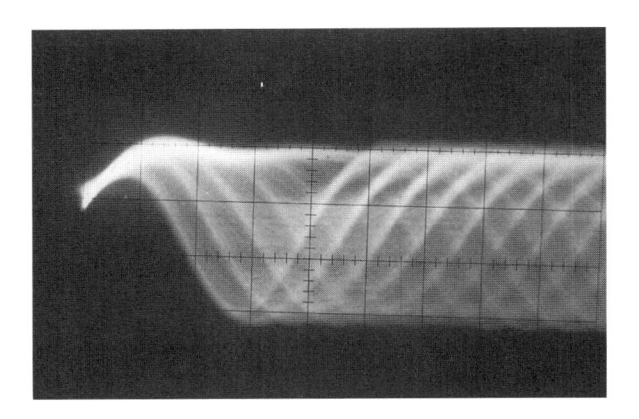

Figure 6.9 *Blurred RF eye pattern waveform*

A preset may be present on the APC board, and if this adjusted to increase the eye pattern level, the laser diode will again most certainly achieve an early demise.

Comparison of RF eye pattern levels

Ideally the eye pattern waveform should be as clean and in focus as possible (as shown in Fig. 6.13). With reference to a wide range of service manuals from different manufacturers, the peak-to-peak levels of the RF eye pattern waveform shown in Table 6.1 may prove a useful guide. All measurements were achieved with the use of a test disc.

The information in Table 6.1 is intended as a guide across a wide range of players, but absolute verification should be obtained from the relevant service manual wherever possible. Whilst most manufacturers quote a tolerance range, the author

Test mode and adjustment procedures

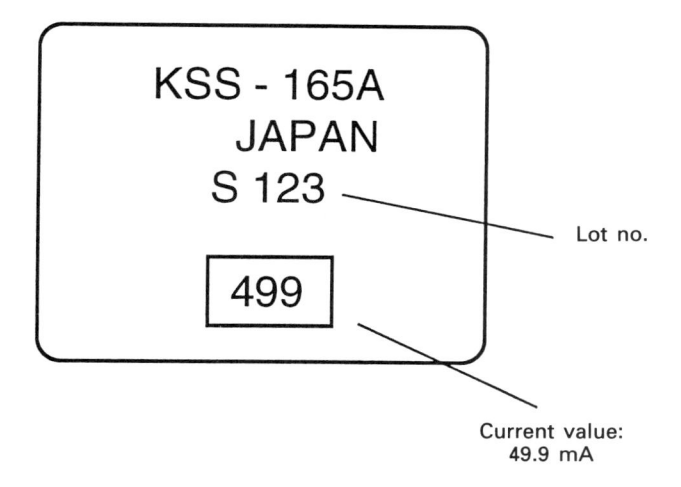

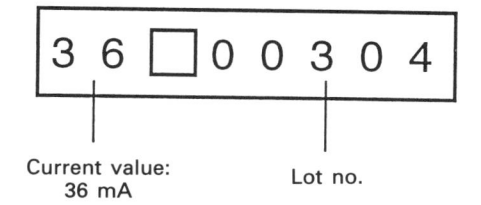

Figure 6.10 *Example of Sony laser current label*

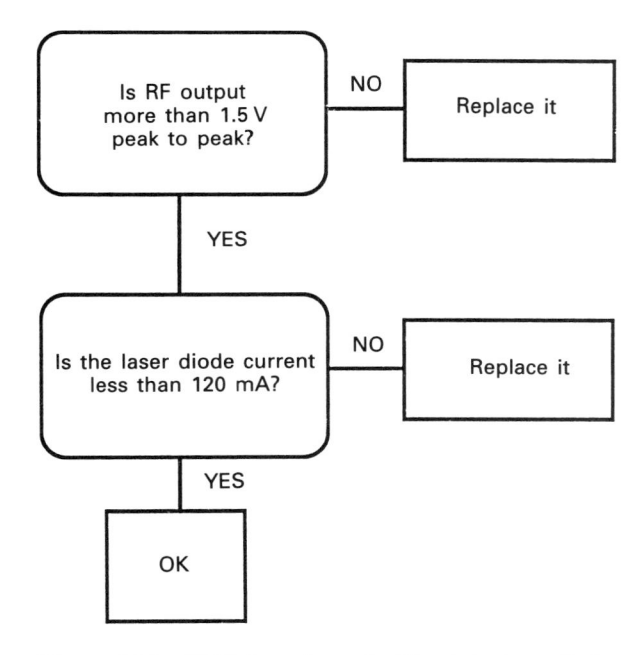

Figure 6.11 *JVC flow chart: the life of the laser diode*

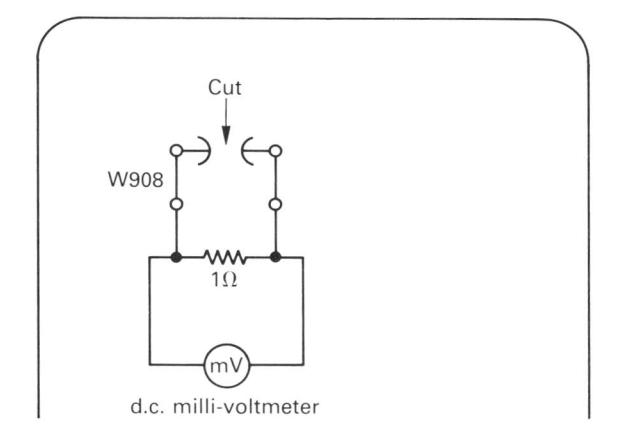

Figure 6.12 *JVC laser diode current measurement*

prefers to maintain the centre of the quoted tolerance wherever possible.

Tracking balance adjustment

Tracking balance adjustment is present in many players and its purpose is to ensure that the gains of the tracking signal amplifiers are identical before the signals are applied to the tracking error amplifier to produce the tracking error signal. In order that this adjustment can be carried out it is necessary to prevent the tracking servo (Fig. 6.4) from fully functioning. There are three methods of achieving this.

1. If the player has a test or service mode facility adjustment is carried out with disc inserted and rotating, with the laser on and the focus servo operating, but with the tracking servo off. The example previously explained relating to the Pioneer players is briefly repeated below, but it

121

Test mode and adjustment procedures

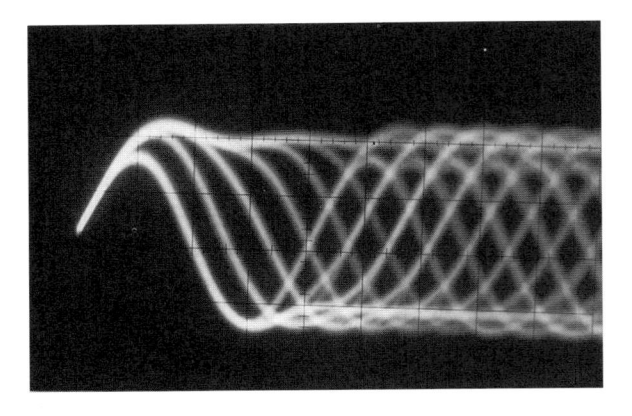

Figure 6.13 *RF eye pattern waveform*

will be necessary to determine the same functions in other players to achieve the same results by referring to the service manual. The Philips and similar players (i.e. with the radial tracking method) do not have the tracking balance facility, as the gain level of each tracking signal line is automatically determined and adjusted as necessary.

1 Select the test mode.
2 Move the optical block to the centre of the playing area of the inserted test disc.
3 Connect the oscilloscope, via a 10:1 probe, to the tracking error test point (TE or TRK ERR) or equivalent.
4 Press 'TRK FWD' (or PGM). The laser and focus servo circuits should become operational.
5 After 3 s press 'PLAY'. The disc should now rotate at a reasonable speed.
6 Adjust the tracking balance preset until the resultant waveform is balanced either side of the zero line.

2. *If no test mode facility is available*, it will be necessary to disrupt the tracking servo circuit.

1 First insert and play the test disc, and select a track near to the centre of the playing area of the disc.
2 Connect the oscilloscope, via a 10:1 probe, to the tracking error test point (TE or TRK ERR) or equivalent.
3 Now turn the tracking gain control to

Table 6.1

	Model	Level (V)	Test point
Aiwa	DX-M77	2.0–2.5	TP102
	DX-M740	2.0–2.5	TP102
Fidelity	CD200	1.2	TP4
	CD202	1.2	HF
Hitachi	DA-W650	1.3	TP1
	DA-6000/6001	1.3	TP1
Philips	CDM2 Mech	1.2–1.5	HF Out
Pioneer	PWY1003/11 (optical blocks)	1.5	TP1 pin 1 (RF)
	PEA 1030 (optical blocks)	1.2	TP1 pin 1 (RF)
	PEA 1179 (optical blocks)	1.2	TP1 pin 1 (RF)
Samsung	CD-35	1.5	HF
	SCM-780	1.5	TP.RF
Sharp	DX-R250	1.3–2.0	HF
	DX-361	0.9–1.5	HFI
	DX-450	1.2–1.8	HF
Sony	CDP-101	1.0–1.5	TP3
	CDP-M20	1.2	TP3
Technics	SL-PJ38A	1.5	HF Out (Pin 3 IC6501)
		1.2	DET (Pin 4 IC6501)
	SL-P111	0.8	TJ 501
Toshiba		0.9–1.0	RF Signal
Yamaha	AST-C30/C10	1.35–2.0	Pin 1 of TP751
	CDX-750	0.92	EFM

Test mode and adjustment procedures

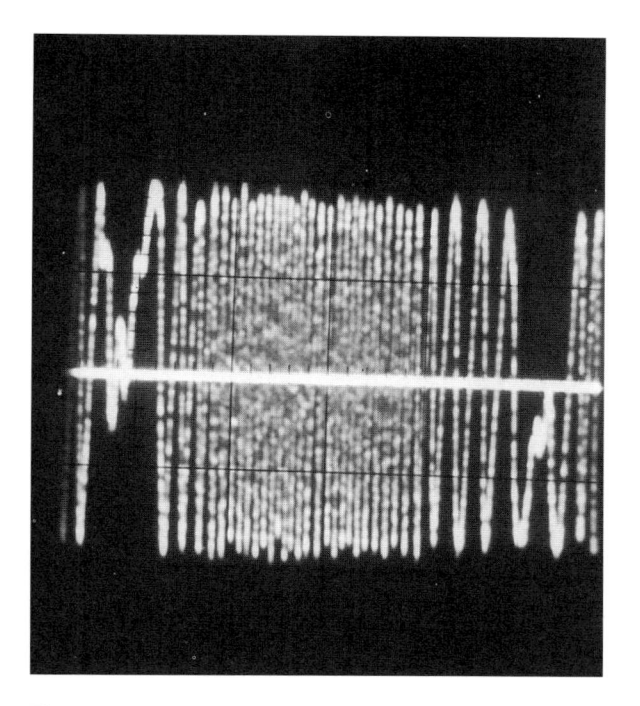

Figure 6.14 *Tracking balance waveform*

minimum, taking note of the *original* position of the control before proceeding.
4 Adjust the tracking balance preset until the resultant waveform is balanced either side of the zero line, as illustrated in Fig. 6.14.

Due to the tracking gain control being turned to minimum, the player may appear to object to this procedure, and it may eventually stop operating, but there is usually sufficient time to carry out this operation successfully.

On completion, restore the tracking gain control to its original position.

3. Alternative method if there is no test mode; this method can also be applied to players with a test mode.

1 First insert and play the test disc, and select a track near to the centre of the playing area of the disc.
2 Connect the oscilloscope, via a 10:1 probe, to the tracking error test point (TE or TRK ERR) or equivalent. The oscilloscope will display the standard tracking noise type of signal.
3 Now press the manual search FWD or manual search REV keys. The lens will gradually track across the disc, and during this process a tracking error waveform will 'ghost' behind the tracking noise waveform.
4 Adjusting the tracking balance control until the 'ghosting' waveform is balanced either side of the tracking noise waveform.

Automatic check of the tracking balance adjustment

A limited number of players have an automatic facility of adjusting the tracking balance, one in particular being the Pioneer PD-8500, which contains a Toshiba TC9220F digital servo processor.

With the player in the test mode, as previously described, and the oscilloscope connected to the tracking error test point, set the oscilloscope to the centre of the screen. There will be a positive potential of approx. 2 V at this point, which amazingly is quite normal.

With a test disc inserted, press 'TRK FWD', followed by 'PLAY'. Three seconds later, the tracking balance waveform will be observed on the oscilloscope.

Now press the 'REPEAT' key; the waveform should balance either side of the centre line.

The purpose of this check is to verify that tracking balance can be achieved, as this is an automatic procedure which is carried out each time a disc is played, and is in fact one of four automatic sequences that are carried out whenever a disc is played, the others being the automatic adjustment of the focus offset and the focus and tracking gains.

Focus balance, focus bias, photo-diode balance adjustments

The above adjustments will appear in various players, but not all in the same player, with the procedure for any of the adjustments being identical even though each of their functions may well be different.

1 Insert and play the test disc, and select a track near to the centre of the playing area of the disc.
2 Connect the oscilloscope, via the 10:1 probe, to the RF test point to monitor the RF eye pattern waveform.
3 Play the test disc, and adjust any of the relevant controls until the eye pattern waveform appears to be at its maximum amplitude, consistent with being as cleanly in focus as possible.

RF level adjustment

RF level adjustment is usually found as a specific preset control in the earlier players, and wherever possible reference should be made to the relevant circuit diagram. However, many later players, such as Pioneer, verified that the RF level was correct by adjusting the laser power (refer to the section on laser power adjustment), whilst playing a recommended test disc, to achieve the recommended level as detailed earlier in this chapter under the heading 'Comparison of RF eye pattern levels'.

Focus and tracking gain adjustments

Some players do not have focus and tracking gain controls, but for those that do there are various methods of carrying out these adjustments. These are described in the relevant service manuals, and can involve additional test equipment, some of it being specialised. Generally these methods will require an external audio frequency source, special filters together with the oscilloscope, and eventually adjustment of the relevant gain control to achieve the required Lissajous waveform. Engineers who wish to pursue this method of gain adjustment should refer to the relevant service manual for the player involved.

Though the adjustment of these controls is relatively critical, there will still be a reasonable margin of tolerance in their adjustment, and if either adjustment proves to be extremely critical then the possibility of a fault condition must be considered.

Many engineers have adopted a method of using a tried and tested 'ever faithful' disc which has fine scratches over the surface as a result of possible mis-use over a period of time. 'If it plays that, it will play anything' is the comment frequently heard, and this does prove to be a method with a fair amount of merit.

Unfortunately players do vary in overall quality, with some having optical blocks that appear to be more noisy than others, thus creating an impression that maybe the focus and tracking gain is high and generating excessive noise, whilst the less noisy optical blocks can create the impression of low levels of gain.

Ideally, neither of the gain controls should be touched; they should be kept in their factory-set positions, which in fact will usually be near to their mid-point position. Therefore if the required test equipment is not available, and in times of servicing expediency, setting both presets at around their mid-point setting or just a fraction less will generally be quite an acceptable measure.

Occasions have occurred where the author has set up all the equipment described in the service manual and followed the procedures by the book, including the filters and the recommended test disc, and where, on completion of the procedure where the required Lissajous waveform was as recommended in the service manual. And where were the two controls? . . . well blow me down . . . they are just before the mid-point position, and it took twenty minutes to discover that!

The author, when visiting a production line where different models of CD players were being manufactured, made an observation of the focus and tracking gain presets after following the factory procedures outlined above to maintain the required specification, which indicated consistently exactly the same position of just before the mid-point. Of course, it was appreciated that these were brand new players of the same manufacturer.

As a result, the author has preferred not to use the involved methods described and has tended to follow a more simple approach, which is not of course intended to detract from any of the methods that are recommended by various manufacturers to achieve optimum effectiveness.

Focus gain adjustment

Play the test disc, preferably with minimal or no scratches, and monitor the focus error (FE) waveform with an oscilloscope via a 10:1 probe.

Select a track near to the central playing area of the disc. As the gain adjustment is increased, so will the noise being monitored on the oscilloscope.

Two possible options can be achieved when the focus gain preset is altered:

1 an overall noise amplitude of no more than 500 mV, or
2 the centre line d.c. level of the noise at about 100 mV.

As the gain is reduced, the d.c. level will tend to increase and the waveform will tend to bounce due to insufficient gain, whilst increasing the gain will certainly increase the noise level.

Both of the above results have been observed in some service manuals, and still maintain the

Test mode and adjustment procedures

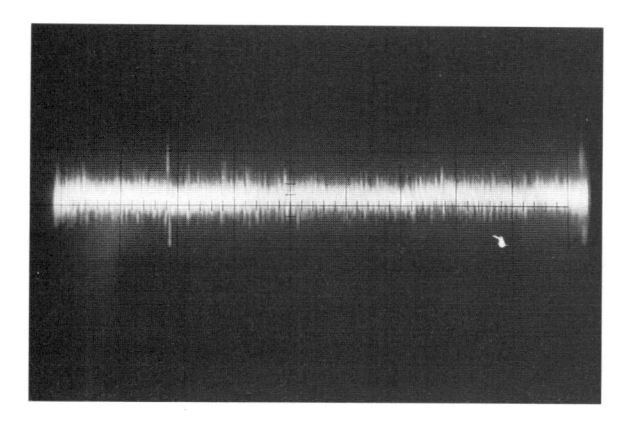

Figure 6.15 *Focus error waveform*

consistency of the control being set to just before the mid-point.

Another option is to listen to the acoustic noise that comes from an optical block whilst a disc is being played. As the gain is increased, the acoustic noise will increase, and will naturally decrease with the reduction of gain. Providing this method is carried out in a quiet environment it is possible to get the 'feel' of different players whilst adjusting the control for an acceptable relatively faint level of noise that still enables the control to be around the mid-point level.

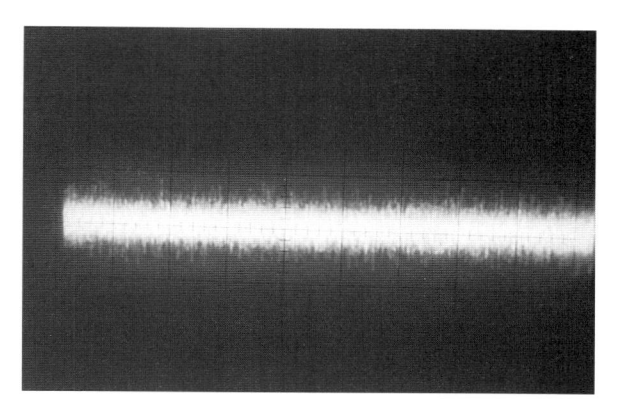

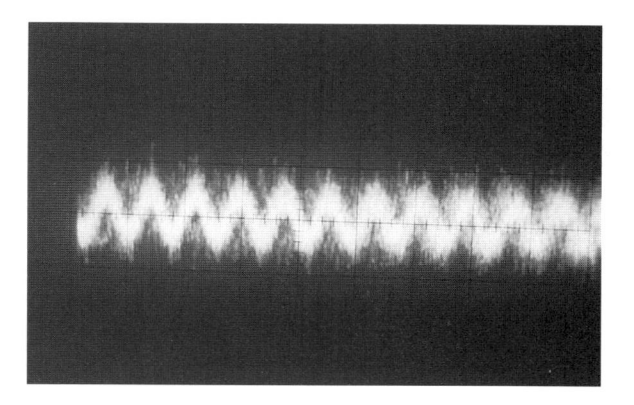

Figure 6.16 *(a) Typical tracking servo noise, normal (b) Excessive tracking gain*

Tracking gain adjustment

Tracking gain adjustment is probably more important than focus gain, especially if the player is susceptible to skipping and jumping, and it is possibly easier to achieve a more effective setting.

1 Monitor the tracking error (TE) waveform (Fig. 6.16) with an oscilloscope, via a 10:1 probe. The noise level will normally be higher than for the focus error, and is usually centred on the zero potential line.

2 Increase the tracking gain until a frequency becomes apparent amongst the noise; this is the natural servo frequency. Note the position of the control at which this frequency becomes apparent.

3 Reduce the gain until the waveform now begins to bounce up and down. Again, note the position of the control when this becomes apparent.

4 Set the control midway between these two points; usually this will be just before the mid-point position.

The procedure outlined regarding acoustic noise for focus gain can also be adopted to achieve similar results. Optical blocks tend to vary in the amount of acoustic noise they emit and thus this method should be treated with caution especially if a particular type of optical block tends to be more noisy when compared to others.

Voltage controlled oscillator (VCO) frequency

If the VCO frequency is not correct, it is possible that the disc will not play, or that it may suffer from intermittent audio output.

It is usually preferable to use a frequency

125

Test mode and adjustment procedures

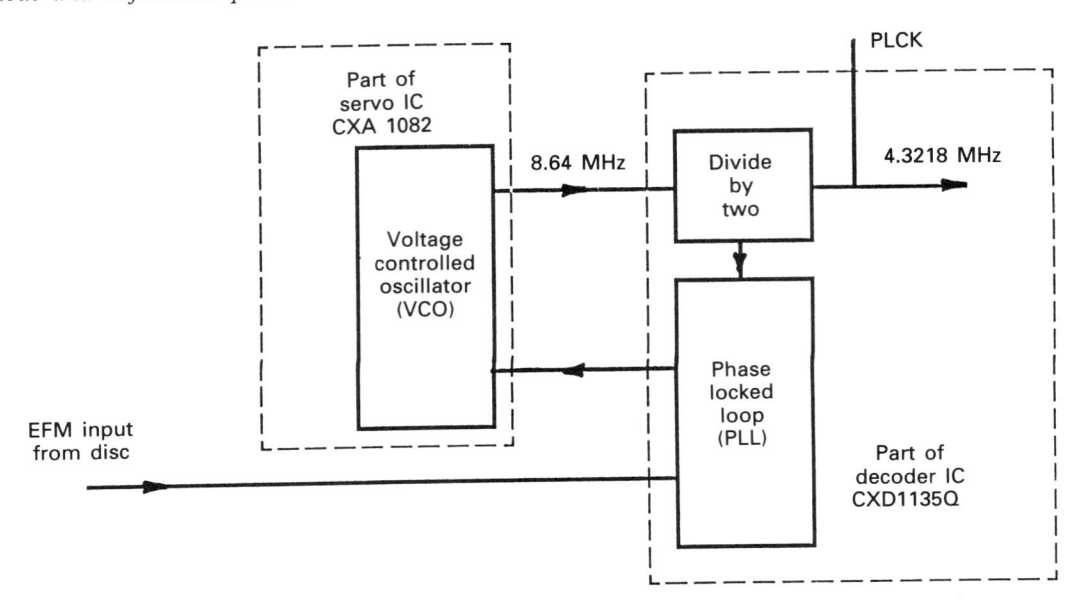

Figure 6.17 *VCO block diagram*

counter, and ensure that the free run VCO frequency is set to the recommended setting, which may be in the region of 4.3 MHz or 8.64 MHz, depending upon where in the circuit the VCO frequency is being monitored (Fig. 6.17).

When a disc is being played correctly, the VCO circuit will usually operate at a frequency of twice the disc clock frequency of 4.31218 MHz, (i.e. 8.62436 MHz), which is divided by two to produce the required resultant clock frequency.

The majority of players will usually refer to the lower frequency, and the author's preference is usually to set the VCO free run frequency at around 4.375 MHz, without inserting a disc, and not in the test mode.

It may prove necessary to verify from the service manual the relevant point to which to connect the frequency counter, usually via a 10:1 probe, and also to ascertain whether the phase locked loop (PLL) has to be disabled by shorting two links, such as in Pioneer players by connecting together the asymmetry signal line (ASY) and ground (GND).

The reason for choosing 4.375 MHz is to ensure that the VCO is running at a frequency higher than the required 4.3218 MHz to assist in a possibly more efficient capture range as the disc runs up to speed at the commencement of play.

Another method, especially if a frequency counter is not available, is to observe the VCO frequency on an oscilloscope, again via a 10:1 probe, whilst a disc is being played, and to adjust the VCO frequency adjustment in one direction until the waveform begins to jitter and then to adjust back in the opposite direction until jitter occurs again; then to set the control to the position midway between the two points where the jitter occurs.

The foregoing completes the majority of electrical adjustments that may be found in CD players, and now it is necessary to consider the relevant mechanical adjustments.

Mechanical adjustments

Generally mechanical adjustments are only necessary when an optical block or spindle motor has been replaced, or some other mechanical servicing has been necessary within the optical assembly.

Many optical blocks do not require any major adjustments when they are replaced, whereas others, such as those in many Pioneer players, require an extensive and involved procedure to be carried out; to the un-initiated this may seem overwhelming, whilst many engineers who have

Test mode and adjustment procedures

trod this path before may consider that the adjustments are more acceptable in the light of experience.

Turntable height

If it has proved necessary to replace a disc, spindle or turntable motor (spindle motor will be used hereinafter), it is essential to ensure that the disc turntable, if it is a removable type, is replaced at the correct height.

Some spindle motors are supplied as an integral component of a complete optical assembly, which is usual with the Philips radial assembly. Shown below are a range of recommended turntable heights with respect to a range of players. If the height information is not available, and it is necessary and possible to replace the motor, it is recommended that the original turntable height is first observed and measured with a suitable gauge to ensure correct reassembly afterwards.

If the turntable height is incorrect it is possible for the disc to 'bind' within the internal mechanism or scrape on the drawer assembly, with possible damage to the disc. Also, the RF eye pattern waveform can be affected by the height of the turntable, and the focus servo will have to adjust the lens position to suit the incorrect height which could cause inefficiency of the servo operation.

Reference to a wide range of service manuals has shown that some spindle motors are supplied complete with the turntable fitted. Access to the motor fixing screws is through holes in the turntable, and the complete assembly is removed sideways from the mounting plate via a slot.

In other examples the turntable is bonded to the spindle, which can make removing it difficult; the bonding material may be softened by the gentle application of heat to the spindle shaft with a soldering iron, and a suitable bonding material is used to bond the turntable onto the replacement motor shaft. The Sanyo MCD-730L is an example of this method.

The Technics SL-P111 method is illustrated in Fig. 6.18. With this method the turntable height is set to the correct height with a 0.9 mm feeler gauge (Technics Part No. RZZ0297), and is then fixed to the shaft by a screw tightened with a 1.27 mm hexagonal wrench.

With Pioneer players a height gauge is supplied

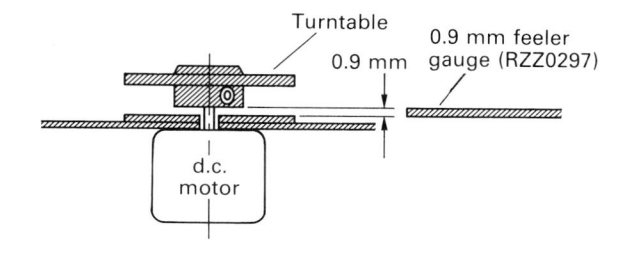

Figure 6.18 *Technics turntable height adjustment*

moulded onto the complete optical assembly unit, as illustrated in Fig. 6.19.

It is necessary to use the correct gauge for the type of player (i.e. single, twin or multi-disc), as the type of turntable and its respective height may well be different for different types of player.

With the later players the gauge moulded onto the optical assembly can be observed to be in effect two gauges in one: a thin inner gauge, usually for the single turntable platter and the multi-disc players, and a thicker outer rim for the single and twin tray players.

Earlier players had a single gauge which had a recess on one side to allow for a small amount of protrusion of the spindle motor through the mounting plate.

It is recommended that before removing the spindle motor, the engineer determines and selects which gauge is required, and checks this against the original turntable height.

Removal of the gauge from the moulding can be achieved with a pair of fine nippers, ensuring that any burrs are removed.

When refitting the turntable, apply the correct gauge, and gently press the turntable onto the motor shaft, preferably with a base support for the motor, until the turntable rests upon the gauge.

Remove the gauge on completion.

Tangential and lateral (radial) adjustments

When an optical block is replaced, some players require mechanical alignment of the optical block to ensure that the laser beam is directed accurately towards the surface of the disc, and in effect the beam is at right angles to the surface of the disc (as illustrated in Fig. 6.20).

If this adjustment is not set correctly, not only

Test mode and adjustment procedures

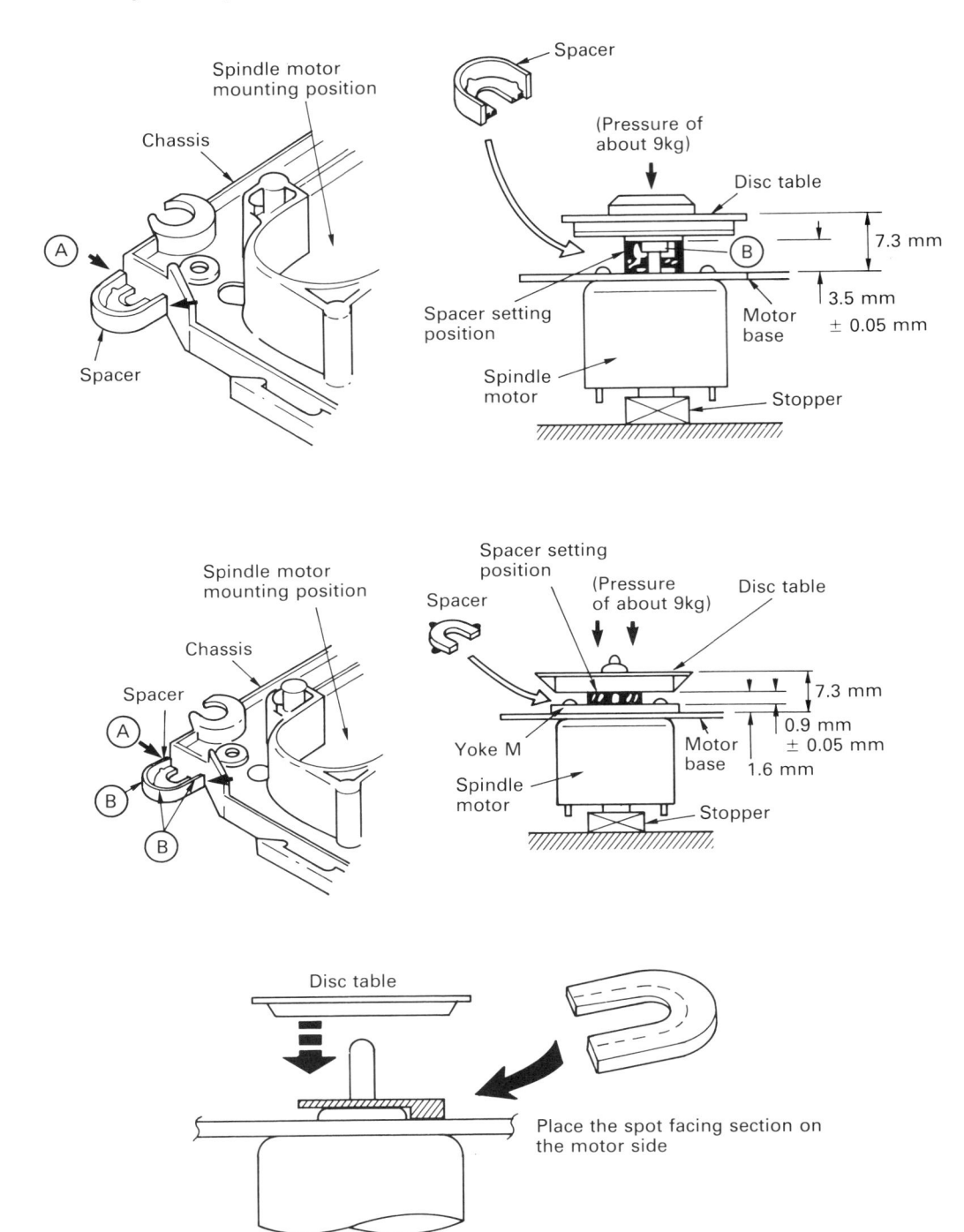

Figure 6.19 *Pioneer turntable height gauge locations*

Test mode and adjustment procedures

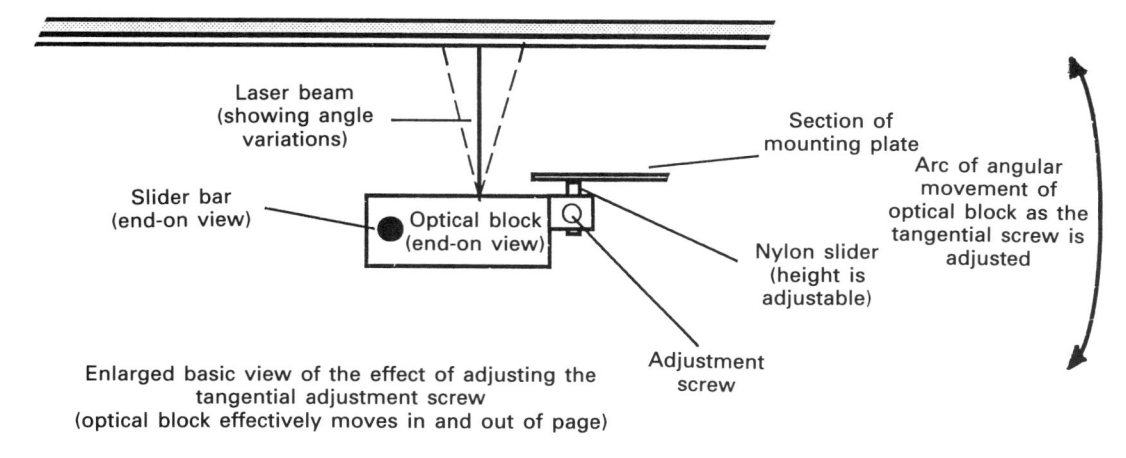

Figure 6.20 *Angle of laser beam to disc surface*

will the RF eye pattern waveform be unsatisfactory, but also the player may be susceptible to skipping and jumping, and with some players the diffraction grating adjustment will prove difficult or impossible to achieve correctly.

Because these adjustments are not necessary on all models of players, when an optical block is replaced, the procedures for the Pioneer players provide an excellent basic outline which can be applied as necessary on those models that may require this adjustment to be implemented.

With earlier Pioneer players, i.e. those using the PWY1011 type optical block (Fig. 6.21), the optical assembly requires only the tangential adjustment, but with the later models, using the PEA1030 and subsequent type optical blocks, tangential and lateral (or radial) adjustments are necessary.

This adjustment can be achieved with a 1.5 mm hexagonal wrench, but because access can prove difficult a 1.5 mm hexagonal ball wrench can prove extremely useful; this is available from RS (Part No. 662–670, which is a pack of four different size wrenches), but engineers may know of alternative sources for this item.

Before fitting a replacement optical block it is recommended that the hexagonal screw is initially adjusted to set the small white nylon runner to its mid-point setting, or if the block is being replaced, adjust the new one to the same point by comparison.

It is necessary for this adjustment to be finally carried out in the test mode as it will be necessary to move the optical block towards the outer edge of the disc until the hexagonal screw can be accessed. By observing the RF eye pattern waveform on an oscilloscope, whilst playing a disc, the hexagonal screw is adjusted until the opimum waveform is achieved (Fig. 6.24).

The recommended procedure is as follows:

1 Set the player to the test mode (refer to the test mode section).
2 Move the optical block until the adjustment screw can be accessed.
3 Insert a recommended test disc.
4 Connect an oscilloscope, via a 10:1 probe, to the RF test point.

Figure 6.21 *Pioneer PWY1011 type optical assembly*

129

Test mode and adjustment procedures

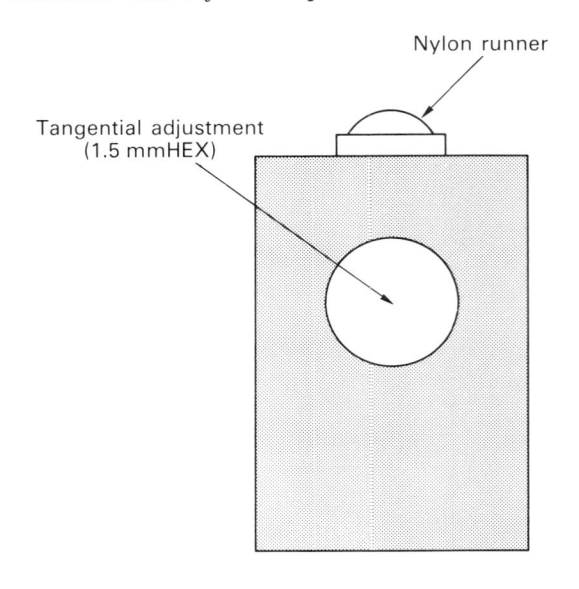

Figure 6.22 *Enlarged section of Pioneer PWY1011 optical block illustrating the tangential adjustment and nylon runner*

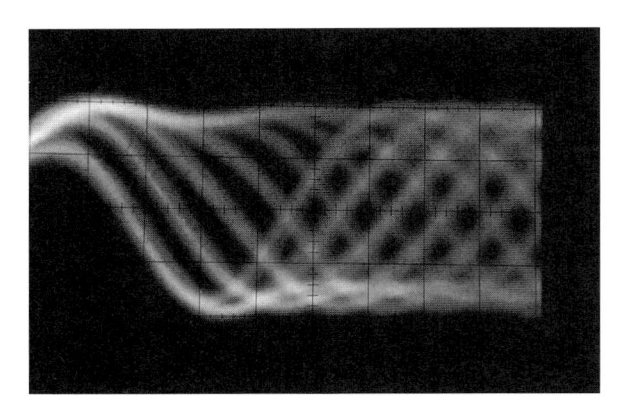

Figure 6.23 *Blurred eye pattern due to poor tangential adjustment*

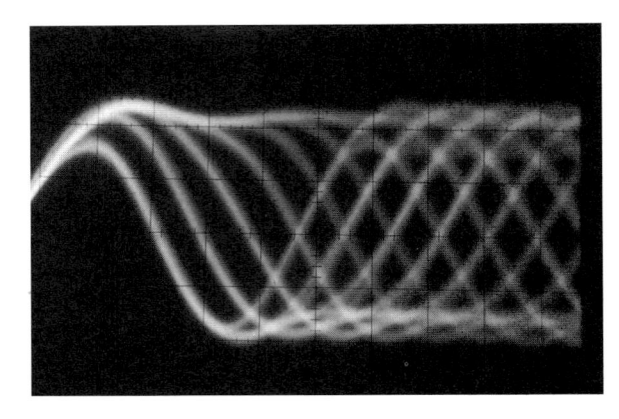

Figure 6.24 *Acceptable eye pattern after tangential adjustment*

5 Press 'Track Forward' (TRK FWD) or 'Programme' (PGM).
6 Press 'PLAY'. The disc will rotate at a reasonable speed.
7 Adjust the tangential screw for maximum amplitude of the waveform.

Note. This waveform should appear as a 'blurred' RF eye pattern. The service manuals will usually illustrate a waveform that can be achieved after the 'PAUSE' control has been pressed when an in focus RF eye pattern will be achieved. This adjustment can frequently 'jar' the optical block whilst it is being attempted, with frustrating effects on the player and engineer alike, especially if the screw is 'way out' of adjustment. The above procedure will minimise these effects, and act as a coarse adjustment; when the 'feel' for this adjustment has been obtained, pressing 'PAUSE' will enable the adjustment to be 'fine tuned', providing the diffraction grating adjustment is correct.

With the later Pioneer players (those that use the PEA1030 optical blocks and subsequent types) there are two adjustments: one for tangential and the other for the lateral or radial adjustment. Access for the relevant screws is much easier, whether the adjustment is necessary in either single, twin or multi-disc players where the optical blocks can be mounted 'lens up' or 'lens down'.

Fig. 6.24 shows the location of the adjustment points for a 'lens up' optical assembly using a 1.5 mm hexagonal wrench. With a 'lens down' assembly the location of the screws will effectively be the same, but as it is now the other end of the hexagonal screws that are to be adjusted, a small Philips screwdriver will be required. The procedure is as previously described, with the exception that both screws are adjusted in turn until the optimum RF eye pattern waveform has been achieved.

References to the tangential and lateral (radial) adjustments have been related to the relevant service manuals, but the perfectionist studying this book may well find that the reference to tangential adjustments in the earlier Pioneer players has become somewhat different for later players. In

Test mode and adjustment procedures

(a)

Lateral (radial) and tangential adjustment screws

Turntable height gauge

(b)

Diffraction grating adjustment

Turntable height gauge

Lateral (radial) and tangential adjustment screws

Figure 6.25 *Pioneer pick-up assembly (PEA1030); (a) top, (b) bottom*

(a)

12 cm disc

Diffraction grating adjustment position 8 cm disc

(b)

Turntable heignt gauge

Diffraction grating adjustment postion

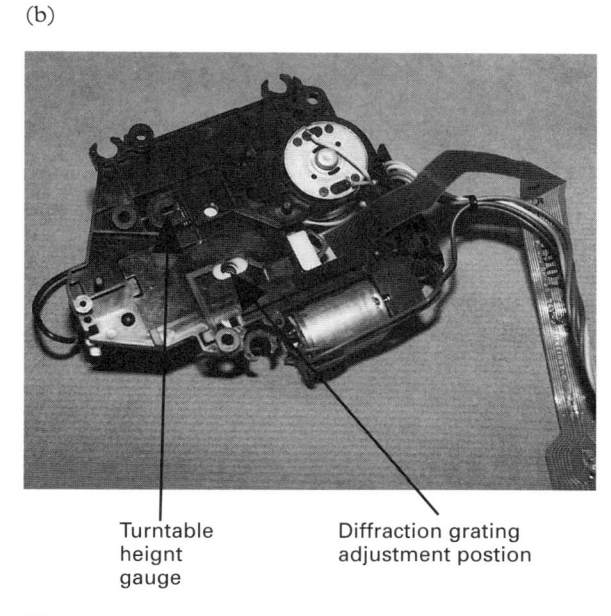

Figure 6.26 *Pioneer pick-up assembly (PWY1011); (a) top, (b) bottom*

Test mode and adjustment procedures

(a)

(b)

(c)

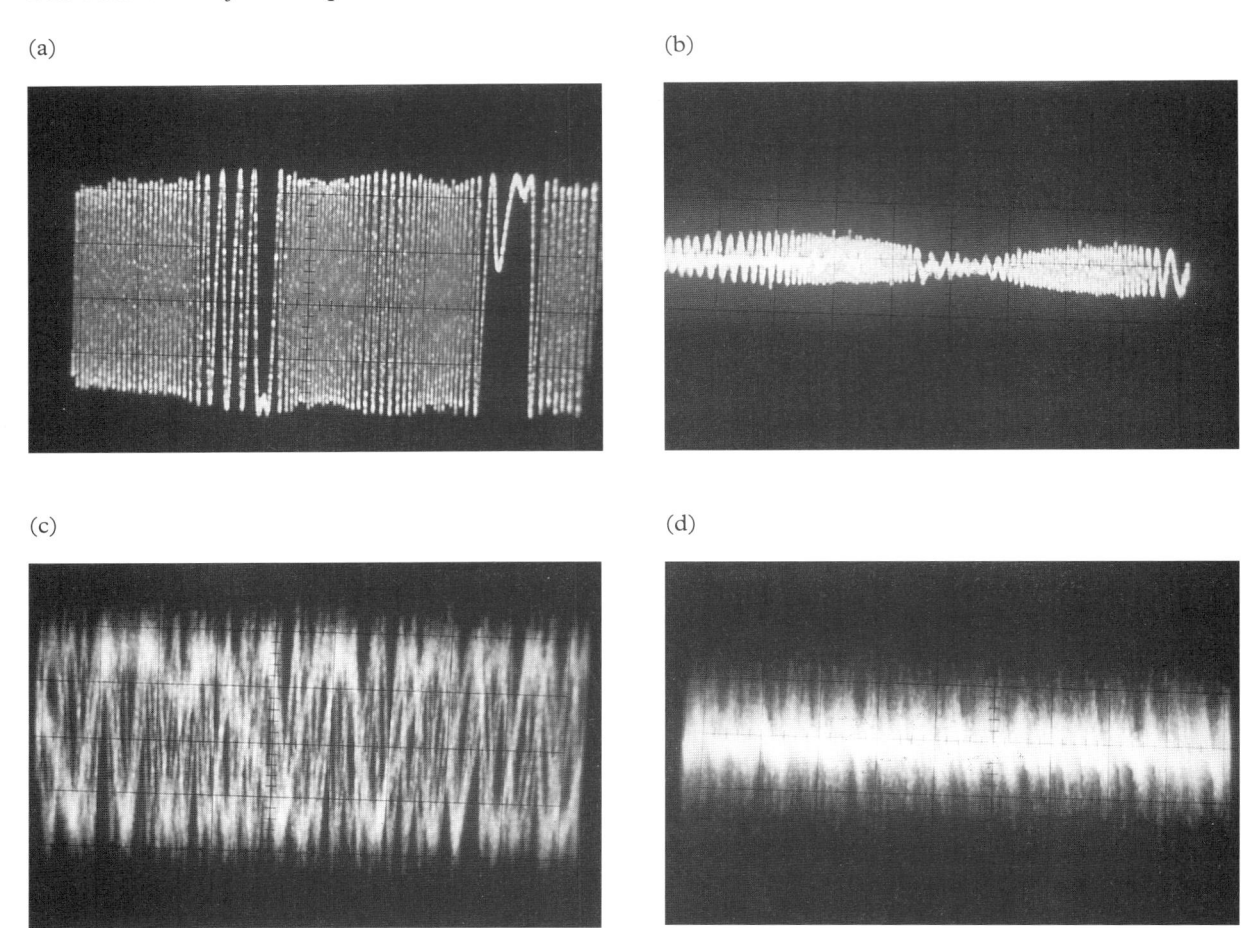

(d)

Figure 6.27 *Tracking error waveform. (a) Maximum with a filter (as referred to in relevant service manuals). Oscilloscope settings: x = 20 ms/cm, y = 0.1 V/cm (x is probe). (b) Minimum with a filter. Oscilloscope settings as in (a). (c) Maximum without filter. Oscilloscope settings: x = 5 ms/cm, y = 0.1 V/cm (x is probe). (d) Minimum without filter. Oscilloscope settings as in (c)*

fact it would seem that the term lateral or radial is more correct when applied to the earlier players, and that tangential adjustment is more correct in the later players. This slight disparity is noted, and the author will leave the reader(s) to draw her, his or even their, own conclusions.

Diffraction grating adjustment

Diffraction grating adjustment, based not only on the author's experience but on discussions with many engineers, can prove to be one of the most frustrating, as well as 'hit and miss', of all the adjustments previously described. Yet, providing a few basic ground rules are followed, there should be no major problems.

Fortunately not all players require adjustment of the diffraction grating as it may be factory set, as is now the case with the latest Pioneer optical block. This will no doubt produce a sigh of relief from the many engineers who endeavour to proceed with this adjustment with some trepidation.

Pioneer players in particular have, over the years, always required diffraction grating adjustment whenever an optical block has required replacement. The author's attempts at this

Test mode and adjustment procedures

adjustment following the relevant service manual have yet to prove successful, and the same applies to the majority of engineers. The main problem is related to obtaining the required minimum tracking error signal, which has always proved nigh impossible. A wide band oscilloscope is probably required in the region of 100–150 MHz, but these do not reside in many service workshops.

However, the following procedure, which is based upon Pioneer players using the PWY1011 and PEA1030 types of optical blocks, will prove successful in the majority of, if not all, cases of diffraction grating adjustment.

Before proceeding with this adjustment it is necessary to obtain the correct sized screwdriver: a standard insulated 2.5 mm flat-bladed, parallel-shaft screwdriver – *no other will do*.

Inserting the screwdriver into the adjustment point and turning the diffraction grating will indicate that there is a very small amount of rotation (no more than 20°) of the screwdriver, and care should be taken otherwise damage can be done to the grating adjustment which may prevent the correct adjustment point being achieved.

1 Set the player to the 'test mode' (refer to the Test Mode section)

Note. It is essential that the tangential and lateral (radial) adjustments have been provisionally set as previously described, otherwise there is the possibility of being unable to successfully complete the diffraction grating adjustment because the laser beam will be looking slightly 'sideways' at the tracks on the disc, preventing the tracking servo from locking onto the track correctly. Also it may well be necessary to check the laser power as described.

2 Connect the oscilloscope, via a 10:1 probe, to the tracking error (TR. ERR.) test point.

3 Move the optical block, by pressing the 'MAN SEARCH FWD'. or 'REV.' keys, until the grating point can be accessed.

Note. With the 'lens down' method of fitting the optical block, i.e. the single disc turntable platter and multi-disc players, the optical block is moved to approximately halfway across the playing area of the disc, at which point the grating adjustment can be accessed through a hole in the mounting plate. With the 'lens up' method, i.e. conventional single and twin disc players, the optical block is moved towards the outer edge of the disc where access can be gained to the grating adjustment. With these particular types of players it frequently proves useful to use an 8 cm disc with a playing time of 20 min, instead of a conventional 12 cm disc.

4 Press 'Track Forward' (TRK. FWD.), or 'Programme' (PGM). Disc 1 in the multi-disc players will be accepted. The laser and focus search sequences will now operate.

5 After 3 s, press 'PLAY'. The disc will now rotate at an acceptable speed.

6 Set the oscilloscope to observe the tracking error waveform ($X = 5$ ms, $Y = 0.1$ V cm^{-1} with a 10:1 probe).

7 Insert the 2.5 mm screwdriver into the grating adjustment, and turn the screwdriver gently fully anticlockwise, taking care not to exert excessive pressure.

8 Now carefully turn the screwdriver fully clockwise, counting the number of maximum responses observed on the oscilloscope.

Note. If the response is a maximum when the screwdriver is turned fully anticlockwise, count this response as the first maximum.

9 Return the screwdriver to the fully anticlockwise position, and now turn the screwdriver to the mid-point maximum position.

Note. If an even number of maximum responses is observed (e.g. 8) then the mid-point could be either the fourth or fifth maximum response, but the fourth would be recommended at this point. If an odd number of maximum responses were observed (e.g. 9) then the fifth would be recommended at this point.

10 Ensure that the grating is adjusted to indicate an absolute maximum response.

11 Remove the screwdriver, press the 'PAUSE' control, and observe whether the player now plays the disc by monitoring the audio output, or checking that the front panel display is showing track and timing variations.

12 If this does not occur, press 'PAUSE' once more as the incorrect grating point has been selected.

If the incorrect grating point has been selected:

13 Re-adjust the grating to one of the maximums either side of the incorrect point, again ensuring that the absolute maximum response has been achieved.

Note. Referring to the examples mentioned in stage 9, with an even number of responses, take the next maximum response (e.g. fifth), and with an odd number of maximum responses, the one either side will suffice.

Test mode and adjustment procedures

In effect, when carrying out this adjustment it can be observed that alternate maximums will enable the player to operate, whilst those in between will not. This is a consistent rule with all grating adjustments experienced by the author, and the reason for choosing the most central maximum that operates is to ensure that the mid-tolerance point is selected.

The actual number of maximums that can be obtained can vary from one optical block to another, but a reasonable average is 9 or 10 for a new block. If the optical block has been abused with frequent unsuccessful adjustments, the actual grating adjustment point can wear and therefore the number of maximum responses can be much less.

This completes the recommended adjustment procedures, but occasions do arise when it may prove necessary to repeat or double check some of the adjustments to ensure that one adjustment has not inadvertently affected another.

7
SYSTEM CONTROL

Most domestic electronics products are controlled by some form of system control, which whilst usually being extremely complex, provides virtually complete control of the operation and functioning of a specific item of electronic equipment.

Whilst the general aim of system control is no doubt to provide a wide range of facilities, the frequent problem that can occur is endeavouring to drive whatever the product may be, as well as achieve a complete understanding of what all the facilities a particular product may provide, and one can wonder if the extensive technology involved is really feasible when maybe only basic functions are required.

This is especially so where the extensive programming facilities that are possible enable a person to record a TV programme which is scheduled to occur n days or weeks away, subject to world and political catastrophes, or even program a CD player capable of playing 6 to 12 discs in a specific order of disc and track selection.

The engineer comes into this when the requirement arises for servicing products which 'appear' to have problems in this area. Many engineers seem to have a love–hate relationship when the subject of system control arises; what does one look for when the system control 'apparently' goes on the blink? Many a tale has been told of the system controller chip being replaced, only to show that the previous one was probably operating satisfactorily after all, as the same symptoms still exist.

Thus this chapter endeavours to approach the subject in a fairly logical manner, in order that a reasonable appreciation may be achieved.

Generally the master system controller (Fig. 7.1) will control the overall functions of the player, and in practice the same concepts can be applied to other involved items such as VCRs, in-car products, etc. Communication between the controller and a controlled item is achieved by an interchange of information, and it is quite usual for an output expander to be included because the output circuit of the controller may not be sufficient to drive the controlled item; this may be more readily illustrated below.

Consider the operation of opening the drawer of the CD player, which will be similar to ejecting the tape from a VCR (Fig. 7.2). This aspect of system control is relatively straightforward, and operating the 'open' or 'eject' control causes data pulses from the system control via the relevant switch to pass back into the controller, which will pass a high or low level to the motor via the output expander. The 'open' or 'close' limit switches serve to indicate to the controller that either the drawer is closed, and therefore able to open, or that the drawer is open and the motor supply is removed.

This arrangement is similar to the optical block 'home' operation; when a CD player is first switched on, it is essential for the optical block to be at the start or 'home' position, and this is detected by the 'home' switch on the optical block.

The high level is usually a positive level to enable the motor to rotate in one direction, whilst the low level, usually a negative level, allows the reverse operation of the motor, and the zero is the 'off' or 'stop' condition.

Frequently a slave system controller is included which will communicate with the master system controller, especially where the control panel operates a unit remote from the main unit(s) of an overall system, and is typical of in-car and a wide range of hi-fi products.

Consider the focus search operation of a CD player when a disc is inserted into a player (Fig. 7.3). The system controller informs the CPU interface to commence the focus search sequence, which is a series of commands sent via the data lines, from the system control to the CPU interface, which then instructs the focus servo, also via data lines, to commence the focus search, whereby the lens is moved up and down a number of times.

System control

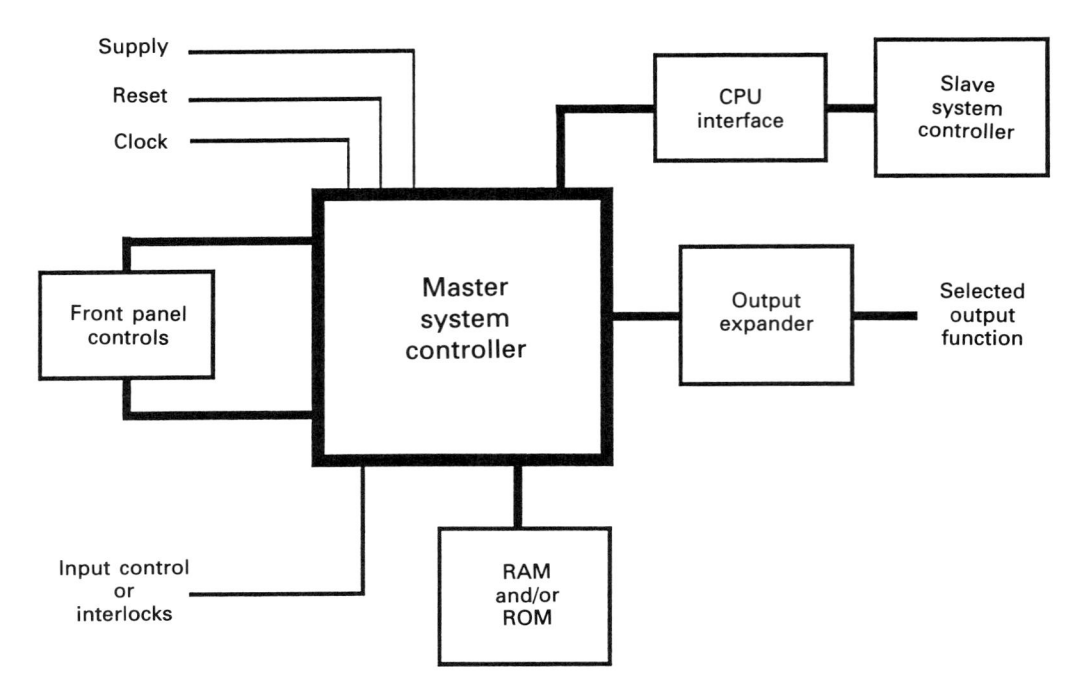

Figure 7.1 *Basic system control arrangement*

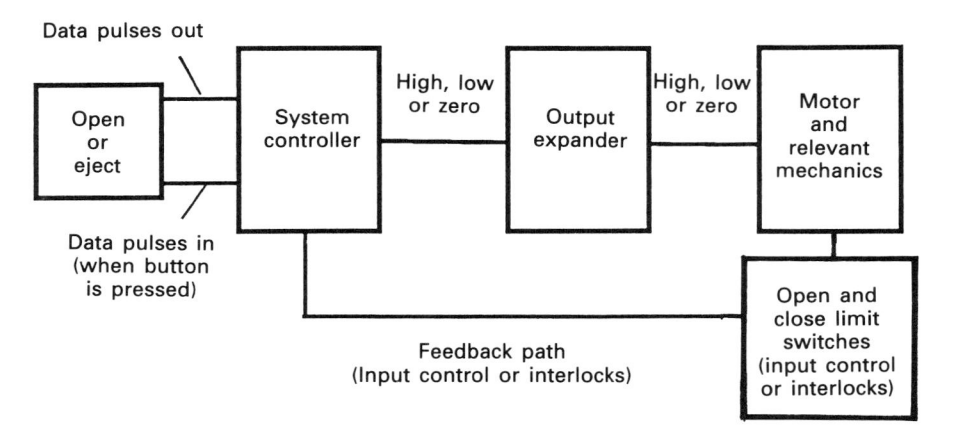

Figure 7.2 *System control and motor drive arrangement*

When focus has been achieved, the laser is assumed to be switched on; a high level will be obtained from the photo-diode array and RF amplifier, which is the focus OK (FOK) signal, which then informs the focus servo and the system control that focus has been achieved.

The system control then instructs the focus servo, via the CPU interface, to stop the search function and close the focus servo loop to ensure that the laser maintains focus on the disc via the focus error signal.

Fig. 7.4 illustrates the system control arrangement that can be found in a wide range of players that utilise the CXD2500Q decoder, which, typical of many decoders, contains the CPU interface.

System control

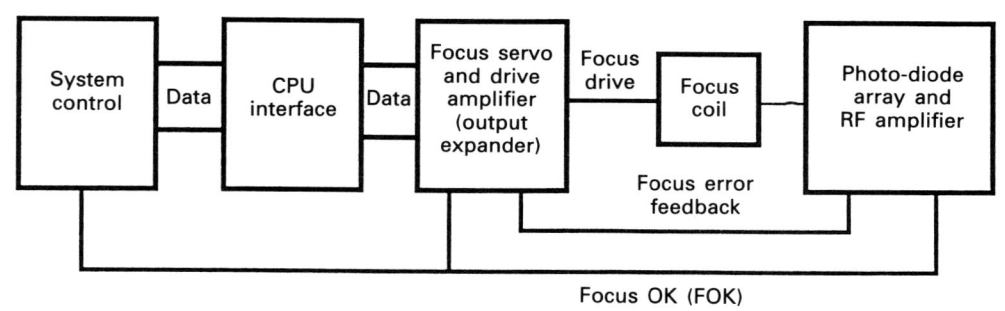

Figure 7.3 *System control and focus servo arrangement*

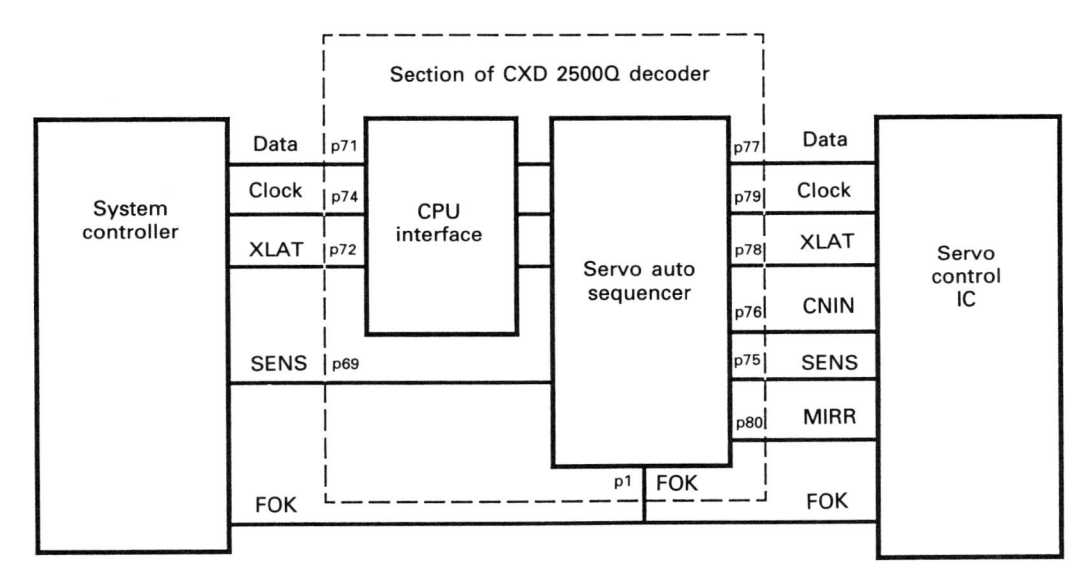

Figure 7.4 *Basic system control and CPU interface*

Whenever a function is required by the system controller in relation to a servo function, a data signal is sent to the CPU interface simultaneously with the clock signal. The data will comprise an address of 4 bits together with up to 16 further bits in relation to the command that has been requested, and the clock signal will be of the same time duration as the data signal. On completion of the data transfer, a latch signal (XLAT) will be sent to inform the CPU interface that the data transfer has been completed (Fig. 7.5).

These commands are transferred in a similar manner from the CPU interface to the servo auto sequencer, which can perform a series of functions without further commands from the system controller. Typical of this would be the series of operations that take place after selecting the 'play' command, which will include automatic focus and track jumping to achieve the correct playing of the disc.

The focus OK (FOK) signal will indicate with a high level that the laser has achieved focus on the disc, and is passed to the auto sequencer of the CPU interface and the system control. If the FOK signal does not go high, the system control will cease any further function of the auto sequencer and shut down.

The mirror signal continually ensures the laser is tracking on the track of the disc, whilst the table of contents (TOC) is being read, and whilst the disc is being played normally. When the TOC has been read it is necessary to move the optical block to the start of the music playing area of the disc by counting the required number of tracks

137

System control

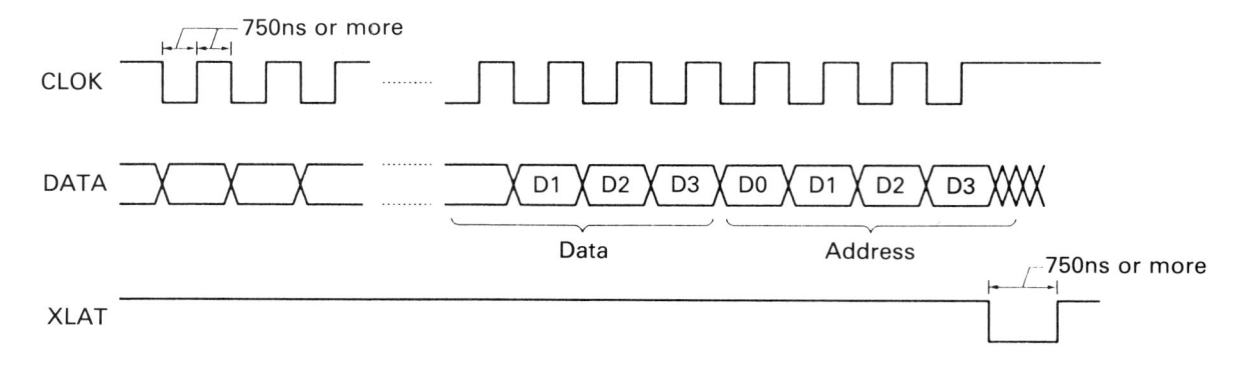

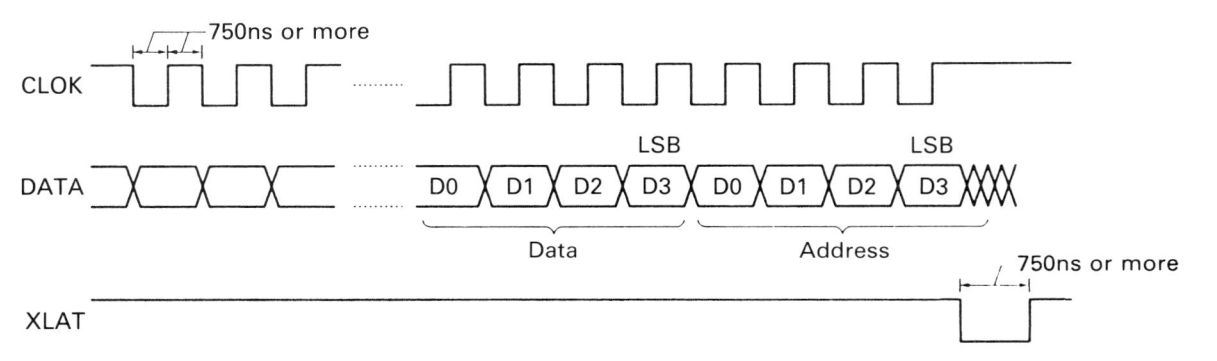

Figure 7.5 *System control clock, data and latch signals*

(CNIN) to the start area to commence playing the disc.

The sense (SENS) signal will provide information on whether various stages of the selected function have been achieved and can be high or low depending upon the specific function that has been selected. Measurement of any of the clock, data and latch signals can prove difficult, but even if the required test equipment is available, the specific data information is rarely provided in the relevant service manuals. The preferred method of monitoring any of these signal lines is to ensure that there is some form of movement present on the oscilloscope, especially when some of the front panel controls are selected.

The circuit shown in Fig. 7.6 is an extract from a typical CD player, showing the system controller, keyboard and display.

The three important items described in the fault diagnosis as the 'magic three' are:

p64	supply (+ 5 V)
p38	reset
p30/31	clock

Communication outputs to the CPU interface are:

p6	CLOK (clock)
p7	DATA
p19	XLAT (latch)

The 'laser on' (LDON) signal is output from p29, with p28 registering the focus OK (FOK) signal.

Opening and closing the disc drawer

When the drawer is selected to 'open', p24 (LOUT) will go high and drive the motor to open the drawer and release the disc. When the drawer is fully open p25 (open) will go low, which in

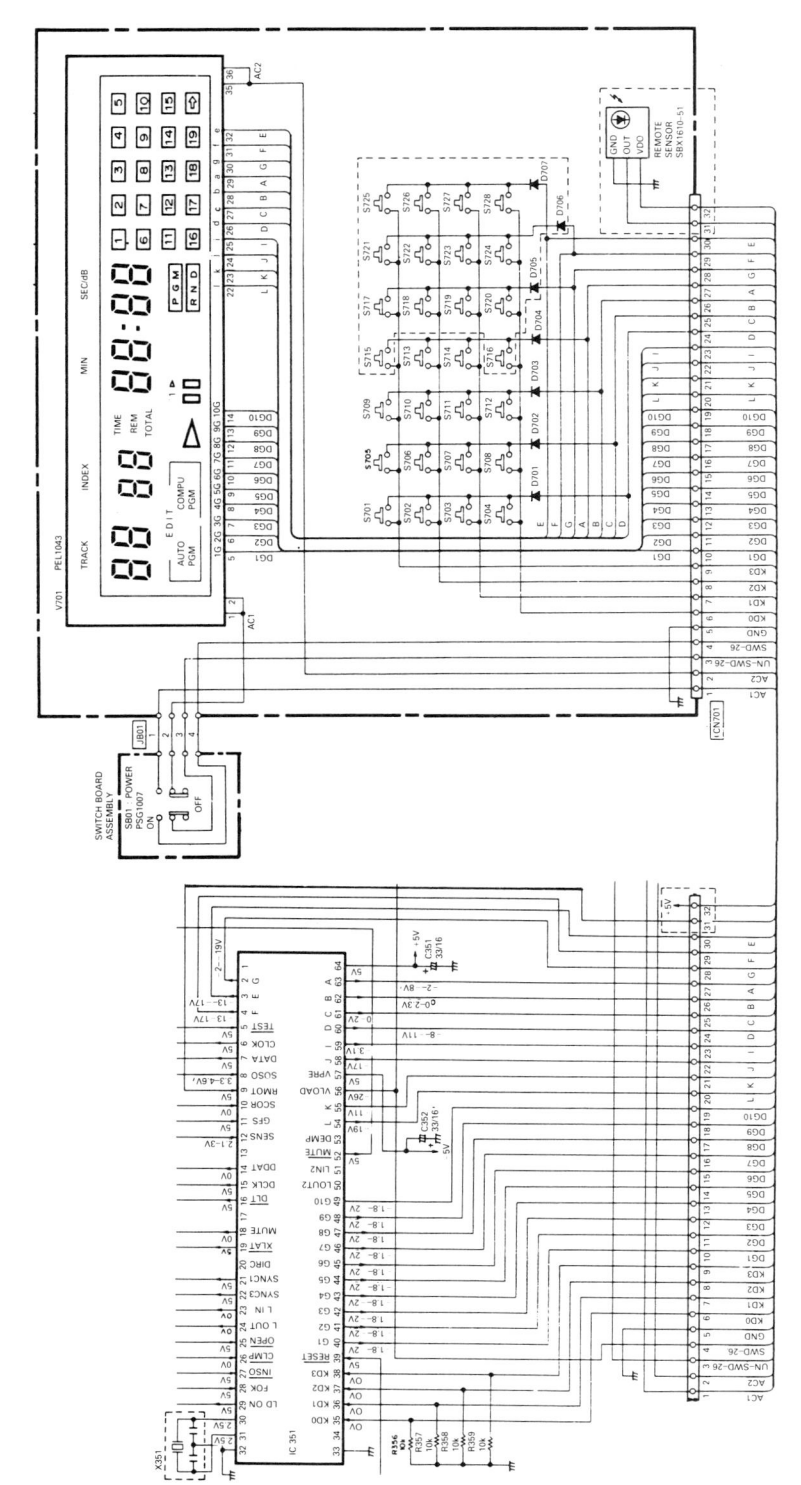

Figure 7.6 Example of a system control circuit in a CD player

System control

turn will cause p24 to go to zero and stop the motor.

Closing the drawer is achieved when p23 (LIN) goes high to drive the motor to close the drawer and clamp the disc. When this is achieved, p26 (CLMP) will go zero to enable the motor to stop.

Sensing whether the optical block is in the home position

In order that the system controller knows that the the optical block is at the start, home, or inside position, to enable the TOC to be read either at switch on, or when a disc is inserted, the optical block is driven by the carriage motor until the inside switch closes and p27 goes low, stopping the carriage motor via data interchange.

Front panel controls and display

The various segments on the display are controlled from pins 40 to 49 (G0–G10), and the A–L lines from pins 1, 2, 3, 54, 55 and 58–63. The A–G lines are also fed to the keyboard matrix through the relevant diodes (D701–D707), with each key making contact with a horizontal line of the matrix to connect to the KD0–KD3 lines, which go to pins 35–38.

Data pulses will appear on the A–G lines, and will appear on one of the KD0–KD3 lines depending upon which key is selected.

Some system control arrangements only provide pulses on the equivalent A–G lines when a relevant key is selected, and therefore in fault diagnosis it is important to ensure that no pulses appear on the KD0–KD3 lines when no keys are depressed.

It is quite often known for keys to be jammed, which effectively 'locks up' the system control, so checking the KD0–KD3 lines can be extremely useful.

Remote control signals are received by the remote sensor and applied to pin 31 (RMDT).

Occasionally some form of memory is included with the system control, in the form of either ROM or RAM, which is used to ensure that the player, or even VCR or other product, carries out specific functions. In the case of programmable players it is necessary to store a sequence or program of tracks on the disc, or discs in the case of multi-disc players, to be played in some specific order.

The information outlined above should enable engineers to achieve a reasonable appreciation of typical system control arrangements.

8

IN-CAR COMPACT DISC PLAYERS

Generally speaking, with the exception of supply voltages, physical size and compactness of in-car CD players, the basic technology is the same as that for the hi-fi range of players. This chapter will therefore be relatively brief with the main intention of considering certain important areas that will hopefully assist engineers who have to service these products.

These players usually appear either as single disc players, which are fitted into the dashboard area of the vehicle, or as multi-disc players which are fitted into the boot of the car (Fig. 8.1).

Single disc players

Single disc players can appear either as an additional unit connected to an existing system which in itself is capable of driving the CD player, or as a virtual stand-alone unit which may also have a tuner and amplifier facility included.

Single disc players in particular can obviously be extremely compact units, and it is important to ensure that the relevant service manual is available in order to determine the correct dismantling, re-assembly and possible adjustment procedures.

The CD is inserted and ejected usually by some form of motor-driven mechanism, and the playing procedure is usually relatively simple in order to minimise distraction to the driver, especially as he/she may be proceeding down the outside lane of a motorway at 70 mph.

Multi-disc players

There are a wide range of multi-disc players that are fitted into the boots of cars, though they have also been known to be fitted under one of the seats; and they are operated by the head unit fitted into the dashboard. A magazine is used in these players, which can accommodate from 6 to 12 discs, enabling an extensive selection of music to be available and providing a total playing time of 5–12 h, depending upon the system installed and the type of discs being played.

With these players two types are generally available with respect to the initial operation when a magazine is first inserted into the unit.

1   No initialisation process takes place when the magazine is first inserted, and the player commences playing discs when requested by the operator.

2   An initialisation process commences the moment the magazine is inserted, where each tray of the magazine is checked to determine which tray contains a disc inserted, and this information is stored in an internal memory; this process is typical of Pioneer players.

Depending upon the type of multi-disc player, either the magazine or the optical assembly is moved vertically to the correct position to enable a specific disc to be selected, clamped into position, and played.

Concerning the installation of these players it is essential that the procedure recommended in either the users' manual or the service manual are accurately followed. Some players can be vertically or horizontally mounted, and with these it is possible that some internal support springs will have to be adjusted to compensate for the weight of the player mechanism. If this is not correct internal damage can result, or excessive skipping, jumping and even total inability to play can result. However, the majority of these players are fitted horizontally, though the author has been aware of players being fitted on their side, or even upside down in the roof of the boot area of the vehicle, which has resulted in some delightful dismantling techniques in order to restore the unit back to the original playing ability. Possibly the

In-car compact disc players

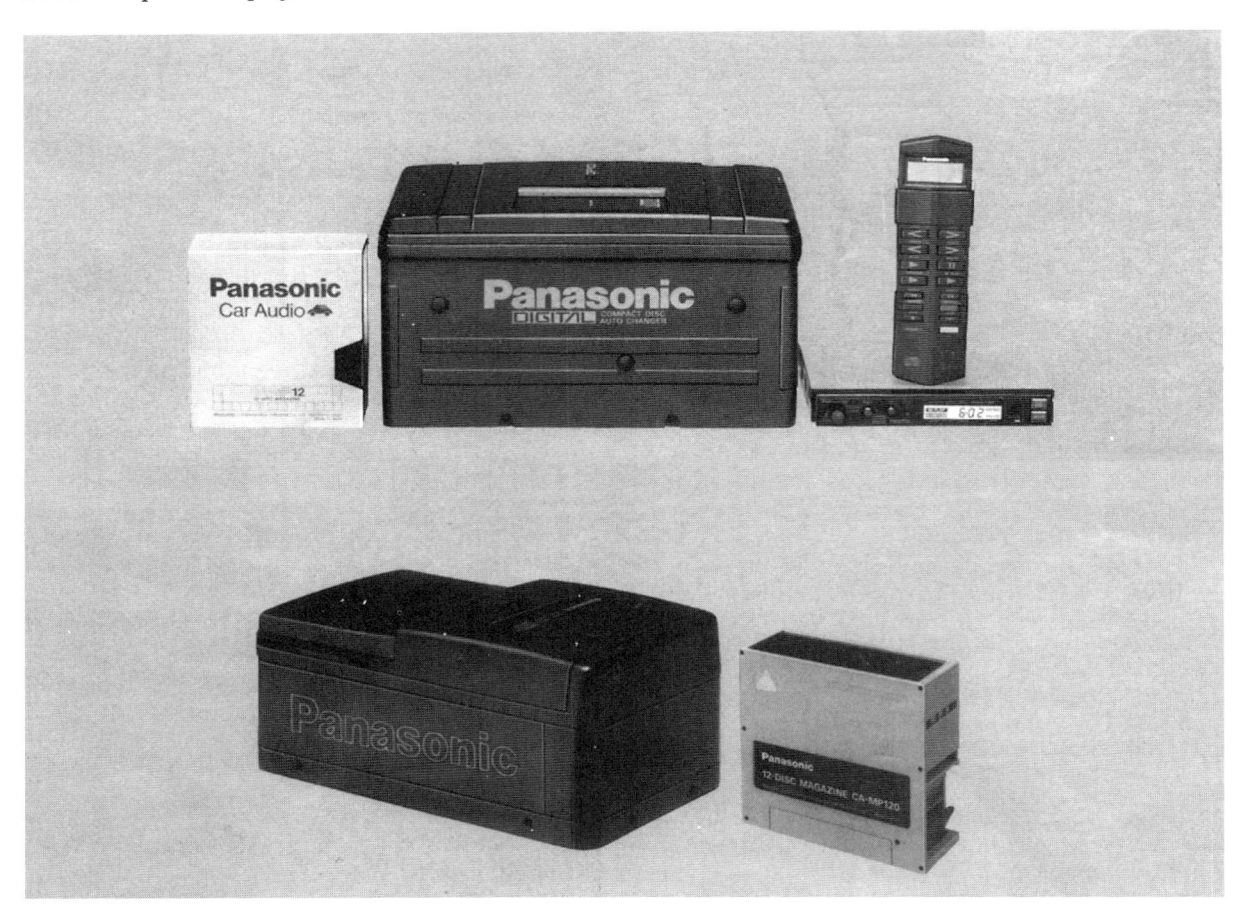

Figure 8.1 *In-car CD players*

'best' incorrect installation observed was into an extremely expensive brand new car, where the owner required the multi-disc unit to be fitted on its side, at a fairly acute angle adjacent to the driver's left leg in the driving area of the car next to the gear change, so that the magazine could also be conveniently changed as required. This installation required extensive cutting of carpeting to enable the mounting to be achieved – and all was well until the first disc was selected!!

With the various types of multi-disc players available, a wide range of magazines are available, a selection of which are shown in Figs 8.2–8.5.

Some magazines have trays which can be drawn open. It is essential to be aware that some magazines contain trays that can only be withdrawn singly, and opening more than one tray at a time when inserting a disc can cause damage to the

Figure 8.2 *Example of magazine with disc inserted (Pioneer)*

In-car compact disc players

Figure 8.3 *Examples of 12 cm and 8 cm disc magazines (Pioneer)*

Figure 8.4 *Six disc magazine (Pioneer)*

Figure 8.5 *Twelve disc magazine (Pioneer)*

tray, which can result in jamming of the player mechanism if more than one tray is inadvertently withdrawn from the player.

With other magazines it is possible to completely remove all the trays when inserting the disc, but as always enough emphasis cannot be placed upon the need to refer to the relevant instructions for the magazines and players concerned.

Multi-disc player mechanism control

When the magazine is inserted into these players, there is usually some switching that can determine the position of either the magazine or the optical assembly, and whether a disc is clamped ready to play or a tray is in or out of the magazine. Also some form of switching will determine that a magazine has been inserted, or that the ejection process has been completed. All the switching information is usually sent to the system controller as either input control or interlock signals.

143

In-car compact disc players

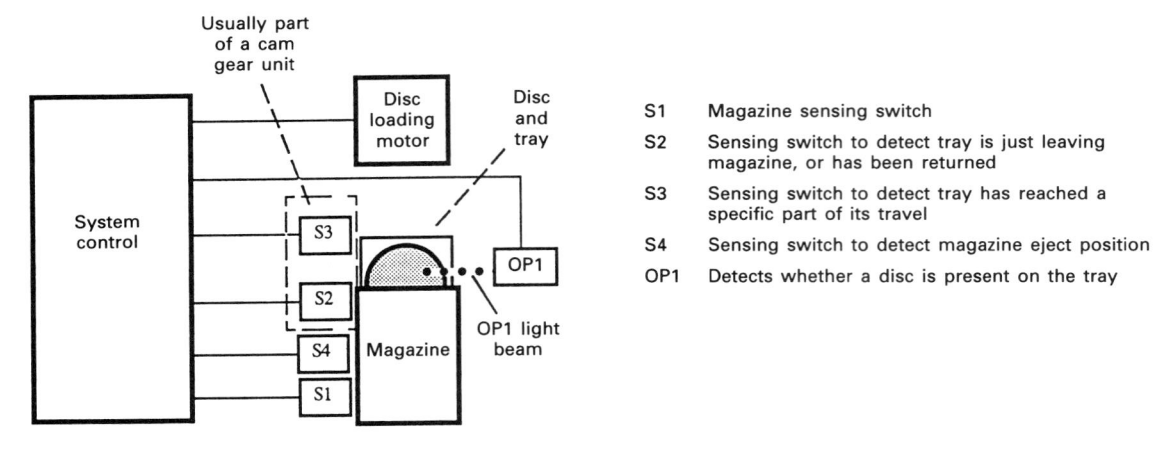

Figure 8.6 *Method of detecting that there is a disc on the tray*

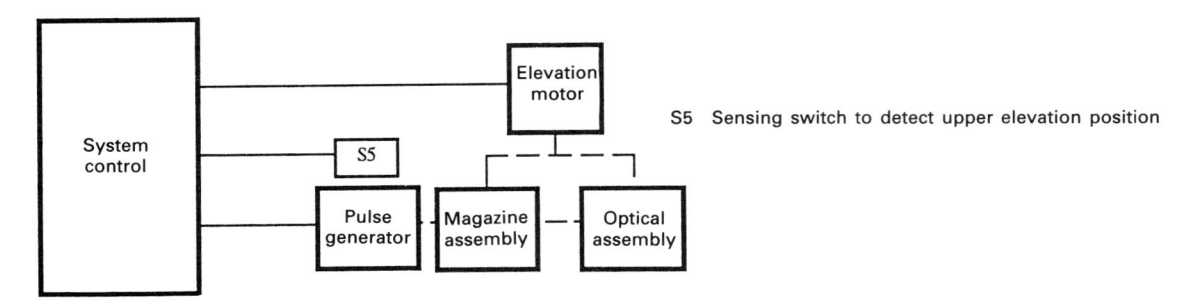

S1   Magazine sensing switch
S2   Sensing switch to detect tray is just leaving magazine, or has been returned
S3   Sensing switch to detect tray has reached a specific part of its travel
S4   Sensing switch to detect magazine eject position
OP1   Detects whether a disc is present on the tray

Figure 8.7 *Detecting the elevation position of the magazine or optical block*

S5   Sensing switch to detect upper elevation position

If the player is of the type that initialises when a magazine is inserted, a fairly complex routine is automatically followed, which involves sensing switches and opto couplers to determine whether a disc is present on the tray (Fig. 8.6) and also the vertical position of the magazine or optical block (Fig. 8.7).

When the magazine is first inserted, and the initialisation process commences on the relevant players, S1 detects that the magazine is inserted and informs the system controller, which usually switches on any applicable power supply.

The disc loading motor now commences to remove the tray from the magazine, and as this is started, S2 detects that the tray is just about to leave the magazine, and may be referred to as the tray position (TRPN) switch. S3 detects when the tray has reached a certain position, and instructs the system control to monitor via the opto coupler, OP1, whether a disc is present on the tray. The tray continues to move slightly further, when S3 changes its state, at which point the loading motor reverses, to return the tray back into the magazine. S3 may be referred to as the 'disc present' (DCPN) switch.

When the tray is back into the magazine, S2 will change its state, and the magazine or optical assembly now moves to the next position in readiness for the next disc.

The pulse generator is usually either a switch or an opto coupler, depending upon the method that is being utilised. As the elevation motor raises or lowers the magazine or optical block, a series of pulses will be generated which will enable the system controller to determine that the next disc, or selected disc in the normal play mode, has been reached. An upper limit switch, S4, may be included in some models to detect that the upper level of elevation has been reached.

In-car compact disc players

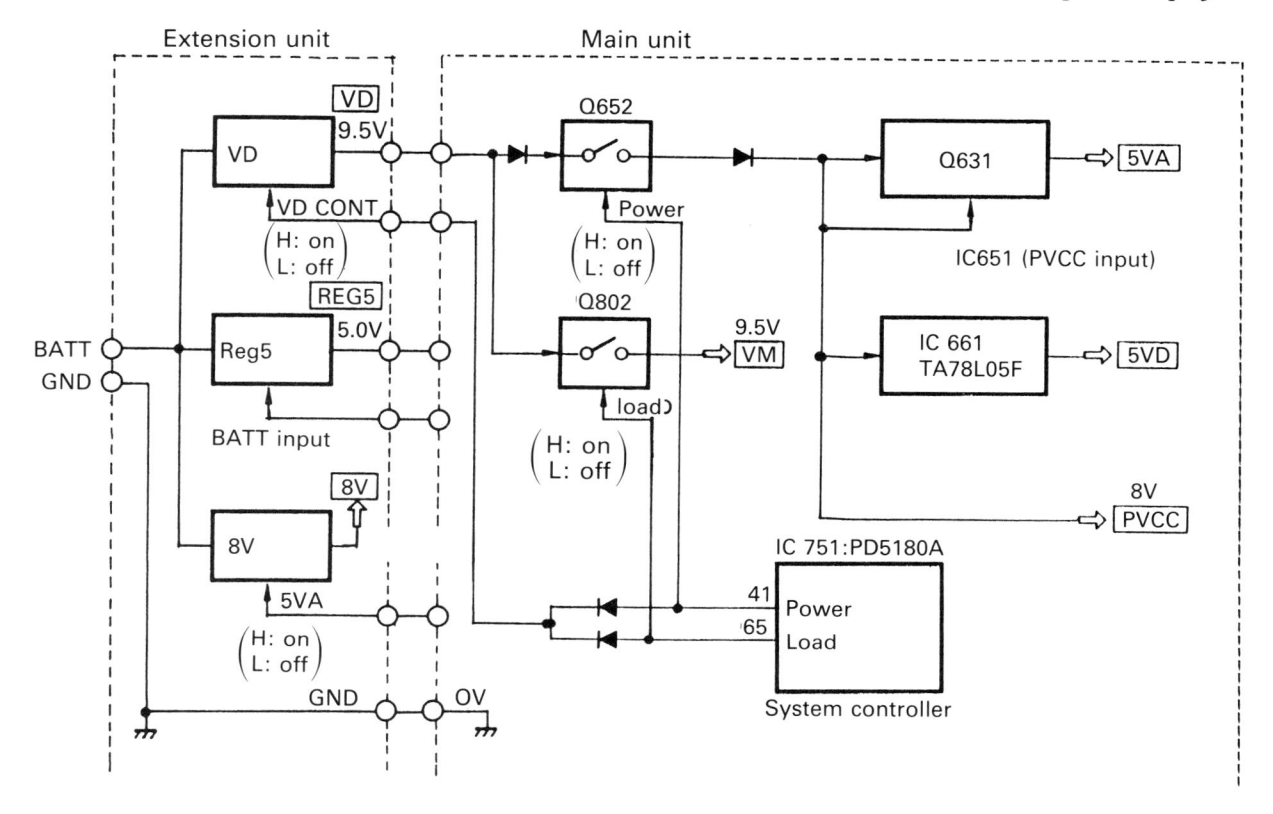

Figure 8.8 *Pioneer CDX-M30 power supply block diagram*

When the magazine is ejected, the elevation motor will restore the magazine or the optical block back to the start position when the magazine was first inserted. Then the loading motor will follow a sequence to release any retaining mechanism to enable the magazine to eject. When this position has been achieved, S5 will change its state and cause the disc loading motor to reverse until S5 changes its state once more when the loading motor will stop to await the next magazine.

The above information will provide an appreciation of this type of mechanism control, but a word of warning to the unwary: the first of the Pioneer multi-disc players, CDX-M100, had a method of reversing the tray back into the magazine immediately a disc was detected during the initialisation process, but if there was no disc present then the tray would continue to be extracted until a limit switch detected that this position had been reached. Therefore the process took slightly longer if there was no disc present on a tray compared to when one was present, and it has been known for engineers to dismantle the mechanism on the quite reasonable belief that the mechanism may be jamming up when a disc is present as the tray travel was not as far. All the later models take exactly the same amount of time whether a disc is present or not.

Power supplies

The power supplies of in-car CD players tend to have a complexity of their own, usually for two reasons: they are frequently distributed throughout the circuit and require a certain amount of time to trace the relevant sources and subsequent paths; also, there is usually a d.c.–d.c. converter, and this comes either in the form of an IC to provide either another voltage level, or more specifically a negative potential to enable servo circuits to perform their work adequately.

145

In-car compact disc players

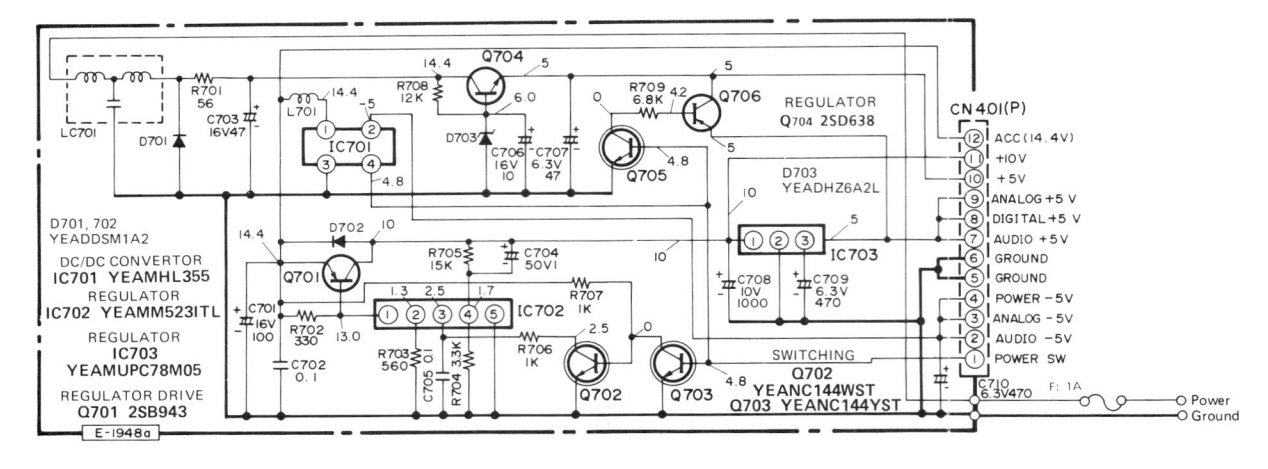

Figure 8.9 *Extract of power supply circuit of the Panasonic CX-DP10EN*

Figs 8.8 and 8.9 are two power supply diagrams, in block or circuit form, which provide an overview of the methods that exist. Essentially the main section of any power supply is controlled by the system controller, though there naturally has to be some form of +5 V supply to maintain a supply to the controller when the in-car unit is switched off.

The battery (BATT) supply is applied to three blocks:

1 Reg 5: to supply a continual +5 V to maintain the Vcc voltage for the system controller.
2 VD: to supply 9.5 V to drive motors and provide other 5 V supplies.
3 8 V: to supply audio pre-amplifier and audio isolator or buffer circuits.

When a magazine is inserted the LOAD output from pin 65 of the system controller will switch on the VD and Q802 circuits, to enable the initialisation process to take place. When 'play' is selected, pin 41 (POWER) and pin 65 (LOAD) will switch on the VD circuit, together with Q802 to provide motor supplies, and also Q652 to provide two +5 V supplies via Q 631 and IC661, as well as the +8 V which is the Vcc supply to a driver IC which operates the focus and tracking coils, and the carriage and spindle motors.

With this circuit there is no negative supply, which is usually required in CD players to enable the focus and tracking coils to move either side of a neutral position, but in the later Pioneer in-car CD players a reference voltage (V Ref.) is derived to provide a +2.5 V supply which is used as the neutral position potential, with +5 V, which is in effect +2.5 V compared to the V. Ref., and ground or 0 V, which is in effect −2.5 V compared to the V. Ref., and it is important to be aware of this fact when adjusting these CD players (see below).

The section of circuit shown in Fig. 8.9 is interesting insomuch that a −5 V potential is derived from IC701. The power supply is applied via the fuse at the bottom right-hand corner of the diagram to be fed to Q704 via LC701, which provides a low current +5 V to the system controller, whilst the player is in the 'off' condition.

The system controller provides the power switch on signal via pin 1 of the connector CN40, which controls Q703, Q705 and IC701 and their associated circuits. The +14.4 V appears when the accessory switch on the ignition is switched on providing ACC via pin 12 of CN40, and is applied to pins 1 of IC701 and IC702, the emitter of Q701, the collector of Q703, and the base of Q702, which will conduct. Q703 collector will go low when switched on, which in turn will now cause Q702 to switch off, with pin 3 of IC702 going high, which via pin 1 will now cause Q701 to conduct as pin 1 goes sufficiently low. The voltage at Q701 emitter provides +10 V to pin 1 of IC703 and pin 12 of CN40, with IC703 providing +5 V for a range of applications via pins 7, 8 and 9 of CN40.

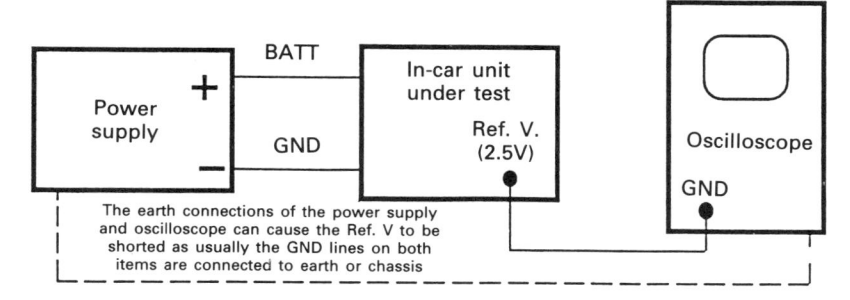

Figure 8.10 *Precautions to observe when adjusting some of the Pioneer in-car CD players*

When Q705 is switched, this causes Q706 to conduct as a result of the +5 V from pin 3 of IC703, and the output from Q706 is in effect a back up +5 V to provide an increased current handling capability for the system controller.

As previously mentioned, when IC701 (which is a d.c.–d.c. converter) switches on, −5 V will be provided at pin 2 to be applied to a range of circuits via pins 2, 3 and 4 of CN40.

Outline of adjustments

When carrying out adjustments of the in-car CD players, many are quite similar to those outlined in Chapter 6, but it is important to refer to the relevant service manuals wherever possible as requirements vary from one manufacturer to another, with possibly the Pioneer in-car players being the most involved; it is with respect to these players that care must be taken when connecting the oscilloscope, as outlined in Fig. 8.10.

Because the Ref. V of 2.5 V is used as the neutral position reference within the player for the operation of the servos, the ground connection of the oscilloscope is required to be connected to this point when carrying out specific adjustments. As it is relatively common practice with some manufacturers of low voltage power supplies to connect the negative potential of the low voltage power supply directly to the chassis or ground of the unit. If the oscilloscope chassis is connected to the earth connection of the mains plug, and likewise for the ground or chassis of the power supply, a situation can be created that when all the connections are made with respect to relevant adjustments, for the +2.5 V Ref. V to be effectively shorted to ground via the earth loop, with the possibility of dire consequences.

To minimise this occurring with some types of power supplies, the relevant service manual suggests that the frame of the measuring instrument (oscilloscope) be changed to a floating status. Whilst some engineers may feel concerned with this procedure, it should be remembered that many oscilloscopes operate in this condition for working on television products, and providing the workshop environment uses isolation transformers and also residual current circuit breakers (RCCB), engineers should feel confident with this procedure; but of course it is recommended that the insulation of the oscilloscope is checked at regular intervals, as recommended in Appendix 2.

Test mode

As a test mode is provided with most Pioneer players (an example of which is shown in Fig. 8.2 where specific keys provide a range of different functions), and as this can vary from one player system to another, complete details are always available in the service manual, but in line with the test mode facilities previously outlined in the chapters related to Adjustments and Fault Diagnosis, the test mode can prove extremely useful to the engineers servicing these particular products.

In-car compact disc players

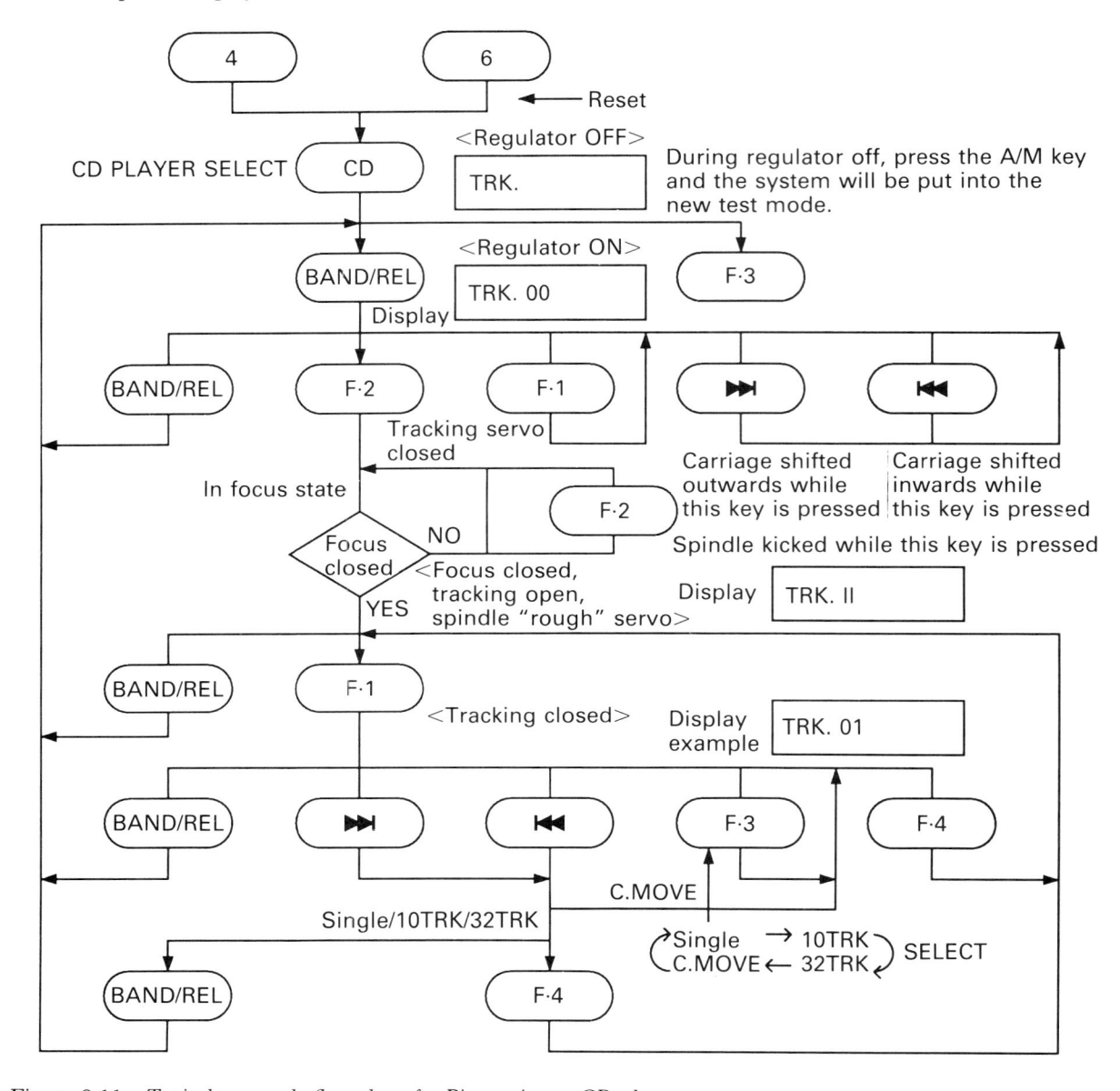

Figure 8.11 *Typical test mode flow chart for Pioneer in-car CD players*

Fault diagnosis

The most common problems that tend to occur are usually related to either power supply problems, or those related to skipping and jumping, especially as the in-car CD player is required to operate in an extremely unfriendly environment. Not only is the player required to operate on demand in varying temperature conditions from the coldest to the hottest (though many players do have a temperature sensing circuit, which switches the player off when extremely high temperatures are encountered), but it is also expected to function no matter what the condition of road surface and thus the player is not expected to give the slightest hint of a skip or jump even when travelling at speed over the bumpiest of surfaces. The installation, as previously mentioned, is important, but cars vary in their suspension characteristics, and therefore it is not unreasonable to expect some form of strain on the system, despite the extremely effective internal suspension techni-ques employed in many players. If skipping and jumping does occur, the relevant areas referred to in Chapter 9 should be considered, with additional consideration being given to the actual installation and suspension of the vehicle concerned.

9
FAULT DIAGNOSIS

With CD players, many faults can be attributed to the mechanical aspects, and actual electronic faults are generally not so common.

The author has never recommended a standard list of CD player faults, on the basis that if the relevant recommendations are implemented with respect to a specific type of player, and if the 'cure' is not achieved, where does one go from here!?

If the optical block or the spindle motor is replaced, because these items are frequently suspected, and the original fault is still present, a waste of time has occurred with respect to the engineer involved, and the eventual bill for the customer may well be unnecessarily high.

Another area where problems can occur is in situations where nothing appears to be happening with the player and the system controller becomes the major suspect, especially as it controls the overall operation of the player. Nothing can be more annoying than replacing a 60 pin (or more) dual in line or four-sided flat pack, only to prove that the component that has been replaced was not defective after all, as the same fault condition still persists.

The test or service mode facilities that are available with some players can greatly assist in fault diagnosis, and reference to these will be included as applicable, though reference to the relevant service manual will provide more specific information for a particular model.

Therefore, with respect to fault diagnosis, a recommended checking sequence is preferred. Whilst it can prove difficult to be absolutely definitive when recommending specific procedures, experience will enable some short cuts to be taken; the following procedures and information have been based on previous encounters with players of various types.

Before really delving into recommended procedures or methods, it is useful at this stage to consider the basic 'start up' routine that most CD players adopt at 'switch on' and when a disc is inserted (Fig. 9.1). Though some players may vary slightly from this routine, it is a useful basis from which to consider whether a player is at least trying to make some attempt to conform to the requirements of playing a disc.

Being aware of the basic initialisation routine of a typical CD player, together with some preliminary observations, can prove extremely useful in helping to determine some initial symptoms with respect to fault diagnosis. If problems are encountered, then the following fault diagnosis procedures will generally assist in determining the faulty area of most CD players.

Suggested check list to assist in determining the possible faulty area, in relation to a specific symptom

The check list given in Fig. 9.2 is not intended to be exhaustive, i.e. the last word in CD player fault diagnosis, but it should at least guide engineers towards resolving most problems, and the following recommended stages (Fig. 9.3) are represented in a logical manner, so that, if 'push comes to shove', following each of the stages and applying some of the relevant suggestions should assist.

Two of the main problems with servicing CD players are inheriting someone else's frustrated attempts, and those nasty obscure faults that all attempts to resolve bear little relevance to the final diagnosis, and which frequently combine together to create what can occasionally become a 'horrendous' experience.

A recent example occurred with a respected engineer who had inherited a Pioneer three-magazine player to service, which totally refused to respond in any manner or form, and certainly gave the final impression of a system control problem. Fortunately previous incumbents of the unit

Fault diagnosis

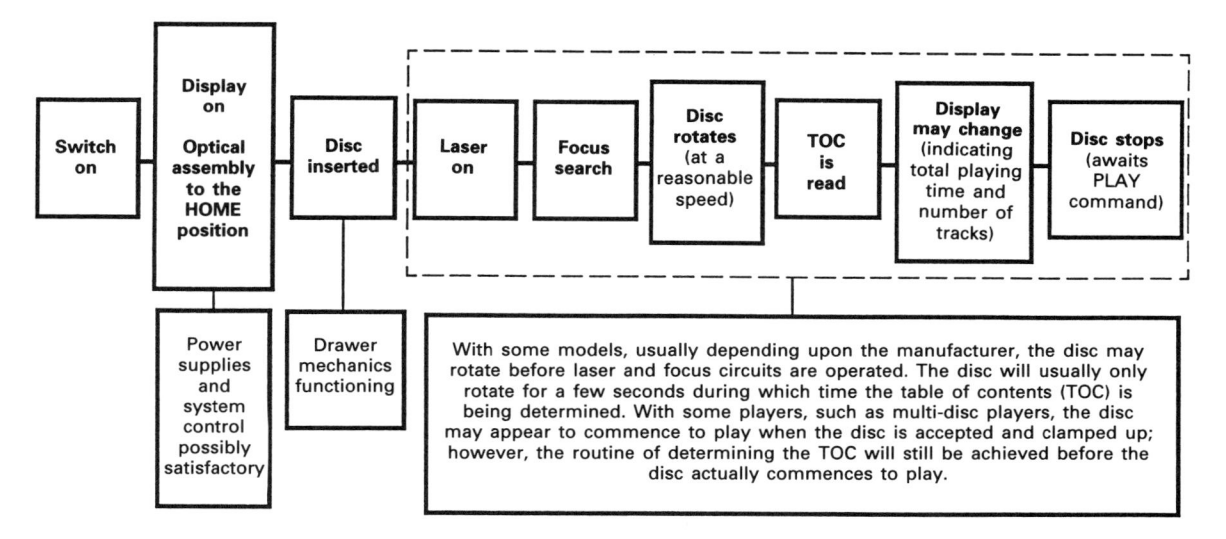

Figure 9.1 *Typical CD player initial operating sequence*

had decided that the unit required a more expert approach, and nothing drastic had taken place. So the engineer reluctantly chose to replace the system controller; as in many cases of this nature, this proved that this item was working quite satisfactorily, as no further progress had been made. It was now a case of back to basics, and a thorough check of all things mechanical was necessary, with the final result being that a small piece of plastic 'swarf' had jammed itself within a couple of the magazine selection gears thereby preventing any further progress or operation.

Many engineers will have experienced the somewhat frustrating problems that can create such catastrophic symptoms, and whilst not endeavouring to bolt the stable door after the horse has bolted, an overall logical approach to fault diagnosis should assist either the uninitiated, or those who may find themselves floundering in the 'slough of despair'.

Stage 1: initial checks

Before proceeding with any major diagnostic thoughts or procedures it is always useful to look round the inside of the player to observe if any strange happenings have occurred, or are taking place (Fig. 9.4).

Has someone been here before?

Unfortunately CD players do tend to be a 'tweakers' paradise', and of course all presets come fitted with a screwdriver slot which is meant for turning... so why not turn it?

Are the preset controls at or around the midpoint settings? If they are at one extreme then the possibility exists that the dreaded twiddler has cast a magic spell, especially with focus and tracking gain controls, as not a lot will happen if they are at the minimum setting.

Many engineers frequently have a 'tweak', but of course the golden rule is to return the preset back to the original position if nothing specific occurs. If this is not done, any fault condition can be further compounded by unnecessary 'adjustment' of the preset controls.

Are all wiring looms still nice and tidy? If they are somewhat dishevelled it is a sure sign that items have been removed and replaced.

It is frequently a pity that engineers are not fitted with X-ray vision in order that the underside of the PCB can be viewed more easily without removal. Some methods and techniques of removing and replacing 80 pin four-sided flat packs can indeed cause trauma of the most undesirable kind, to engineer and customer alike, especially by way of the final bill.

Fault diagnosis

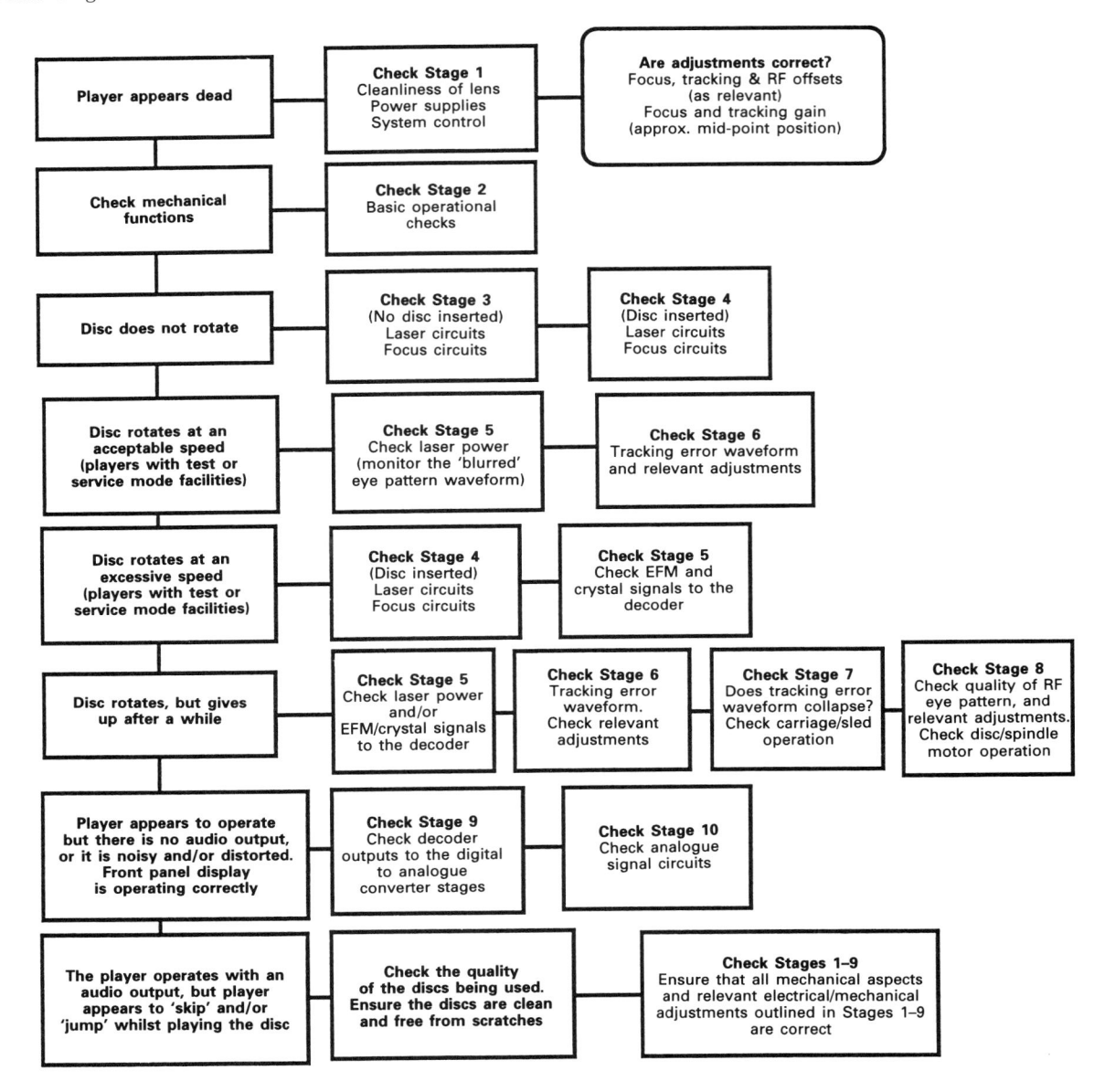

```
┌─────────────────┐     ┌──────────────────┐     ╭──────────────────────────╮
│ Player appears  │────▶│ Check Stage 1    │────▶│ Are adjustments correct? │
│     dead        │     │ Cleanliness of   │     │ Focus, tracking & RF     │
└─────────────────┘     │ lens             │     │ offsets (as relevant)    │
                        │ Power supplies   │     │ Focus and tracking gain  │
                        │ System control   │     │ (approx. mid-point       │
                        └──────────────────┘     │ position)                │
                                                 ╰──────────────────────────╯

┌─────────────────┐     ┌──────────────────┐
│ Check mechanical│────▶│ Check Stage 2    │
│   functions     │     │ Basic operational│
└─────────────────┘     │ checks           │
                        └──────────────────┘

┌─────────────────┐     ┌──────────────────┐     ┌──────────────────┐
│ Disc does not   │────▶│ Check Stage 3    │────▶│ Check Stage 4    │
│     rotate      │     │ (No disc         │     │ (Disc inserted)  │
└─────────────────┘     │ inserted)        │     │ Laser circuits   │
                        │ Laser circuits   │     │ Focus circuits   │
                        │ Focus circuits   │     └──────────────────┘
                        └──────────────────┘

┌─────────────────┐     ┌──────────────────┐     ┌──────────────────┐
│ Disc rotates at │────▶│ Check Stage 5    │────▶│ Check Stage 6    │
│ an acceptable   │     │ Check laser      │     │ Tracking error   │
│ speed (players  │     │ power (monitor   │     │ waveform and     │
│ with test or    │     │ the 'blurred'    │     │ relevant         │
│ service mode    │     │ eye pattern      │     │ adjustments      │
│ facilities)     │     │ waveform)        │     └──────────────────┘
└─────────────────┘     └──────────────────┘

┌─────────────────┐     ┌──────────────────┐     ┌──────────────────┐
│ Disc rotates at │────▶│ Check Stage 4    │────▶│ Check Stage 5    │
│ an excessive    │     │ (Disc inserted)  │     │ Check EFM and    │
│ speed (players  │     │ Laser circuits   │     │ crystal signals  │
│ with test or    │     │ Focus circuits   │     │ to the decoder   │
│ service mode    │     └──────────────────┘     └──────────────────┘
│ facilities)     │
└─────────────────┘

┌─────────────────┐     ┌──────────────┐  ┌──────────────┐  ┌──────────────┐  ┌──────────────┐
│ Disc rotates,   │────▶│Check Stage 5 │─▶│Check Stage 6 │─▶│Check Stage 7 │─▶│Check Stage 8 │
│ but gives up    │     │Check laser   │  │Tracking error│  │Does tracking │  │Check quality │
│ after a while   │     │power and/or  │  │waveform.     │  │error waveform│  │of RF eye     │
└─────────────────┘     │EFM/crystal   │  │Check relevant│  │collapse?     │  │pattern, and  │
                        │signals to the│  │adjustments   │  │Check carriage│  │relevant      │
                        │decoder       │  │              │  │/sled         │  │adjustments.  │
                        └──────────────┘  └──────────────┘  │operation     │  │Check disc/   │
                                                            └──────────────┘  │spindle motor │
                                                                              │operation     │
                                                                              └──────────────┘

┌─────────────────┐     ┌──────────────────┐     ┌──────────────────┐
│ Player appears  │────▶│ Check Stage 9    │────▶│ Check Stage 10   │
│ to operate but  │     │ Check decoder    │     │ Check analogue   │
│ there is no     │     │ outputs to the   │     │ signal circuits  │
│ audio output,   │     │ digital to       │     └──────────────────┘
│ or it is noisy  │     │ analogue         │
│ and/or          │     │ converter stages │
│ distorted.      │     └──────────────────┘
│ Front panel     │
│ display is      │
│ operating       │
│ correctly       │
└─────────────────┘

┌─────────────────┐     ┌──────────────────┐     ┌──────────────────┐
│ The player      │────▶│ Check the        │────▶│ Check Stages 1-9 │
│ operates with   │     │ quality of the   │     │ Ensure that all  │
│ an audio        │     │ discs being      │     │ mechanical       │
│ output, but     │     │ used. Ensure     │     │ aspects and      │
│ player appears  │     │ the discs are    │     │ relevant         │
│ to 'skip'       │     │ clean and free   │     │ electrical/      │
│ and/or 'jump'   │     │ from scratches   │     │ mechanical       │
│ whilst playing  │     └──────────────────┘     │ adjustments      │
│ the disc        │                              │ outlined in      │
└─────────────────┘                              │ Stages 1-9 are   │
                                                 │ correct          │
                                                 └──────────────────┘
```

Figure 9.2 *CD player fault symptom check list*

A list of stages which if applied in a logical sequence as relevant to individual players will assist engineers in determining where, in a CD player, a specific fault may be occurring. Each stage is subsequently expanded to provide further guidance where possible.

Stage 1: Initial checks

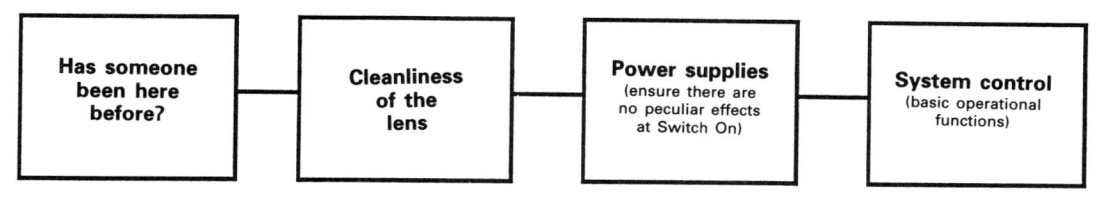

- Has someone been here before?
- Cleanliness of the lens
- Power supplies (ensure there are no peculiar effects at Switch On)
- System control (basic operational functions)

Stage 2: Basic operational checks

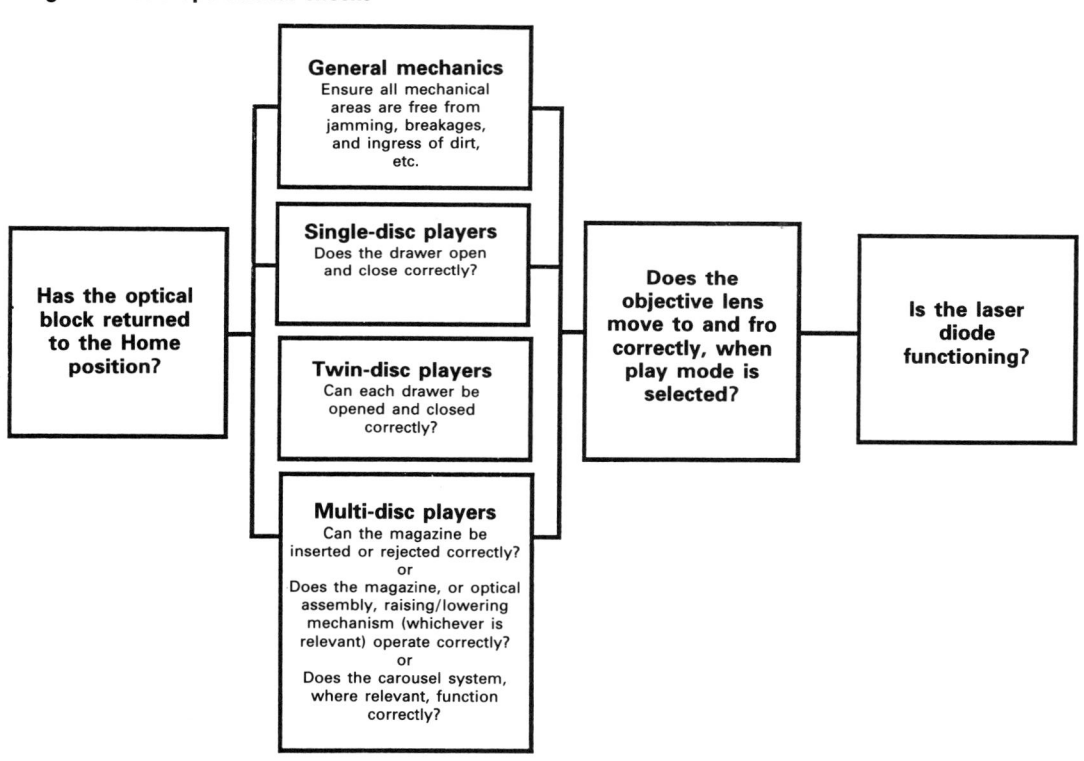

- Has the optical block returned to the Home position?
  - **General mechanics** — Ensure all mechanical areas are free from jamming, breakages, and ingress of dirt, etc.
  - **Single-disc players** — Does the drawer open and close correctly?
  - **Twin-disc players** — Can each drawer be opened and closed correctly?
  - **Multi-disc players** — Can the magazine be inserted or rejected correctly? or Does the magazine, or optical assembly, raising/lowering mechanism (whichever is relevant) operate correctly? or Does the carousel system, where relevant, function correctly?
- Does the objective lens move to and fro correctly, when play mode is selected?
- Is the laser diode functioning?

Stage 3: Players with the test or service mode facilities selected (with no disc inserted) (refer to Adjustments – Test or Service Mode section)
Philips players: Select the service mode; servicing position '0'
Pioneer players: Select the test mode

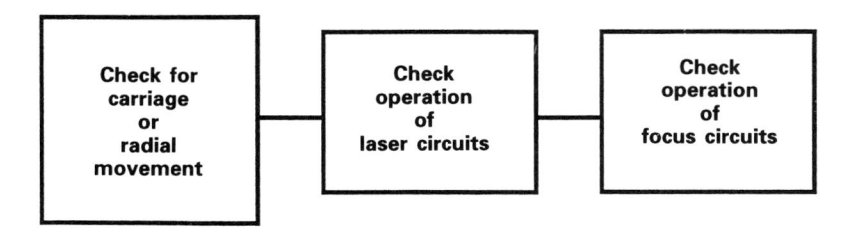

- Check for carriage or radial movement
- Check operation of laser circuits
- Check operation of focus circuits

Figure 9.3 *Suggested stages in fault diagnosis – Stages 1–3*

Fault diagnosis

Stage 4: **Players with the test or service mode facilities selected** (with a disc inserted) (refer to Adjustments – Test or Service Mode section)
Philips players: Operate the NEXT key; Servicing position '1'
Pioneer players: Operate track forward (TRK FWD) or Programme (PGM)

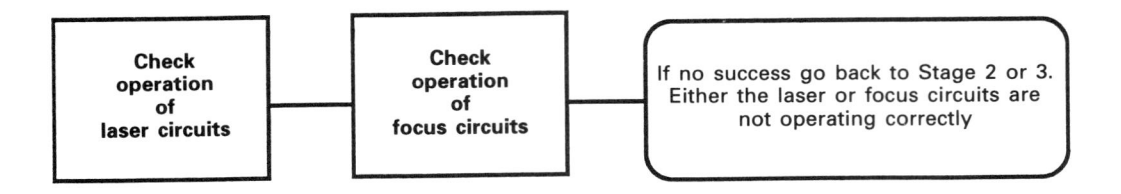

```
┌─────────────┐    ┌─────────────┐    ╭──────────────────────────────╮
│   Check     │    │   Check     │    │ If no success go back to     │
│ operation   │────│ operation   │────│ Stage 2 or 3. Either the     │
│    of       │    │    of       │    │ laser or focus circuits are  │
│laser circuits│   │focus circuits│   │ not operating correctly      │
└─────────────┘    └─────────────┘    ╰──────────────────────────────╯
```

Stage 5: **Players with the test or service mode facilities selected** (with a disc inserted) (refer to Adjustments – Test or Service Mode section)
Philips players: Operate NEXT again, Servicing position '2'
Pioneer players: Operate PLAY

Note: Always wait 3 s after operating the relevant functions outlined in Stage 4 before operating any of the above keys, otherwise disc 'runaway' is guaranteed

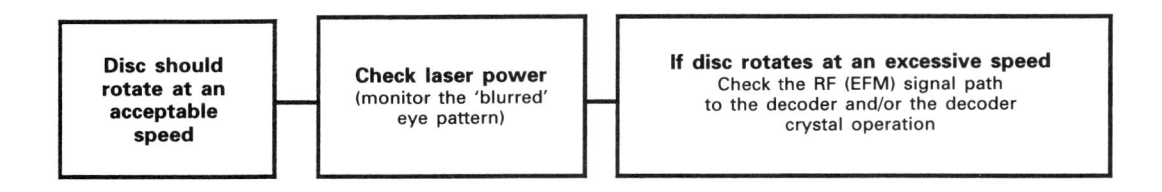

```
┌─────────────┐    ┌─────────────┐    ┌──────────────────────────────┐
│Disc should  │    │Check laser  │    │ If disc rotates at an        │
│rotate at an │────│power        │────│ excessive speed              │
│acceptable   │    │(monitor the │    │ Check the RF (EFM) signal    │
│speed        │    │'blurred'    │    │ path to the decoder and/or   │
│             │    │eye pattern) │    │ the decoder crystal operation│
└─────────────┘    └─────────────┘    └──────────────────────────────┘
```

Stage 6: **Players with the test or service mode facilities selected as in Stage 5**
This stage can apply particularly to Pioneer players or any players that require adjustment of the diffraction grating. The disc will be rotating at an acceptable speed

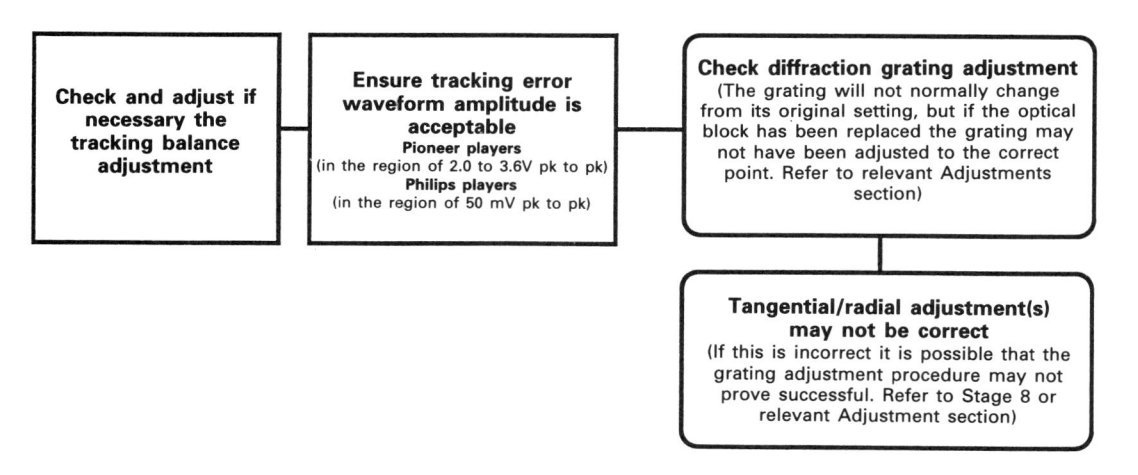

```
┌─────────────┐    ┌─────────────────┐    ╭──────────────────────────────╮
│Check and    │    │Ensure tracking  │    │ Check diffraction grating    │
│adjust if    │    │error waveform   │    │ adjustment                   │
│necessary the│────│amplitude is     │────│ (The grating will not        │
│tracking     │    │acceptable       │    │ normally change from its     │
│balance      │    │Pioneer players  │    │ original setting, but if the │
│adjustment   │    │(in the region of│    │ optical block has been       │
│             │    │2.0 to 3.6V pk to│    │ replaced the grating may not │
│             │    │pk)              │    │ have been adjusted to the    │
│             │    │Philips players  │    │ correct point. Refer to      │
│             │    │(in the region of│    │ relevant Adjustments section)│
│             │    │50 mV pk to pk)  │    ╰──────────────────────────────╯
└─────────────┘    └─────────────────┘                  │
                                        ╭──────────────────────────────╮
                                        │ Tangential/radial            │
                                        │ adjustment(s) may not be     │
                                        │ correct                      │
                                        │ (If this is incorrect it is  │
                                        │ possible that the grating    │
                                        │ adjustment procedure may not │
                                        │ prove successful. Refer to   │
                                        │ Stage 8 or relevant          │
                                        │ Adjustment section)          │
                                        ╰──────────────────────────────╯
```

Figure 9.3 *(Continued) Suggested stages in fault diagnosis – Stages 4–6*

Fault diagnosis

Stage 7: Players with the test or service mode facilities selected as in Stage 5
This stage can apply particularly to those players that have the facility of 'opening' and 'closing' the tracking servo

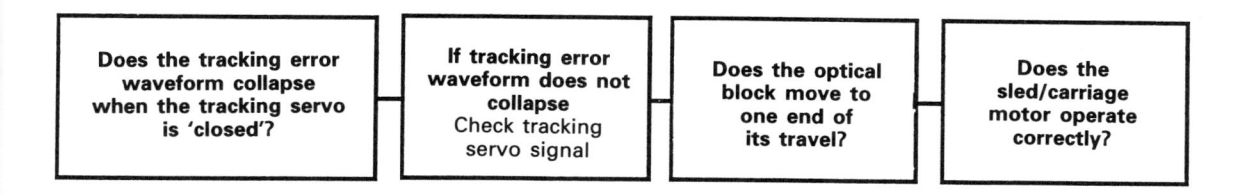

```
┌──────────────────┐   ┌──────────────────┐   ┌──────────────────┐   ┌──────────────────┐
│ Does the tracking│   │ If tracking error│   │ Does the optical │   │ Does the         │
│  error waveform  │───│ waveform does not│───│  block move to   │───│  sled/carriage   │
│ collapse when the│   │     collapse     │   │   one end of     │   │  motor operate   │
│  tracking servo  │   │  Check tracking  │   │    its travel?   │   │    correctly?    │
│    is 'closed'?  │   │   servo signal   │   │                  │   │                  │
└──────────────────┘   └──────────────────┘   └──────────────────┘   └──────────────────┘
```

Stage 8: Players without the test or service mode facility and/or Players with the test or service mode facilities selected as in Stage 5
This stage can apply particularly to those players that have the facility of 'opening' and 'closing' the tracking servo

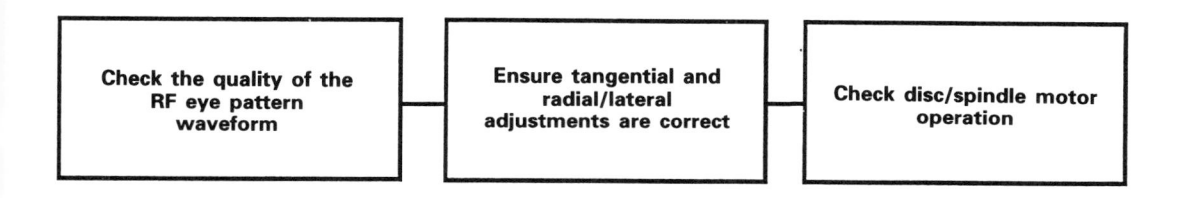

```
┌──────────────────┐   ┌──────────────────────┐   ┌──────────────────┐
│ Check the quality│   │ Ensure tangential and│   │ Check disc/spindle│
│   of the RF eye  │───│   radial/lateral     │───│  motor operation │
│ pattern waveform │   │ adjustments are correct│ │                  │
└──────────────────┘   └──────────────────────┘   └──────────────────┘
```

Stage 9: Player operating in the normal Play mode 1

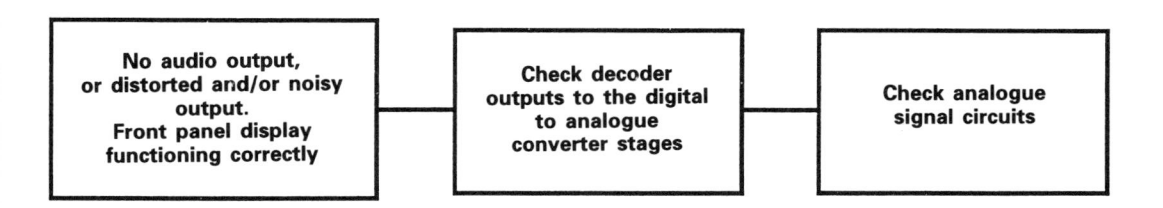

```
┌──────────────────┐   ┌──────────────────┐   ┌──────────────────┐
│  No audio output,│   │  Check decoder   │   │                  │
│   or distorted   │   │  outputs to the  │   │  Check analogue  │
│   and/or noisy   │───│   digital to     │───│  signal circuits │
│     output.      │   │   analogue       │   │                  │
│  Front panel     │   │  converter stages│   │                  │
│ display function-│   │                  │   │                  │
│  ing correctly   │   │                  │   │                  │
└──────────────────┘   └──────────────────┘   └──────────────────┘
```

Stage 10: Player operating in the normal play mode 2

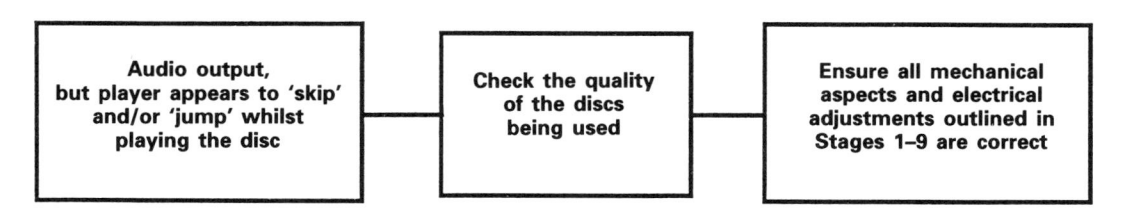

```
┌──────────────────┐   ┌──────────────────┐   ┌──────────────────────┐
│  Audio output,   │   │                  │   │ Ensure all mechanical│
│ but player appears│  │ Check the quality│   │  aspects and electrical│
│  to 'skip' and/or│───│  of the discs    │───│  adjustments outlined│
│ 'jump' whilst    │   │   being used     │   │  in Stages 1–9 are   │
│ playing the disc │   │                  │   │      correct         │
└──────────────────┘   └──────────────────┘   └──────────────────────┘
```

Figure 9.3 *(Continued) Suggested stages in fault diagnosis – Stages 7–10*

Fault diagnosis

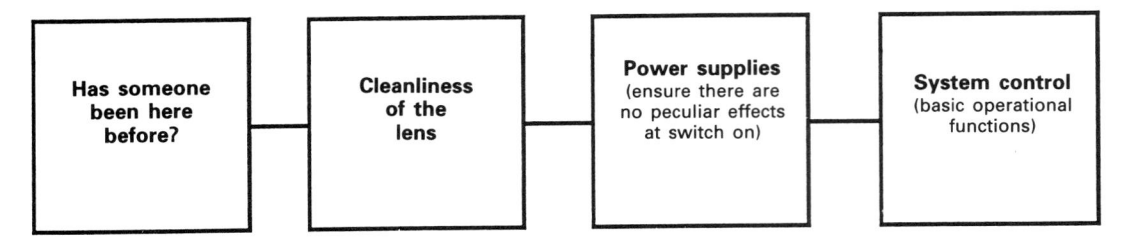

```
┌─────────────┐   ┌─────────────┐   ┌─────────────────┐   ┌─────────────────┐
│ Has someone │   │ Cleanliness │   │  Power supplies │   │  System control │
│  been here  │───│   of the    │───│(ensure there are│───│(basic operational│
│   before?   │   │    lens     │   │ no peculiar     │   │   functions)    │
│             │   │             │   │ effects at      │   │                 │
│             │   │             │   │ switch on)      │   │                 │
└─────────────┘   └─────────────┘   └─────────────────┘   └─────────────────┘
```

Figure 9.4 *Initial checks*

On some players, such as Pioneer, it is possible to determine whether the spindle motor has been changed, because the disc table spacer may be missing from the original location on the optical assembly.

Cleanliness of the lens

It is essential to ensure that the lens is absolutely clean, as many a player has fallen by the wayside as the result of a dirty lens.

Many of the hi-fi players, which just have a drawer or magazine for inserting the disc, are very good at minimising the ingress of dirt into the player, but many of the portable 'mean machine' and 'ghetto blaster' types, which have a lift-up lid to insert the disc, prove extremely good at allowing dirt to find its way onto the lens of the optical block.

Also some of the hi-fi midi systems have ventilation slots in the cover or back of the complete unit. So, by virtue of the natural ventilation of the system, it is possible for dirt and dust to find its way into the unit, and this naturally always seems to accumulate on the lens of the optical block.

In fact there may be a good reason as to why this usually happens; it may be due to the fact that the lens on the optical block is usually composed of a plastic material, and sitting just above the lens will be a CD rotating at anything from 500 to 200 rpm, or thereabouts. What better combination is required for a static generator, where the lens can attract dirt, dust, smoke or pollution from the surrounding environment to desposit itself upon the surface of the lens, thereby effectively inhibiting the laser beam from doing its necessary job of extracting the information from the disc?

Cleaning the lens should be carried out with care, and because some lenses have a special coating on them, chemical-based cleaning agents are not generally recommended. The author's tried and tested method is to gently rotate a 'moistened' cotton bud over the surface of the lens, and gently polish up the lens with a dry cotton bud if necessary.

There are some special lens-cleaning discs available which are inserted into the player. The actual process of inserting these special discs is intended to clean the lens, a process which sounds somewhat traumatic as far as the lens is concerned, and this is a method which is not generally recommended, though engineers through their own experience will be aware of the virtues of this method of cleaning the lens.

Power supplies

Ensure there are no peculiar effects at switch on. The majority of CD players, especially of the hi-fi type, incorporate a balanced power supply which provides positive and negative supplies to drive the electro-mechanical items such as focus and tracking coils, and relevant motors, as well as other areas of the circuits. It can prove useful to ensure the supplies and circuit protectors where fitted are all correct and operational, especially if there are problems when first switching on, and motors are observed to be either running or buzzing because they may be stalled due to some mechanical aspect being at one end of its travel. Even observing the focus lens can prove interesting, as usually it will be in a neutral position, especially with no disc inserted. If it is either sucked right in to the optical block or trying to climb out of its bed this is usually sufficient to indicate that all may not be well with the power supply and related drive circuits.

With the majority of CD players, the positive and negative 8/16 V supplies are used to drive the electro-mechanical components, such as focus and tracking coils and any associated motors, with their

Fault diagnosis

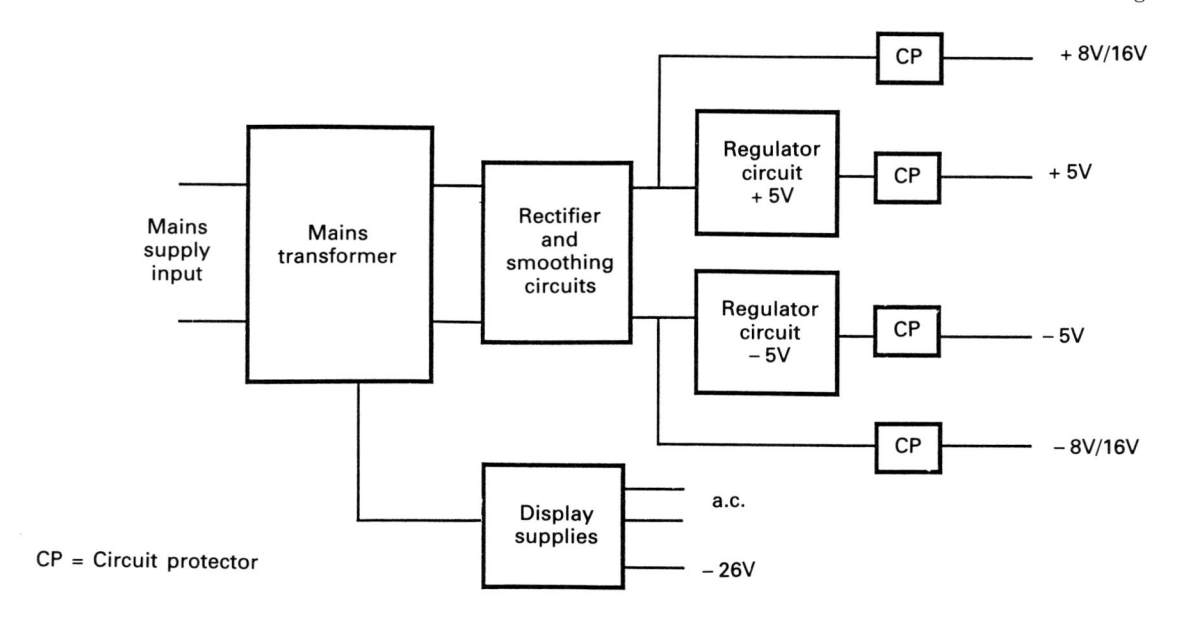

Figure 9.5 *Typical power supply block diagram*

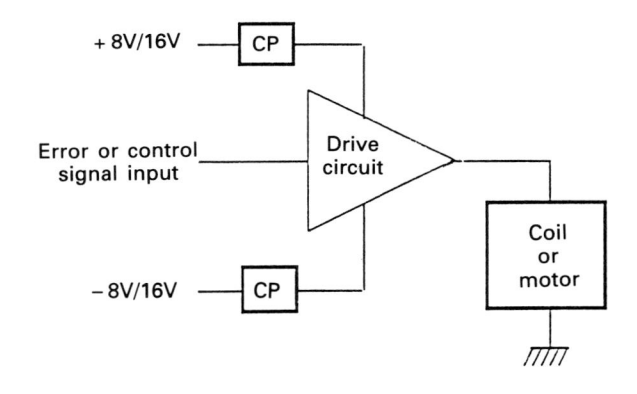

Figure 9.6 *Typical servo drive circuit*

relevant direction of movement being related to the applied polarity derived via the necessary drive circuits (Fig. 9.6).

When the coils are in their neutral position, or the motors are stopped, this is achieved by applying a neutral or zero signal to the drive circuit, which in turn will provide a zero output to effectively switch off the supply to the relevant device. The principle of operation of the circuit shown in Fig. 9.6 is that with a zero input, a zero output will be achieved, and therefore the coil or motor will not operate. If the input is positive or negative, the resultant output will cause the coil or motor to operate in one direction or the other, depending upon the polarity of the input.

Circuit protectors are frequently included in many of the supply lines, though not necessarily all of them. Should one of the circuit protectors become defective, causing one of the supplies to be missing, the result will be an inbalance in the operation of the drive circuit, which in turn will cause a voltage to appear at the output which will cause the coil or motor to function.

System control – basic operational functions

Amongst the wide range of sophisticated domestic electronic products that service engineers are called upon to service in these modern times, there is usually lurking somewhere within the inner depths a multi-pin pack package, either dual in line or four-sided flat pack, that goes under the identification of the microprocessor or system controller.

Service engineers 'love' them, as these mysterious and sometimes magical 'black boxes' appear to assume virtual total control of whatever the product may be, and unfortunately, by the nature of their complexity, they are blamed for a multitude of sins of which they are usually quite innocent.

157

Fault diagnosis

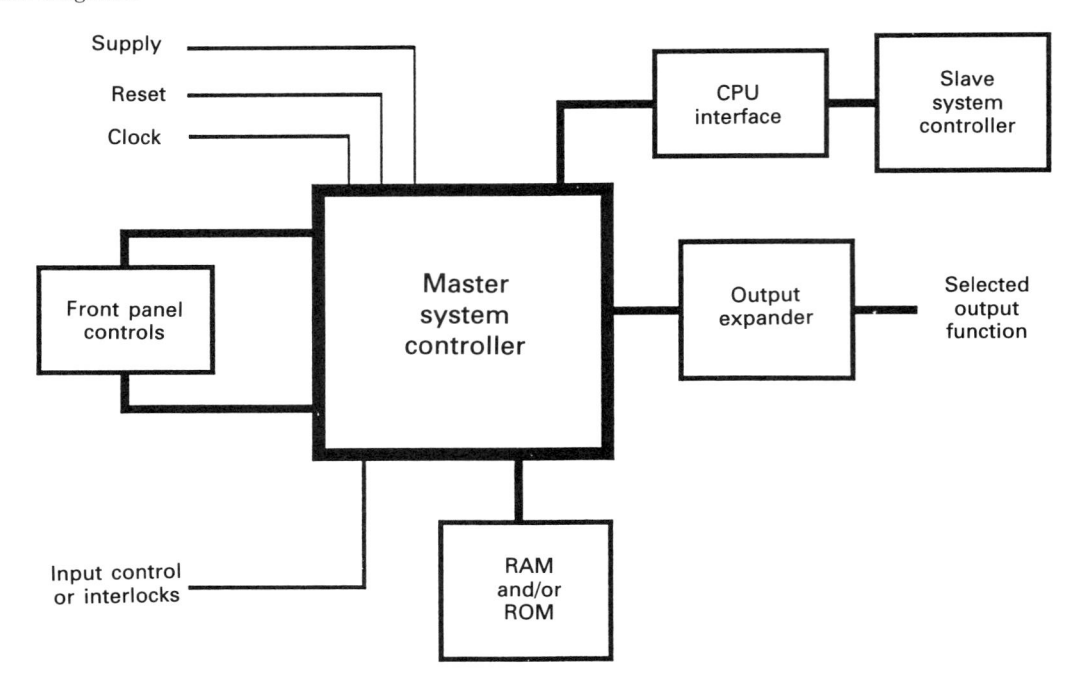

Figure 9.7 *Basic system control arrangement*

As a result, these items are frequently replaced, and because they are usually found as the dual in line or four-sided flat pack type, and are therefore an absolute 'delight' to change, there can be nothing more frustrating than having replaced one of these 'beasts' only to find that the existing fault is still apparent, or even worse, extra symptoms have occurred as a result of some dubious soldering, broken print or other unspeakable niceties that can occur.

Experience has shown all too frequently that it is usually some straightforward and rather mundane aspect of the circuit that has been the real cause of the fault condition, thereby preventing the system controller from performing the function that has been selected.

Whilst Chapter 7 covers the system control aspect, it is worthwhile repeating certain diagrams as necessary in order to maintain consistency when considering to which area of the CD player the fault symptom relates, especially the basic outline of the concept of the system control as shown in Fig. 9.7.

When considering a recommended procedure in order to determine whether the system controller may or may not be at fault, it is essential as a preliminary process of elimination to check the '*magic three*':

1 **Supplies to the 'beast'** (to make it work). Usually the supply to the system controller is + 5 V; occasions have been known when this supply has been as low as only + 4.5 V or + 4 V and the system control has refused point blank to function, and therefore it can prove useful to check that this supply is correct.

2 **Reset** (to make it work correctly). As previously mentioned, the reset is necessary to ensure that the system controller is in the correct logic state at switch on, to ensure the various internal sequences and functions are in their required state(s) to enable correct operation to be achieved.

If the reset function is incorrect, some players can give the impression of being totally 'dead', with no display indications, possibly pointing to a power supply problem, and it certainly can prove worthwhile to ensure all supplies are correct.

It is also possible for random display functions to appear, with the possibility of some totally illogical functions taking place, though some other system control malfunctions can create functioning problems which will be discussed later.

Fault diagnosis

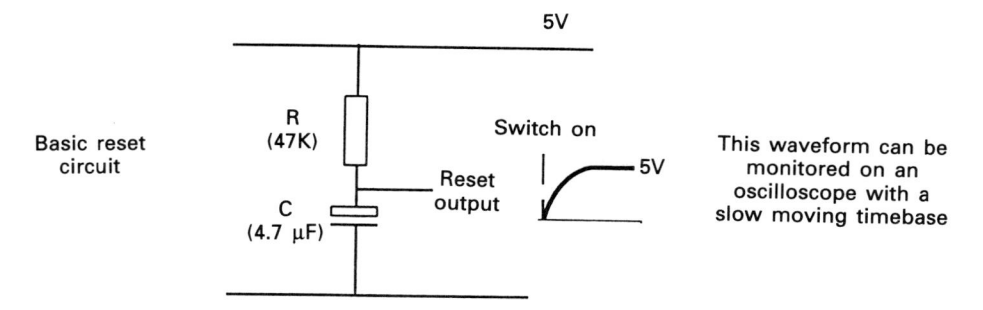

Figure 9.8 *Basic reset circuit*

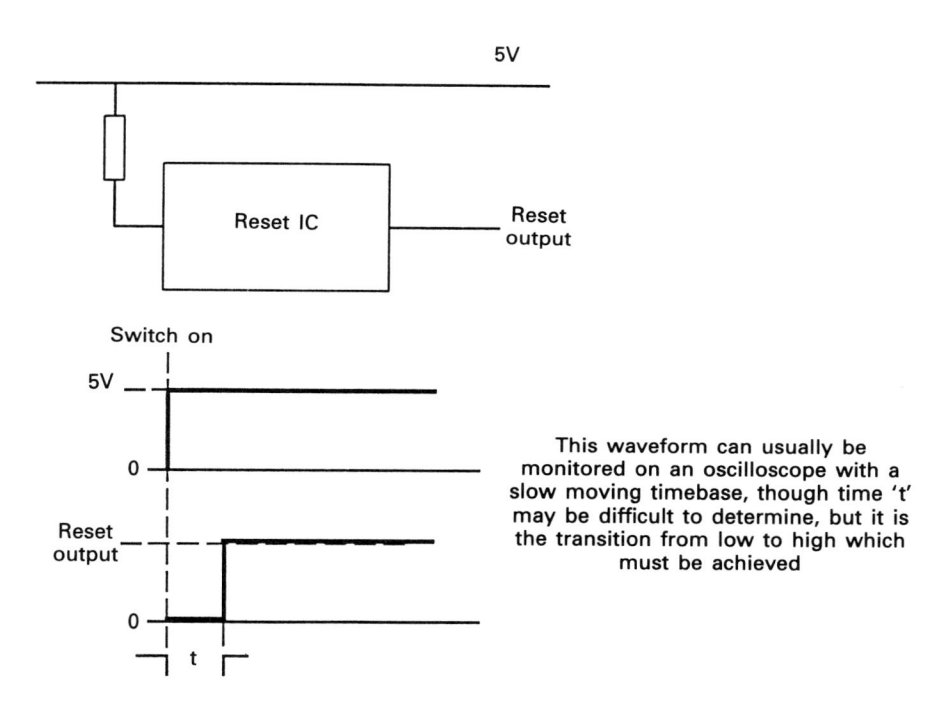

Figure 9.9 *Alternative reset circuit*

It may be necessary to refer to the relevant service manual in order to determine the correct reset output requirements, but the information given in Fig. 9.8 may prove useful. Basically the reset pulse should complete its required function a brief period of time after switching on the player or associated equipment. The component values shown in Fig. 9.8 are an example from a particular player, and may vary from one player or manufacturer to another.

The information given in Fig. 9.9 should help engineers in determining whether the reset circuit is functioning correctly, but it is also useful to appreciate that some system control circuits are known not to provide a clock signal if the reset circuit is not operating correctly.

3 Clock (to help it along or perform the necessary functions). This is usually derived from a crystal oscillator within the system control IC, and will operate at a frequency ranging from 4 MHz to greater than 8 MHz, depending upon manufacturer and type of player.

Measurement of the clock frequency can be readily achieved with an oscilloscope, but it is

159

Fault diagnosis

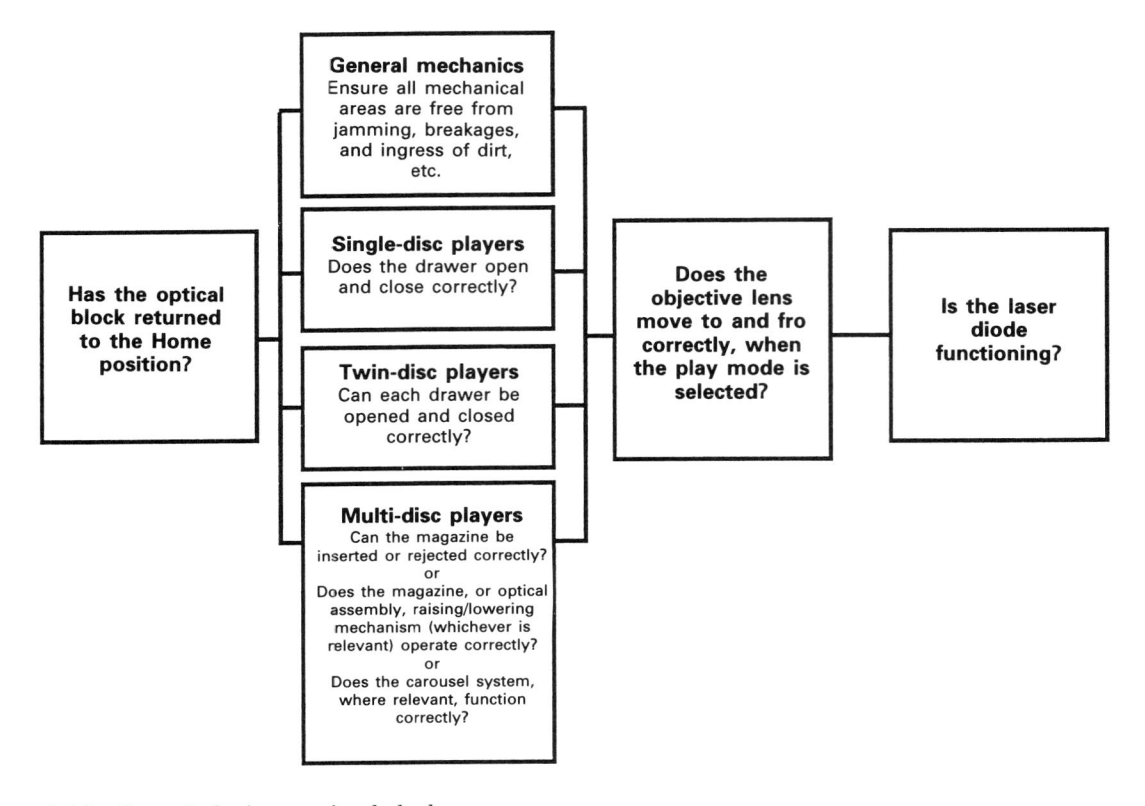

Figure 9.10 *Stage 2: basic operational checks*

essential that the ×10 probe is used. If the a.c. setting is also used, ensure that the end of the probe is first discharged to ground, as otherwise any inadvertently stored d.c. potential within the input circuits of the oscilloscope can prove somewhat detrimental to the system controller.

The author prefers wherever possible to use the d.c. facility of oscilloscopes, having observed a few circuits suffer from a quick discharge from the oscilloscope when the a.c. input is selected.

Stage 2: basic operational checks

Without inserting a CD, check the following (Fig. 9.10).

Has the optical block returned to the home position?

When a CD player is first switched on, it is usual for the optical block to return to the home or start position, but if the block has refused to do this there is some form of electrical or mechanical problem or failure that is causing this to occur.

Electrical checks

These can be carried out in relation to the procedure explained in Stage 3: can the carriage, sled or slider be moved?

Mechanical checks

These should be carried out on the optical unit mechanical drive gear or thread assembly for cleanliness, and for any possible breakages, especially any plastic or nylon gears or linkages. At this stage it is important to ensure that the rubber drive belt present on many units is correctly fitted or intact, as not much will happen if this is missing!

Check the operation of general mechanics, drawers opening and closing, or magazines (where relevant) being inserted and rejected correctly

It is essential to ensure that all mechanical aspects are functioning correctly, as it is quite possible for a player to appear to be unable to function and to create the impression of a severe system control problem, when at the end of the day it is quite possible that some form of mechanical jamming may have occurred.

With few exceptions, the majority of players have some form of mechanical system for accepting the disc into the unit, such as a drawer, or drawers in the case of the twin-disc players, or a magazine for six or more discs, or, more recently, with the advent of the carousel method, a large drawer capable of handling up to five discs.

Regarding the various drawer methods, the correct alignment of relevant gears is obviously essential. The twin-tray type has been a source of frustration for many engineers, because if the assembly achieves a mis-alignment by only one tooth, all sorts of fun can occcur. This problem is usually caused by unsuspecting users pushing the actual drawer to close it rather than pressing the relevant control key on the front panel of the player.

The magazines used in the multi-disc players can also be a source of trouble, especially the six disc type, as the manufacturers' recommendations are to only withdraw one tray at any one time when inserting discs. Indeed within the magazine itself there is usually a locking system comprising a series of ball bearings which prevents more than one tray being withdrawn when inserted into the mechanism. But unfortunately when it comes to some users, three shredded wheat and anything is possible; frequently more than one tray is withdrawn whilst inserting the discs, resulting in some of the locking ball bearings being inadvertently removed. When the magazine is now inserted into the player it is possible for more than one tray to be withdrawn due to the selected tray effectively dragging out the tray above or below the selected one, which will result in quite a spectacular jamming up of the mechanism.

Frequently this situation can be patiently resolved by the engineer with a partial strip down and 'jiggling' with suitable tools. The offending tray(s) can be eased back into the magazine, but unfortunately many a user has aggravated the situation by adopting the attitude of attacking the system with the 'kitchen knife' only to cause further problems, usually resulting in serious mechanical damage, as well as not being too kind to the disc(s) that may have become inadvertently involved.

Does the objective lens move to and fro correctly?

Objective lens facing upwards or forwards

With the type of players which have the objective lens facing upwards, i.e. the majority of single disc, twin-tray and carousel players, it is usually quite easy to observe the lens when no disc is inserted. Pressing the 'play' button, or closing the drawer, will usually activate the 'laser on and focus search' process, and at this stage it should be possible to observe the gradual movement of the lens.

Some players may have some form of sensing or interlocking system, especially the portable type of CD players which usually require the drawer or lid to be fully closed, which may require overriding, as otherwise the 'laser on and focus search' sequence may not be activated. Naturally, care must be taken to ensure that unnecessary exposure to the laser beam does not occur, and that any overriding is removed on completion of any servicing procedures.

Please refer to the relevant references concerning care when the possibility of exposure to the laser beam may occur.

The number of times that the lens moves to and fro during the focus search sequence varies from one manufacturer to another; there will usually be a minimum of two searches but there may be up to 16 searches in some Philips players when in the service mode.

An important factor to consider when observing the lens is that it should move reasonably smoothly and not 'jump up and down', or 'in and out', which can be caused by the focus search capacitor being defective or 'dry jointed'. When this occurs it is also possible to hear the effect of this somewhat erratic movement as the lens hits the 'stops' at each end of its travel.

Usually the lens should adopt a neutral position when the player is switched on, before any search sequences are activated. If the lens does

Fault diagnosis

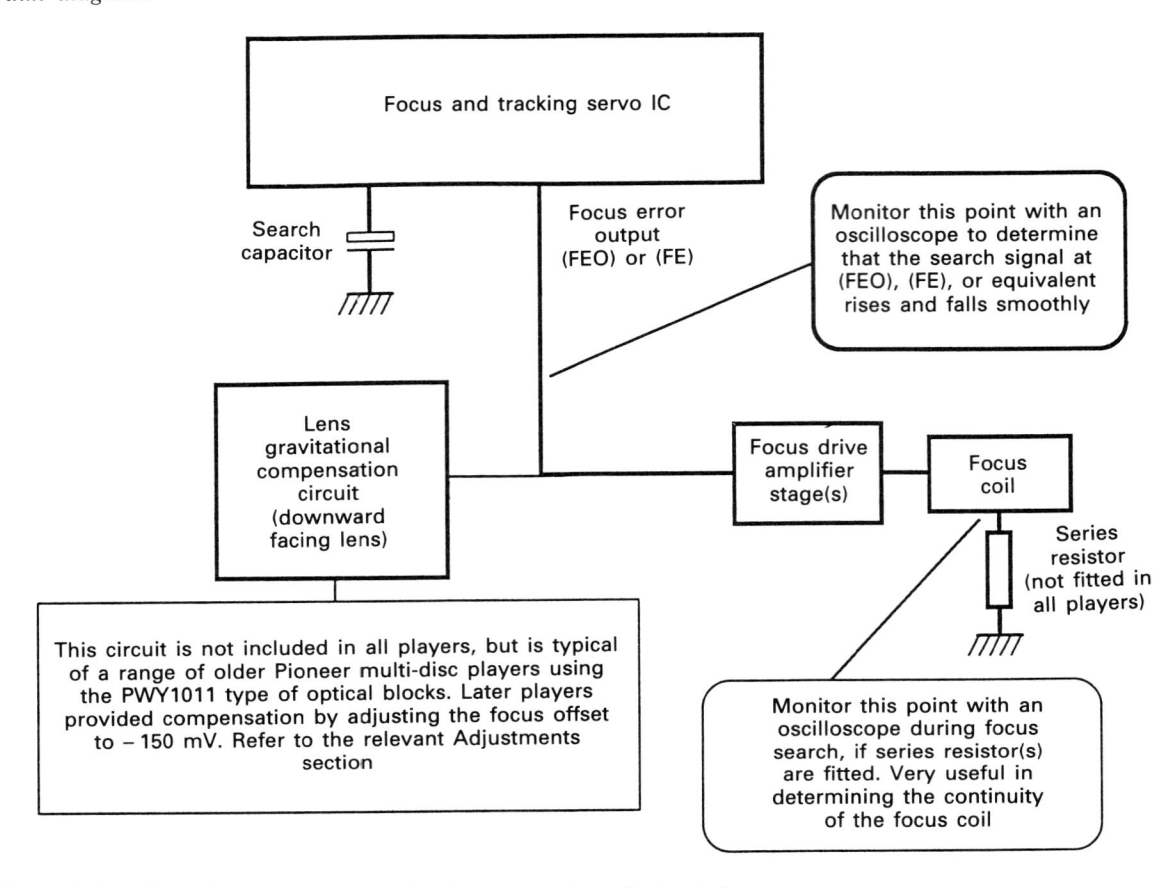

Figure 9.11 *Basic focus servo system, showing suggested monitoring points*

appear to be excessively high or low then voltages within the focus servo, especially relevant positive and negative supplies, may prove to be correct, but also it may prove prudent to ensure that the lens is not mechanically jammed for some reason, preventing correct operation.

Objective lens facing downwards

The optical assemblies that are fitted into players with the lens facing downwards, such as the Pioneer multi-disc and turntable platter players, can make it difficult to observe the movement of the lens, and it may be considered logical to remove the relevant fixing screws for the complete optical assembly and simply turn the unit over to observe the lens movement – take care not to fall into a nasty little trap!!!

With this type of player it is possible that some form of correction circuit may be incorporated in the focus circuits which will provide compensation for gravitational effects upon the lens to enable it to maintain its neutral position. If the unit is turned over to observe the lens movement it will be in effect withdrawn further into the block by gravity, enabling the correction circuit to 'suck' the lens inwards, preventing any hope of monitoring whether the lens is moving correctly or not.

The preferred method of observing the lens in this situation is to remove the relevant screws and just lift the unit vertically a small amount, looking under the unit to check that the lens movement is acceptable. Whichever way the lens may be facing, the majority of players will develop a focus search signal in a manner similar to the block arrangement shown in Fig. 9.11, which also includes the possible circuit arrangement to provide any gravitational compensation requirements that may be applicable to relevant players.

Fault diagnosis

Procedure for monitoring the focus search sequence

Players without test or service mode facilities

1 Without inserting a disc, determine the focus error point, (FEO) or (FE), where possible.
2 Connect the oscilloscope to the relevant point. Set to d.c., 1 V cm^{-1}, fast timebase to provide a continuous line.
3 Switch on and select the 'play' mode (override interlocks if necessary).
4 The FEO should rise and fall sedately; the number of times will vary depending upon the manufacturer.

Players with test or service mode facilities

1 Without inserting a disc, determine the focus error point, (FEO) or (FE), where possible.
2 Connect the oscilloscope to the relevant point. Set to d.c., 1 V cm^{-1}, fast timebase to provide a continuous line.
3 Select the relevant test or service mode facility controls and switch on (see the relevant section in Chapter 6).
4 Depress the relevant control to commence the search sequence: 'NEXT' for Philips players; 'Track forward' (TRK FWD) or 'Programme' (PGM) for Pioneer players.
5 (FEO) and (FE) should rise and fall sedately; the number of times will vary depending generally upon the manufacturer of the player.

If the lens movement is not satisfactory, i.e. it tends not to move smoothly or sedately, it is worthwhile checking the search capacitor and the relevant circuitry for intermittent or dry joints. Those circuits that have a series resistor included in the ground or return side of the focus coil provide an additional benefit insomuch that monitoring across the resistor with the oscilloscope during the focus search sequence should indicate a rising and falling of the trace, and though of a smaller value, can prove invaluable in determining continuity of the focus coil. If it not possible to obtain this measurement, checks of the focus coil, flexi-print and adjacent circuitry should be carried out.

Is the laser diode functioning?

Whilst a laser power meter can provide a useful method of determining whether the laser diode is actually functioning or not, this item of test equipment does not usually appear on the list of required test equipment for many service workshops. This is understandable since access to the laser diode in some types of players can occasionally prove difficult without the need for a certain amount of dismantling; also, many manufacturers recommend the final measurement for laser power being achieved with the use of an oscilloscope monitoring the RF eye pattern waveform.

As a result many engineers are known to check visually whether the laser diode is functioning. Though the author has never recommended that anyone should visually check the laser diode, he is, on the other hand, quite ready to undertake this task personally. It is essential to ensure that direct viewing of the laser beam through the objective lens does not occur. Warning labels are placed on all CD players at manufacture, and warning information is provided in the relevant service manuals recommending the avoidance of exposure to the laser beam. Though it is possible to observe an amount of faint red light if the laser beam is viewed, this is in fact only a small amount of the laser power; the majority of the laser power is invisible, and it is this that can cause damage to the retina if viewed too closely and for too long. If it is considered necessary to view the laser beam, a few basic guidelines should be followed:

1 Ensure the laser beam is of a low power category; domestic CD players come within this category.
2 View the lens at a distance of *no nearer* than 40–50 cm.
3 View the lens at an angle of ~45°.
4 *Do not view for longer than 10 s* – after all, for how long does a person have to look at a conventional lamp to determine whether it is on or not?

It is, however, possible to carry out certain electrical checks to verify that the laser circuit is endeavouring to operate, and the basic block arrangement is shown in Fig. 9.12.

The block arrangement outlined in Fig. 9.12 can be applied to a wide range of players as a basis for typical operation. The LD ON from the system controller will provide a high or low, depending upon the player concerned, though it is usually a high condition and this will in effect switch on the APC circuit.

The laser control transistor will either be a PNP or NPN type, usually depending upon the manufacturer. As a result, the laser diode switch on

Fault diagnosis

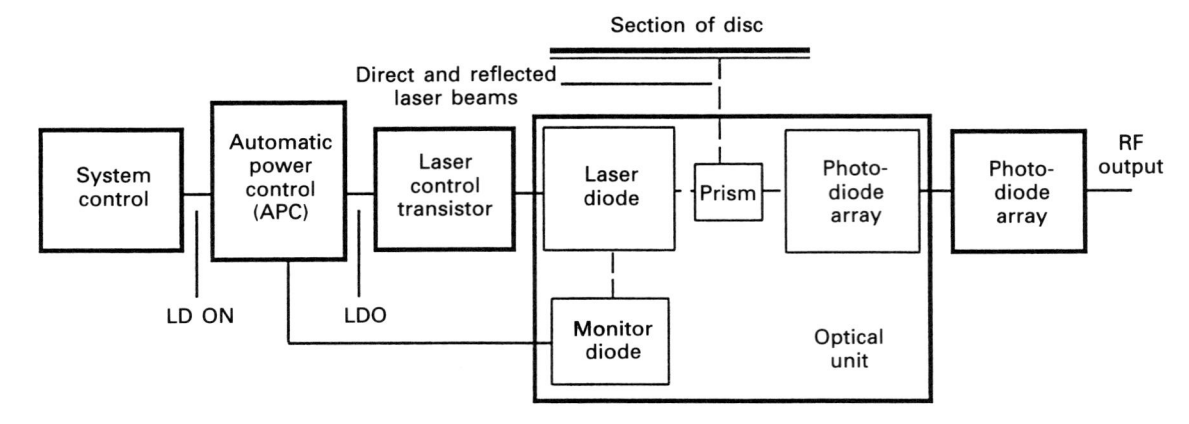

Figure 9.12 *Laser control circuit and monitoring points*

(LDO) signal will be low for PNP and high for NPN, which will enable the laser diode, which is usually connected in the emitter circuit, to switch on, therefore resulting in a high level at the emitter, assuming there is continuity through the laser diode.

Even though the high may be present and is being applied to the laser diode, and even if there is continuity through the laser diode, it is still possible for there to be no laser light output, due to a lack of emission ability from the laser diode.

It is possible to check whether the laser diode is endeavouring to operate without the necessity of viewing the laser beam, or using a laser power meter, by following the procedure outlined below:

1 Connect an oscilloscope to the RF output test point (the point where the RF eye pattern is monitored).
2 Set the oscilloscope to display a continuous line, and the input to d.c. 0.1 V cm^{-1}, using a ×10 probe.
3 Insert a CD.
4 Select the 'play' mode.
5 As the player endeavours to 'run up', the oscilloscope indicator should go high with the possibility of an eye pattern appearing.
6 If this does not occur, ensure that the relevant laser diode ON signals are present throughout the relevant sections of the circuit, as shown in Fig. 9.12.
7 This procedure is easier with players that have the facility of the test or service mode, and by selecting the relevant keys that enable the laser diode to operate, the switching signals, as well as the high at the RF test point, should be achieved with a correctly functioning laser diode.

If during this operation, and whilst monitoring the RF eye pattern, the eye pattern endeavours to appear, or there is excessive movement on the oscilloscope due to the appearance of a 'strong' signal, it is possible that the laser control transistor is faulty with a shorted base emitter, or collector emitter connections, either keeping the laser diode ON permanently or causing excessive current to flow when the laser diode is switched on.

As previously outlined in the section concerning adjustment of laser diodes (Chapter 6), excessive current through the laser diode can indicate a faulty diode, especially if the RF eye pattern waveform is low compared to the manufacturer's recommended level.

Stage 3: players with the test or service mode facilities selected (with no disc inserted)

(See the test or service mode section of Chapter 6.)
See Fig. 9.13.

— Philips players: select service mode
— Pioneer players: select test mode

Fault diagnosis

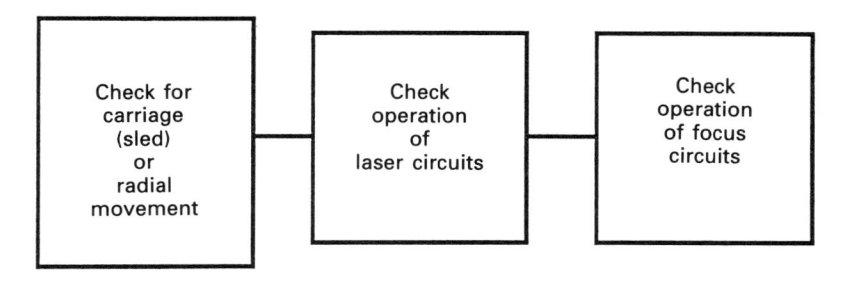

Figure 9.13 *Stage 3: carriage, laser and focus checks*

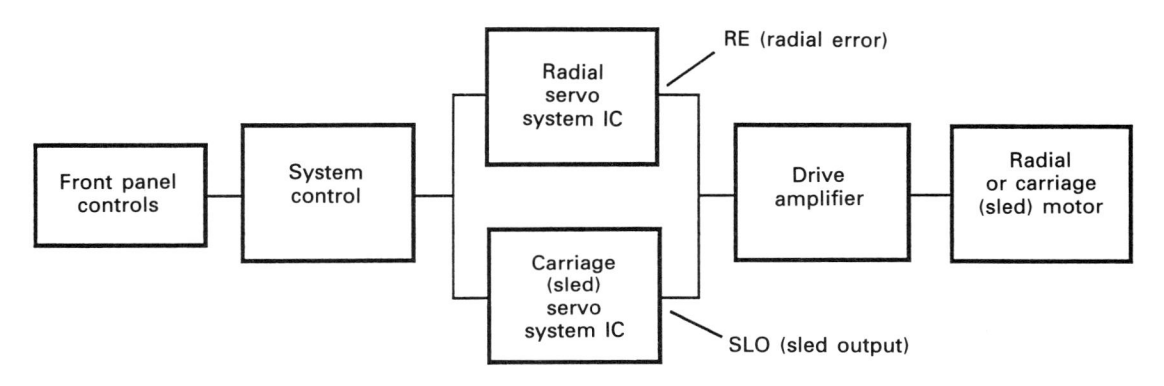

Figure 9.14 *Radial/carriage control*

Check for carriage (sled) or radial movement

Despite the fact that similar front panel functions can enable the movement of the carriage or radial assembly, the related circuit arrangements are quite different when considering the carriage (sled) operation for the majority of the Far Eastern players, such as Kenwood, Pioneer and Sony to name just three, in comparison to the players that use the Philips radial system. However, Fig. 9.14 endeavours to offer a compromise when it comes to applying some thoughts concerning fault diagnosis.

By referring to the service manual for the player concerned it should be possible to determine the controls that will enable the radial assembly or carriage (sled) motor to move in each direction if the player is capable of operating in the service or test mode.

— Philips players:
  standby mode, or
  operate the NEXT key (servicing position 1)
  operate SEARCH FORW. or SEARCH REV.
— Pioneer players:
  operate MAN. SEARCH FWD. or MAN. SEARCH REV.

If the radial assembly or carriage (sled) motor fails to operate the following checks should prove useful. By monitoring either the RE (radial error) or SLO (sled output) with an oscilloscope set to d.c. 0.1 V cm^{-1}, using a ×10 probe, with a straight line displayed, it should be possible to observe either a high or low level, when compared to zero, when the relevant keys are pressed to move either the radial assembly or carriage (sled) in one direction or the other. These checks should determine whether an electrical fault exists.

Check operation of laser circuits and focus circuits

With the facility of the service or test mode it is possible to check the operation of laser and focus

165

Fault diagnosis

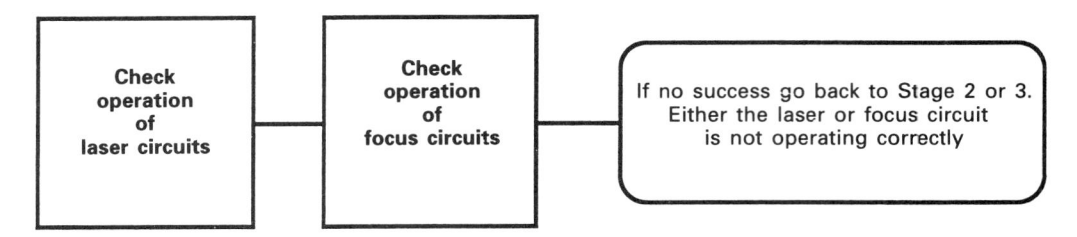

Figure 9.15 *Stage 4: laser and focus checks*

circuits without the need to switch the player off and on in order that the start-up sequence can be repeated, by checking the majority of the procedures outlined in Stage 2 with the operation of selected front panel key functions.

— Philips players:
  operate the 'NEXT' key (servicing position 1)
— Pioneer players:
  operate 'Track forward' (TRK FWD) or 'Programme' (PGM)

When either of these key functions are selected, the laser diode and focus circuits will function for a brief period of time. As previously mentioned, it is essential that the correct precautions are observed with respect to observing laser beams.

Stage 4: players with the test or service mode facilities selected (with a recommended test disc inserted)

(See the Test or Service Mode section in Chapter 6.)
See Fig. 9.15.

— Philips players:
  operate the 'NEXT' key (servicing position 1)
— Pioneer players:
  operate 'Track forward' (TRK FWD) or 'Programme' (PGM)

Operating the specific keys for the relevant player, with a disc inserted, should cause the laser to focus onto the playing surface of the disc.

With single disc and twin-tray players it may appear that not much has actually happened at this stage, but if the disc is gently rotated by hand, a 'squeaking' noise may be heard, which is an acoustic noise from the optical block. This is an excellent method of verifying that the laser and focus circuits are functioning correctly.

With the Pioneer multi-disc magazine players, when the 'TRK. FWD' or 'PGM' keys are operated, the disc from tray 1 will be accepted, and if it remains accepted in the mechanism, this again verifies the correct functioning of the laser and focus circuits.

If no success is achieved at this stage it will be necessary to carry out the checks detailed in Stages 2 and/or 3.

Stage 5: players with the test or service mode facilities selected (with a recommended test disc inserted)

(See the Test or Service Mode section in Chapter 6.)
See Fig. 9.16.

— Philips players:
  operate 'NEXT' again (servicing position 2)
— Pioneer players:
  operate 'PLAY'

Note: Always wait 3 s after operating the relevant functions outlined in Stage 4 before operating any of the above keys, as otherwise disc 'runaway' is guaranteed.

Disc should rotate at an acceptable speed: check laser power

At this point it is extremely useful to verify that the laser power is correct, by monitoring what

Fault diagnosis

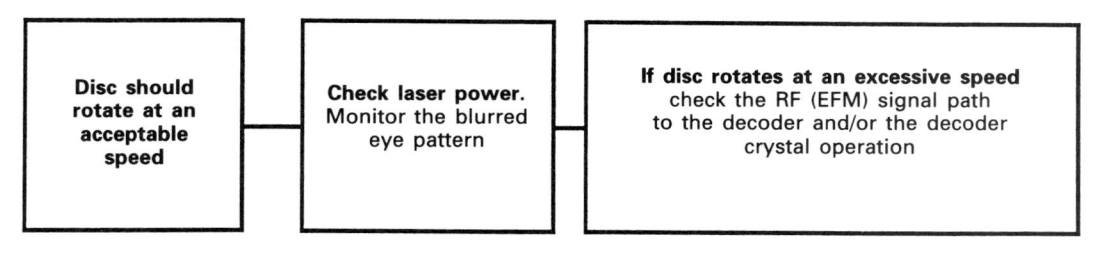

```
┌─────────────┐    ┌─────────────┐    ┌─────────────────────────┐
│ Disc should │    │Check laser  │    │If disc rotates at an    │
│ rotate at an│───▶│power.       │───▶│excessive speed          │
│ acceptable  │    │Monitor the  │    │check the RF (EFM) signal│
│ speed       │    │blurred      │    │path to the decoder and/ │
│             │    │eye pattern  │    │or the decoder crystal   │
│             │    │             │    │operation                │
└─────────────┘    └─────────────┘    └─────────────────────────┘
```

Figure 9.16 *Stage 5: checking laser power and initial disc speed*

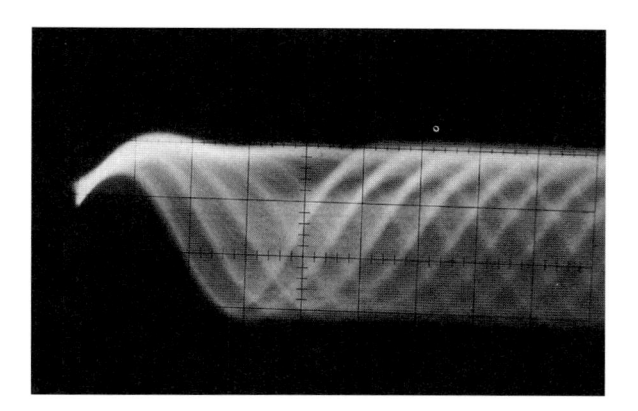

Figure 9.17 *Blurred eye pattern waveform*

is in effect the 'blurred' eye pattern waveform (Fig. 9.17). This is possibly more relevant to the Pioneer players, because if the diffraction grating is not correctly adjusted, it will usually prove impossible to obtain the correct eye pattern waveform, with no sound output or excessive 'skipping' when trying to operate the player in the normal 'play' mode. By monitoring the RF eye pattern waveform test point, the following waveform should be observed, with the oscilloscope set to either d.c. or a.c., 0.1 or 0.05 V cm^{-1}, input, using a ×10 probe, with the horizontal controls set to 0.5 μs cm^{-1}.

The amplitude of the waveform should be in the region of 1.2 V to 1.5 V peak to peak, though absolute verification should be obtained from the relevant service manual. Providing this waveform is similar to that shown in Fig. 9.17, it should indicate that the laser diode circuits are operating satisfactorily, and it is always possible to obtain this waveform irrespective of the setting of the diffraction grating, especially with Pioneer players.

If the disc rotates at an excessive speed

Providing that the laser and focus circuits have been proved to be operational, when the 'play' control on Pioneer players has been selected and the disc rotates at an excessive speed it is necessary to ensure that the EFM and/or crystal clock signals are reaching the decoder circuits, a basic arrangement of which is shown in Fig. 9.18.

The crystal input can either be an actual crystal circuit connected to the decoder or, as in later players which include digital filtering, the crystal circuit may be connected to a different IC with an eventual connection into the decoder.

The presence of a frequency input to the decoder can be monitored with an oscilloscope, using a ×10 probe, though for accuracy a frequency counter can be used to determine the actual frequency input.

The RF input to the RF amplifier can also be checked with an oscilloscope, but of course if the disc is excessive or running away, this can prove somewhat difficult, and some form of ingenuity may be required to ensure that a signal from the disc is present.

Typical methods include the following:

1 Connect the oscilloscope to the RF test point, and with a disc inserted select the 'play' function. Some form of waveform should be present as the player runs up, which will disappear as either the disc reaches excessive speed or the player shuts down because no input is being received by the decoder.

2 With the player in the test or service mode, and laser and focus circuits operating, gentle movement of the disc by hand should indicate that some form of signal from the disc is present.

The EFM output can next be checked usually at the output of the RF amplifier, as the decoder is

Fault diagnosis

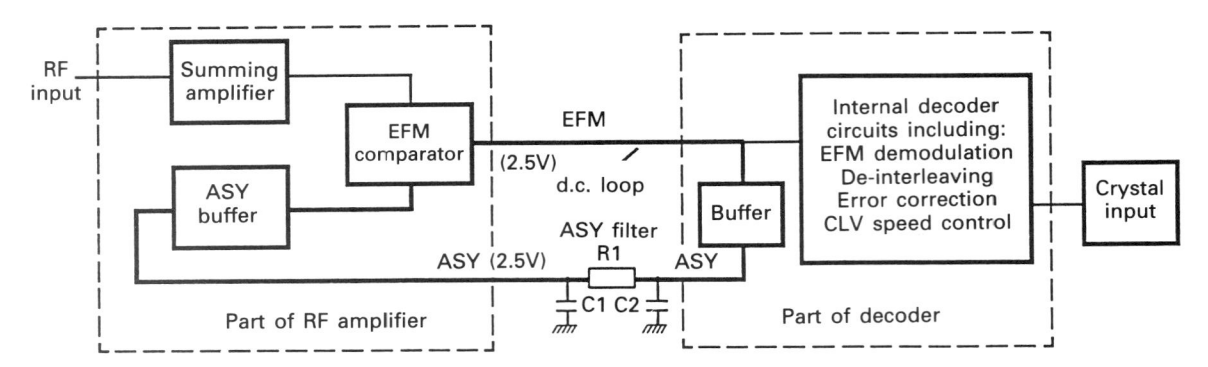

Figure 9.18 *ASY circuit arrangement*

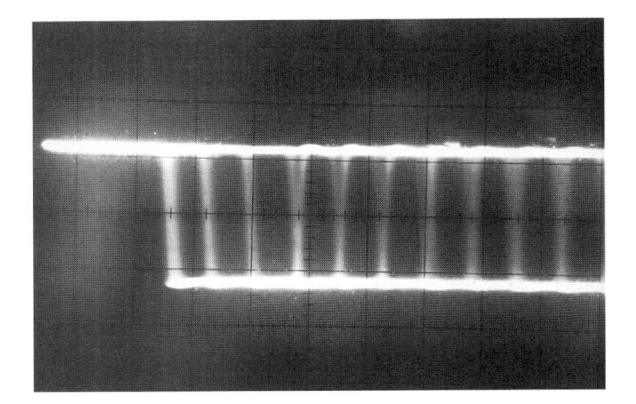

Figure 9.19 *EFM waveform*

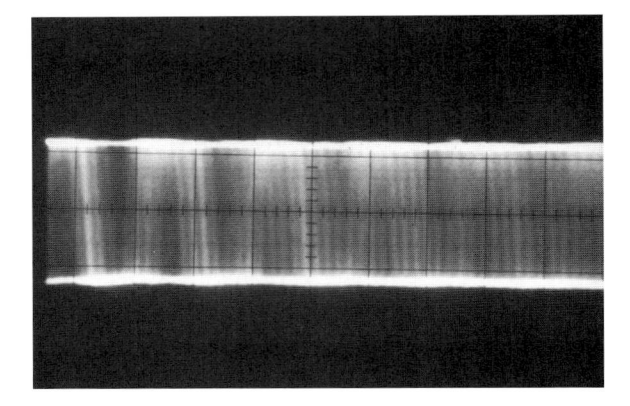

Figure 9.20 *Unlocked VCO waveform from decoder*

usually fitted underneath the printed circuit board, and a typical signal under normal working conditions is shown in Fig. 9.19.

A lack of signal here does not necessarily indicate a faulty RF amplifier and the d.c. loop, denoted by the thicker line in Fig. 9.18, should be checked, and must be complete for the purpose of the ASY function. Generally speaking, 2.5 V should be present at all the relevant points comprising the EFM/ASY circuit.

Occasions can arise when the ASY filter may cause a problem, preventing the 2.5 V reaching the ASY buffer or EFM comparator, effectively switching either or both off and thereby preventing any EFM output. An interesting feature when monitoring the EFM point, with the player just switched on and no functions selected, is that some form of high frequency can be present, even though the disc is not functioning, which is due to the VCO frequency which will not be locked being fed back from the decoder towards the EFM output of the RF amplifier (Fig. 9.20).

Stage 6: players with the test or service mode facilities selected as in Stage 5

This stage can apply particularly to Pioneer players or any players that require adjustment of the diffraction grating. The disc will be rotating at an acceptable speed (see Fig. 9.21).

Check and adjust if necessary the tracking balance adjustment

(Refer to relevant Adjustments section in Chapter 6.) If this adjustment is incorrect, the tracking

Fault diagnosis

[Flowchart:]
- Check and adjust if necessary the tracking balance adjustment
- Ensure tracking error waveform amplitude is acceptable
  Pioneer players: In the region of 2.0 to 3.6V pk to pk
  Philips players: In the region of 50 mV pk to pk
- Check diffraction grating adjustment
  (The grating will not normally change from its original setting, but if the optical block has been replaced the grating may not have been adjusted to the correct point. Refer to the relevant Adjustments section)
- Tangential/radial adjustment(s) may not be correct
  (If this is incorrect it is possible that the grating adjustment procedure may not prove successful. Refer to Stage 8 or the relevant Adjustments section)

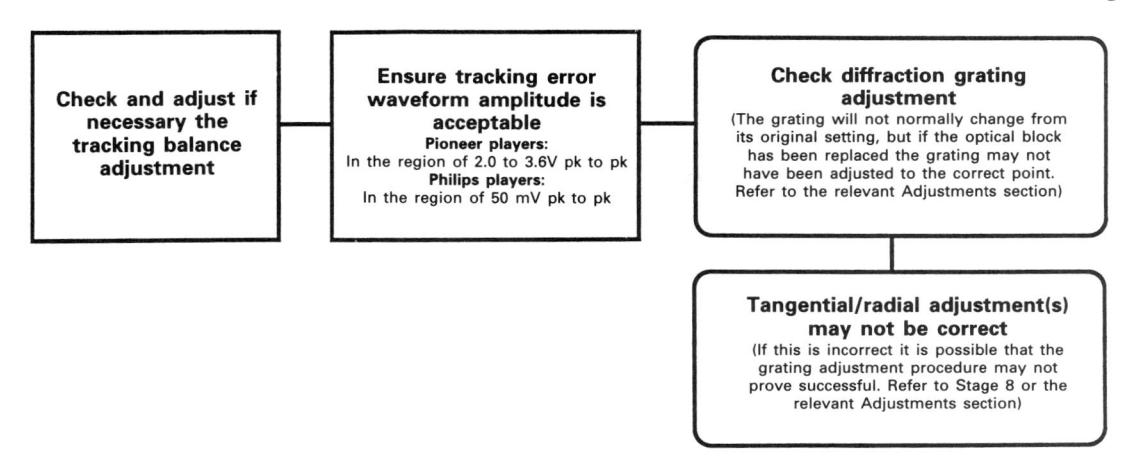

Figure 9.21 *Stage 6: checking tracking error waveform and diffraction grating*

servo will not operate correctly, and can result in 'skipping' and 'jumping' problems. By monitoring the tracking error waveform it is possible to determine how the tracking servo is performing.

Examples of correct and incorrect adjustments are shown in Fig. 9.22.

Stage 7: players without the test or service mode facility (and/or players with the test or service mode facilities selected as in Stage 5)

This stage can apply particularly to those players that have the facility of 'opening' and 'closing' the tracking servo (see Fig. 9.23).

Does the tracking error waveform collapse when the tracking servo is 'closed'?

Some method is required to ensure that the tracking servo is endeavouring to fulfil its function in life. This can occasionally be difficult to prove with some types of player, but naturally it is essential for the tracking servo to operate as required, and if the player does not read the TOC at the start of the disc, then a typical symptom can be that the disc tries to run-up and then give up after a brief period (but note that this symptom can also be related to other problems within the player).

Players that have the service or test mode facility can prove easier in determining the tracking servo operation, but it is also possible to briefly verify its operation in the more basic type of players.

Basically the intention is to determine whether the tracking error waveform collapses, which is usually an indication that the tracking servo is endeavouring to operate, and monitoring this waveform can provide the required indication. If the player has the service or test mode facility, follow the procedure outlined in Stages 5 and 6 in order to monitor the radial or tracking error waveform as shown in Fig. 9.24. Providing the waveform is acceptable, press the key that will now close or operate the tracking servo:

— Philips players:
  operate 'NEXT' again (servicing position 3)
— Pioneer players:
  operate 'PAUSE'

The waveform should now collapse to indicate a noise waveform, which is usually an indication that the servo is working.

If the tracking error waveform does not collapse

If the waveform does not collapse, it is usually an indication that the servo is not functioning

Fault diagnosis

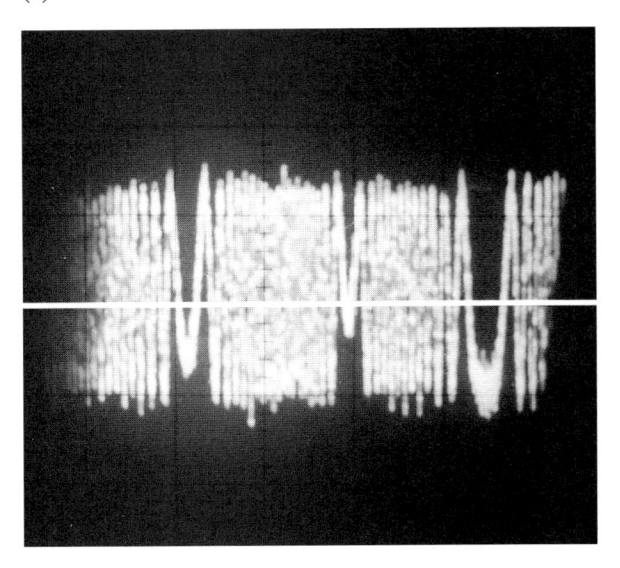

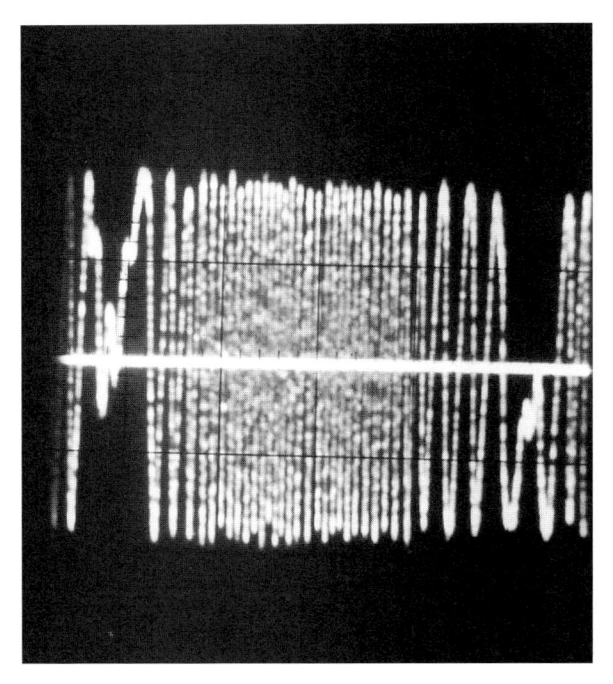

Figure 9.22 *Examples of tracking balance waveforms; (a) out of adjustment and off centre, (b) correctly adjusted*

correctly, and it should now be possible to signal trace the waveform, or something similar to the waveform, as a result of further processing in later stages to determine where the fault may be.

Possible causes may be defective radial/tracking coil, broken print on the PCB or flexi print, or defective amplifier stages.

With many players, the waveform is initially monitored at a point before the tracking gain control, if one is present, and it has not been unknown for this control to be set at the minimum setting, which will very effectively prevent the error signal reaching its objective – that phantom twiddler again!

Players that do not possess the service or test mode facility require a slightly different approach, and it should be confirmed that the laser, focus and disc/spindle motor circuits are functioning.

Determine a point where the tracking error signal, or its equivalent, can be monitored, and connect the oscilloscope. With a test disc inserted, set the player to the normal 'play' operation. Usually, as the player runs up, it is possible to observe the tracking error waveform momentarily before the tracking servo comes into operation, at which point the waveform should collapse. If this waveform does not collapse or appears to be present for an excessively long period before the player throws in the towel, then there may well be a problem with the tracking servo circuits which will need further investigation.

Does the optical block move unexpectedly to one end of its travel?

Occasions have been known when the disc appears to run up, and even indicate that the TOC has been determined, but when the 'play' function has been selected, the player refuses to respond accordingly. It has also been known for the optical block, for no apparent reason, to take off towards the outer edge of the disc at the commencement of playing the disc. On such occasions it has proved useful to determine if there is anything wrong with the tracking coil drive and carriage/sled drive stages.

It has been mentioned previously that it is a normal function of the player to ensure that the optical block is returned to the 'home' position, which is usually a function of the carriage/sled motor circuits, which provide a relevant output

Fault diagnosis

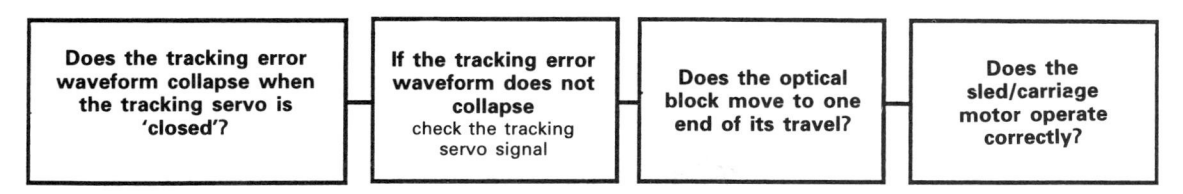

Figure 9.23 *Stage 7: checking tracking and carriage operation*

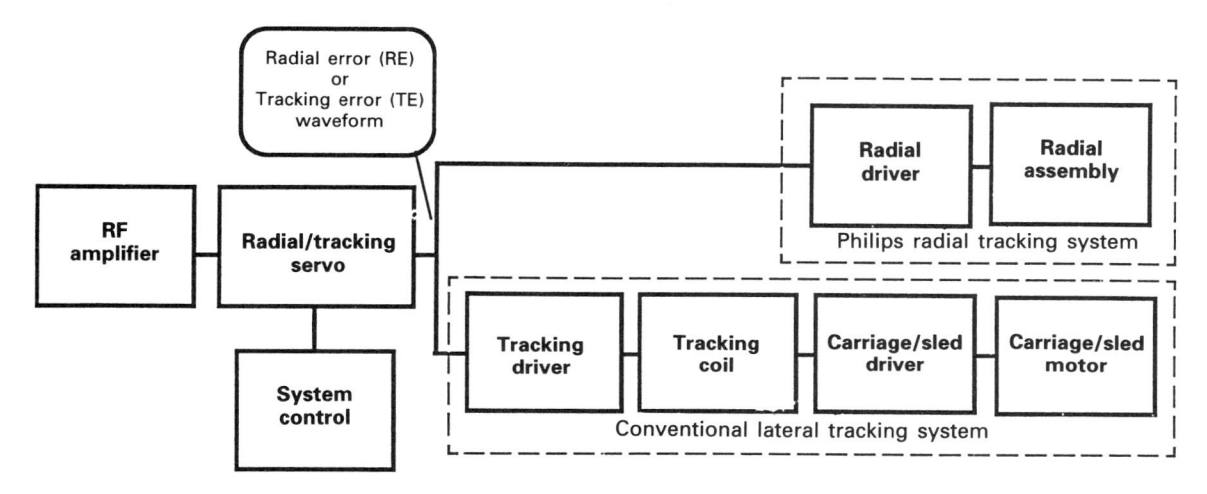

Figure 9.24 *Basic radial/tracking servo system block diagram*

to bring the optical block to the start position. When in this position the tracking coil can track the initial tracks on the disc to read the TOC without the need to move the optical block.

However, whilst it may be possible to move the optical block in one direction or the other, by operating selected key functions in the service or test mode (refer to Stage 3 – Check for Carriage or Radial movement), it is possible that a fault may lie between the tracking coil and carriage drive circuits.

With players that have the service or test mode facility, it is recommended that the optical block is moved to the midway position, though this is not usually possible with the radial tracking assemblies. By going through the process outlined above when determining whether the tracking error waveform collapses, if the optical block goes to one end of its travel when 'pause' is selected, this normally indicates that a d.c. offset fault may well have occurred within one of the relevant drive ICs. Checks for correct and/or incorrect highs and lows along the servo signal lines, i.e. TE or SLO, should be able to determine where this particular fault may be occurring.

Does the sled/carriage motor operate correctly?

Whilst it may be possible with the previous check to determine that the carriage motor is operating correctly, in order that the optical block can travel from the centre of the disc towards the outer edge, it is essential to determine that the carriage motor can in fact operate correctly whilst a disc is being played.

As the tracking coil gradually tracks the laser beam across the surface of the disc, it is important that the optical block is moved a small amount in the same direction, before the tracking coil reaches the limit of its travel, as otherwise 'skipping' will occur, creating an effect similar to being stuck in a groove.

By monitoring with an oscilloscope, the supply to the carriage motor, with the oscilloscope set to

Fault diagnosis

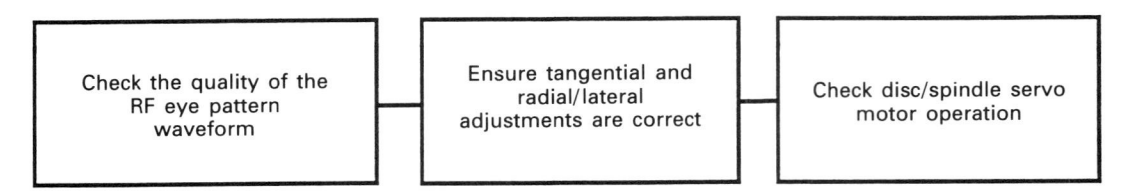

Figure 9.25 *Stage 8: checking the RF eye pattern waveform*

Y = 20 mV/cm and X = 0.2 ms/cm to 2 µs/cm, the resultant trace should appear to be bouncing up and down when a disc is being played. This bouncing should have a tendency to move in a negative direction, but can depend upon the player. As the bouncing moves progressively in the one direction, a point will be reached when the carriage motor will operate, and thus fractionally move the optical block. The bouncing level will now reduce in voltage terms, and then start to increase once again, this process continuing all the time that the disc is being played.

If the sound starts to 'skip' before the motor moves, then the carriage motor, or possibly the relevant drive circuits, may prove to be defective, but it is also worthwhile checking that there is no unnecessary friction or loading of the carriage mechanics, which would lead to the need for a higher potential to produce sufficient torque to operate the motor.

Stage 8: players without the test or service mode facility (and/or players with the test or service mode facilities selected as in Stage 5)

This stage can apply particularly to those players that have the facility of 'opening' and 'closing' the tracking servo. See Fig. 9.25.

Check the quality of the RF eye pattern waveform

Ideally the quality of the RF eye pattern waveform, whilst the disc is being played, should be similar to the waveform shown in Fig. 6.13.

If the amplitude of the waveform is low compared to the manufacturer's recommendation, then the laser power may be low, and should be checked as described in Stage 5 (or refer to the relevant Adjustments section in Chapter 6).

Ensure tangential and radial/lateral adjustments are correct

If either of these adjustments, where applicable, are incorrect this can cause the RF eye pattern waveform to appear low and also out of focus. Refer to the relevant Adjustments section in Chapter 6 to achieve optimum results for these adjustments.

If the waveform appears distorted, especially when using a known and reliable test disc, and providing all relevant adjustments are correct, then it is possible that the photo-diode array within the optical block may be faulty, and the optical block may require replacing.

Check disc/spindle motor servo operation

Again by monitoring the RF eye pattern waveform, especially as the disc runs up to speed in the normal 'play' mode, the waveform should effectively 'lock in' almost immediately, but if the waveform appears to be struggling to achieve its normal response, or if there is excessive sideways jitter at the right-hand side of the waveform, then it is possible that either the VCO phase locked loop is out of adjustment (refer to the relevant Adjustments section), or that the disc/spindle motor may be faulty.

Other areas worthy of being checked

1 Are there any mechanical problems in relation to the optical block, i.e. any dried out grease or dirt on the slide bar(s) or thread drive? Check

Fault diagnosis

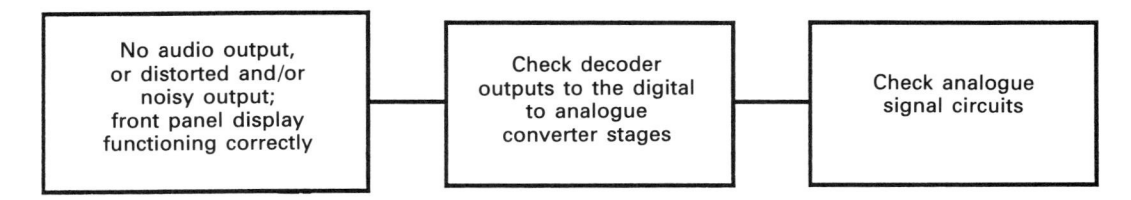

Figure 9.26 *Stage 9: checking the audio output*

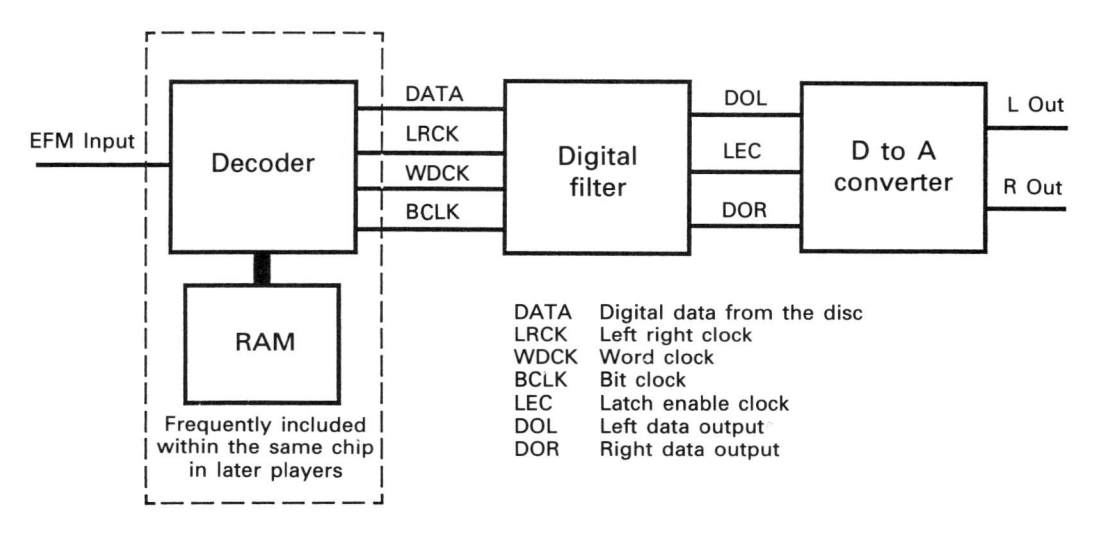

Figure 9.27 *Checking the decoder and D to A converter stages*

that the drive spring (if applicable) is correctly fitted.
2 Is the diffraction grating (if applicable) correct (refer to the relevant Adjustments section)?
3 Is the turntable height correct (refer to the relevant Adjustments section)?
4 Check for incorrect electrical adjustments (refer to the relevant Adjustments section).

Stage 9: player operating in the normal play mode – 1

No audio output, or distorted and/or noisy output with front panel display functioning correctly

With this particular type of symptom (Fig. 9.26), especially the noisy or distorted output, it is the digital to analogue stages that have sometimes been suspected to be the problem – often incorrectly. The important indication is that the disc is effectively playing, and that the front panel display is providing all the required indications of track and time information; from this it may be assumed that the greater part of the player is operational, but that the required output is somewhat lacking.

To determine faults within this area, a test disc with a left-only and right-only track can be beneficial.

Check the decoder outputs to the digital to analogue converter stages. Fig. 9.27 may assist in determining which direction to take when this particular symptom occurs.

The outputs from the decoder can be measured with an oscilloscope using a test disc, preferably one that includes a left-only and a right-only track, which will prove to be extremely useful in determining possible faulty areas. It is important that the four outputs, DATA, LRCK, WDCK and BCLK, are present as shown though WDCK may not be used with players that use the digital filtering stage.

173

Fault diagnosis

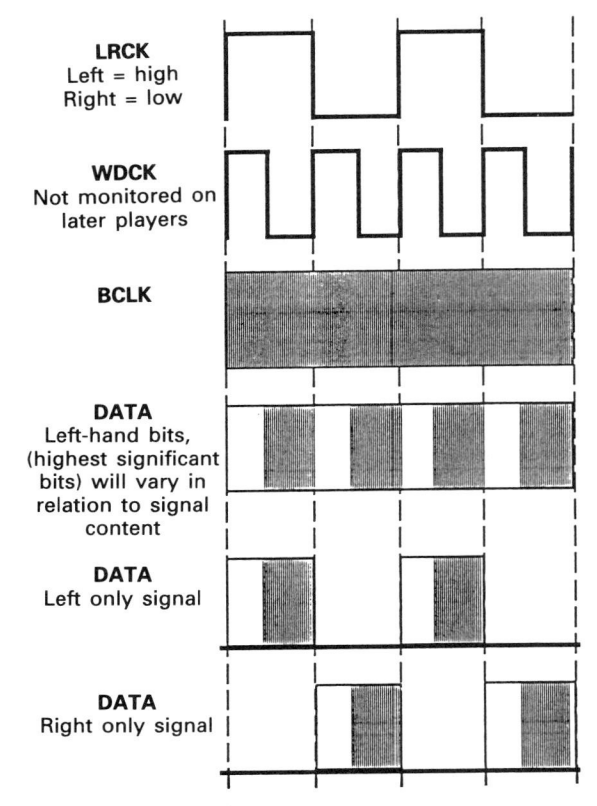

Figure 9.28 *Decoder outputs*

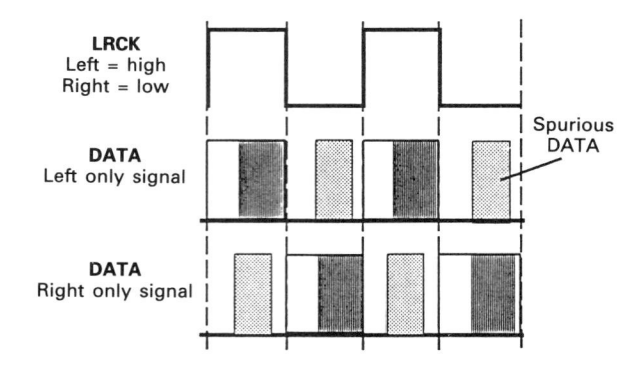

Figure 9.29 *Spurious data (noise) from decoder*

The waveforms that can be monitored are basically illustrated in Fig. 9.28, together with actual photographic examples (Fig. 9.30), and all waveforms are illustrated with the oscilloscope being triggered by the left right clock (LRCK).

Initially it is important to determine that the four outputs from the decoder (if applicable) are present, as otherwise the subsequent decoding stages cannot function correctly.

By monitoring these signals, it can be determined whether the output signals from the decoder are correct, and by using a test disc with the facility of either a left-only or right-only track it is possible to determine at least in which direction to proceed if the audio output is distorted or noisy.

The appearance of spurious data in the spaces between left-only or right-only data will usually indicate either a RAM or decoder fault. If the RAM is a separate IC it is usually recommended to replace this first, but later decoders usually have the RAM incorporated internally (Fig. 9.29).

Further checks at the output of the digital filter enable the following waveforms to be observed which are obtained from a player that has the facility of four times oversampling (4Fs); hence the apparently extra groups of data words when compared to the earlier types of player.

The latch enable clock (LEC), can be either a left right clock (LRCK) or possibly a word clock output (WCKO), depending upon the type of player and which ICs are being used in the data processing stages. The purpose of the LEC is to enable the D to A converter to commence its operation correctly (Fig. 9.30(a)–(f)).

If there is no audio output or a distorted audio output

Providing the previous waveforms, whichever are applicable, are correct and are being applied to the initial AF pre-amplifier stages, further monitoring should enable the faulty area to be determined, and by using a tone from the test disc it is also possible to determine whether distortion is being caused by these stages.

It is possible that either the muting or de-emphasis circuits are causing a problem. The DEMP and MUTE control signals usually originate from the system control circuit, and whether they are high or low will obviously relate to the type of circuit in which they may be operating.

The DEMP (de-emphasis) is not relevant with all discs, and it may be necessary to determine whether this circuit is operating or not by using a test disc that has tracks with emphasis included; these may be determined from the notes usually supplied with the disc (Fig. 9.31).

Fault diagnosis

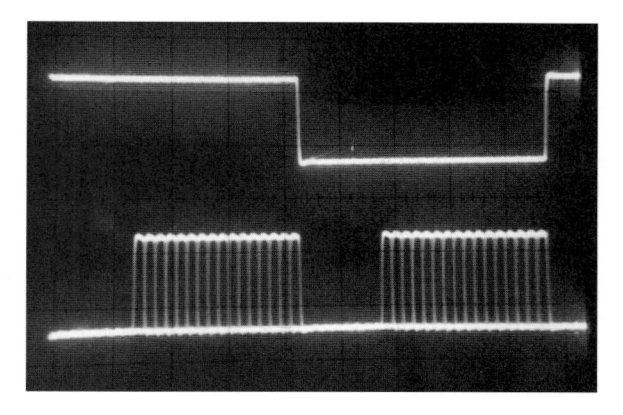

Figure 9.30(a) *Waveform illustrating the signal derived from the decoder on players without digital filtering. Top waveform: LRCK. Bottom waveform: left and right data symbols*

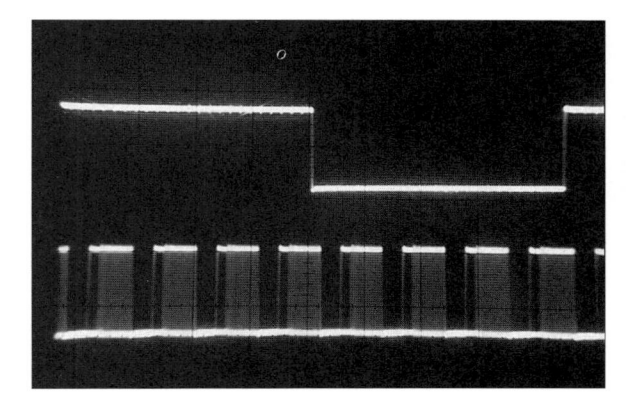

Figure 9.30(b) *Waveform obtained from a player with digital (4Fs) filtering. Top waveform: LRCK. Bottom waveform: four times oversampling (4Fs), left and right data words*

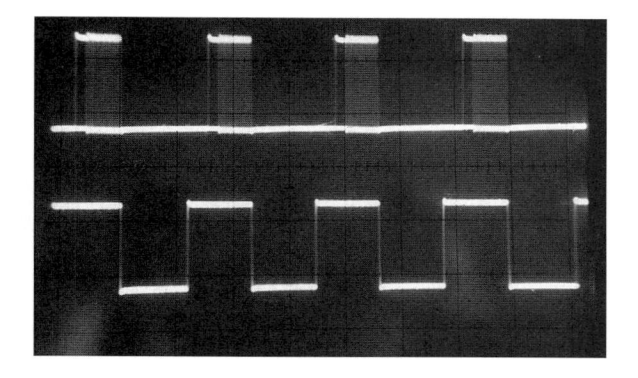

Figure 9.30(c) *Waveform obtained from a player with digital (4Fs) filtering. Top waveform right-only data words. Bottom waveform latch enable clock (LEC)*

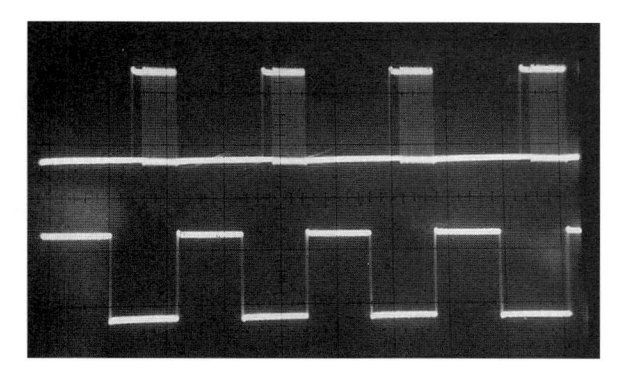

Figure 9.30(d) *As above amended. Top waveform left-only data words. Bottom waveform latch enable clock (LEC)*

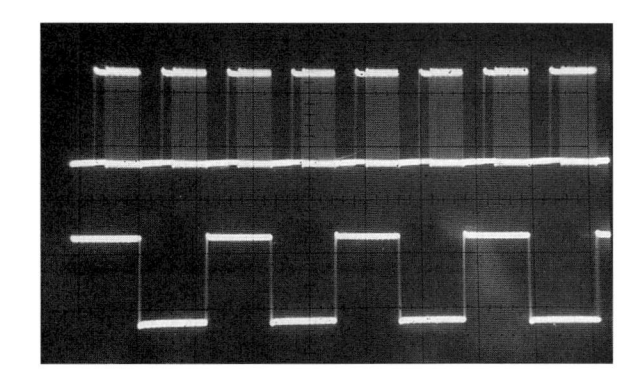

Figure 9.30 (e) *As above amended. Top waveform left and right data words. Bottom waveform latch enable clock (LEC)*

Fault diagnosis

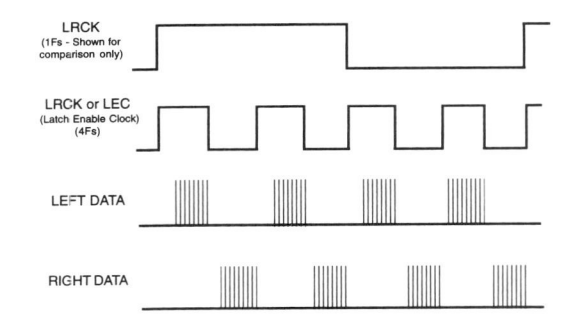

Figure 9.30(f) *Examples of waveforms from a 4 times oversampling (4Fs) decoder*

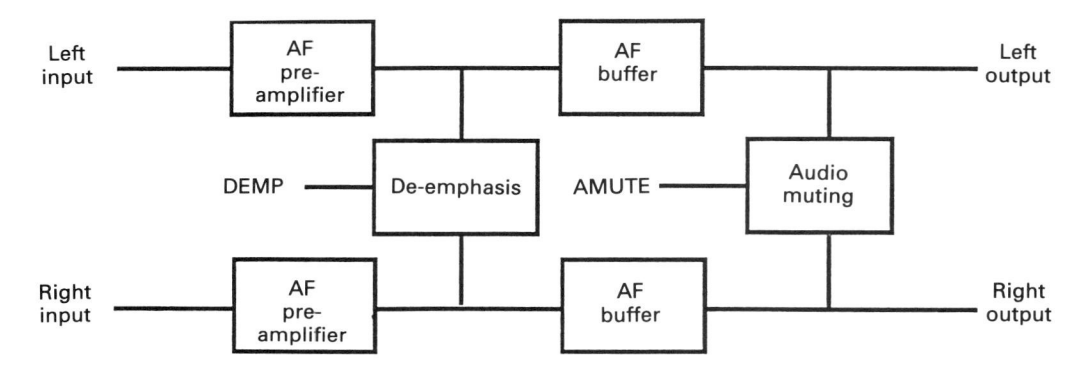

Figure 9.31 *De-emphasis and mute control*

The MUTE control will usually switch on the relevant transistor circuits to prevent unnecessary audio output, and checking of these stages should prove relatively straightforward.

Stage 10: player operating in the normal play mode – 2

There is audio output, but the player appears to 'skip' and/or 'jump' whilst playing the disc

Of the many and various faults that can occur with CD players, the 'skipping' and 'jumping' problems tend to be amongst the most common, and of all the faults that can occur these particular problems can frequently prove to be the most difficult to determine.

Many engineers will have experienced a customer who is emphatic that whilst the discs 'skip' and 'jump' on his or her own player, they naturally operate quite satisfactorily on any other player – especially on the one next door!

The initial recommendation is that when the player comes in to the service centre for investigation, ensure that a selection of the offending discs is also supplied. It is also useful, if the facilities are available, to allow the customer to demonstrate the specific problem.

Check the quality of the disc being used

Those early television programmes displaying the virtues of the CD with the dreaded raspberry jam demonstrations still have a lot to answer for, and whilst the CD is an excellent format in many ways, it should, ideally, be handled in the same manner as the ever faithful vinyl record. Frequently, CDs that customers expect to use in their players appear to have had alternative existences as either coffee, tea or beer mats, as well as having

Fault diagnosis

```
┌─────────────────────┐    ┌─────────────────┐    ┌──────────────────────┐
│  Audio output, but  │    │ Check the quality│   │ Ensure all mechanical│
│ player appears to   │───▶│  of the discs   │──▶│ aspects and electrical│
│ 'skip' and/or 'jump'│    │     being       │   │ adjustments outlined in│
│ whilst playing the  │    │      used       │   │ Stages 1-9 are correct│
│        disc         │    │                 │   │                      │
└─────────────────────┘    └─────────────────┘    └──────────────────────┘
```

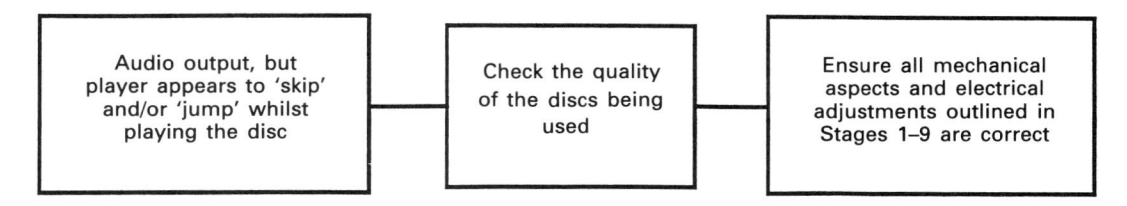

Figure 9.32 *Stage 10: checking for skipping and jumping*

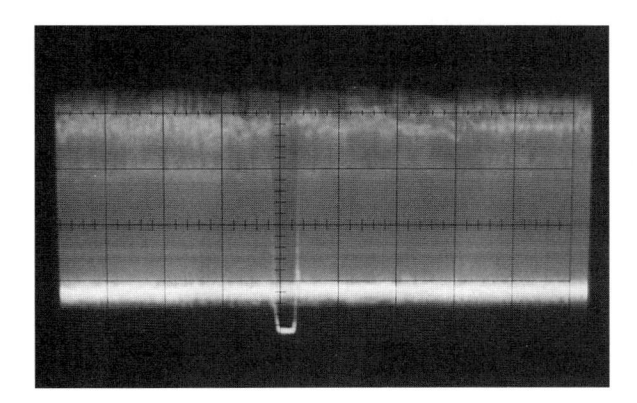

Figure 9.33 *Compressed eye pattern waveform showing excessive scratches on disc. Smaller scratches (not shown) can also be observed with this method*

been used as a 'Frisbee'. When a comment is made with regard to the quality of the disc, the reply is frequently along the lines of: 'You can do anything with a compact disc... they said so'. In response to the further enquiry as to who 'they' might be, there is usually a reference to the dreaded TV programme and raspberry jam.

If a player does appear to be playing a disc as though it is stuck in a groove, and especially if it is consistent on a specific disc at a particular section or time, then it usually goes without saying that the disc must be checked first. If it is just a question of dirt, then gently cleaning the disc with warm soapy water will cure most ills, and gently polishing off with a clean soft duster should restore normal operation. A quick lick and polish up on the front of a shirt is not really to be recommended... it has been observed!

If the disc is scratched, more drastic action is needed depending upon the severity of the scratch. Whilst success can be achieved in most cases, a severely scratched disc can still create problems, even if the scratch appears to have been removed and is not visible to the naked eye.

Many engineers have their own favourite method of removing scratches, often including light abrasive creams such as perspex polish, soft toothpaste (not your denture cleaning type), and silver polish; whilst Brasso is excellent for the more desperate, car polish and the buffing wheel are definitely not recommended.

It is important when polishing out any scratches that a technique is developed of using either small circular strokes, or strokes that are at right angles to the track. This is because the action of polishing the disc in this manner in itself creates extremely fine scratches, which providing they are at right angles to the track can be compensated for by the error correction process within the decoder. But the ideal is to get the disc back to its original condition, which may prove extremely difficult. If the disc is cleaned in large circular strokes, fine scratches will occur that will in part be parallel to the track which in turn will cause tracking problems when the disc is played.

On completion of any scratch removal process, it is again recommended that the disc be gently washed in warm soapy water to remove any residue from the cleaning agent, which if it is oily or greasy in nature can affect the optical characteristics of the disc, and create continued problems.

It is also useful to monitor the RF eye pattern waveform to determine disc quality, and this will be explained in the following section.

Ensure all mechanical aspects and electrical adjustments outlined in Stages 1-9 are correct

Assuming that the disc quality is proved to be satisfactory, various aspects of the adjustments that are applicable to a particular player must

177

Fault diagnosis

now be considered. Some that can create 'skipping' and 'jumping' problems are highlighted as follows:

Mechanical

1 Ensure all slide bars are clean and free from dried-out grease.
2 Check the disc turntable is set to the correct height.
3 Check that any drive springs that may be fitted are correctly located.
4 Check that the optical block flexi-print is correctly formed, and not creating any restriction of movement.
5 Ensure that the complete optical assembly is correctly located, and any resilient mounts are properly fitted.
6 Tangential and lateral/radial adjustments (if applicable).
7 Incorrect grating adjustment (if applicable).
8 Check for a defective disc/spindle motor.
9 Check for a defective carriage/sled motor.

Electrical

1 Check for an incorrect RF level (eye pattern waveform), i.e. not too low or high.
2 Check for incorrect focus, tracking and RF offsets (if applicable).
3 Check for incorrect tracking balance adjustment.
4 Check for a wrongly set VCO.
5 The focus and tracking gain may be incorrect.
6 Check the quality of the RF eye pattern waveform.

Occasions have been known where everything has seemed to be absolutely satisfactory, both mechanically and electrically, and the last resort has been to replace the optical block, but from the author's point of view this should be the last resort, as the previously outlined recommendations resolve the majority of the 'jumping' and 'skipping' problems.

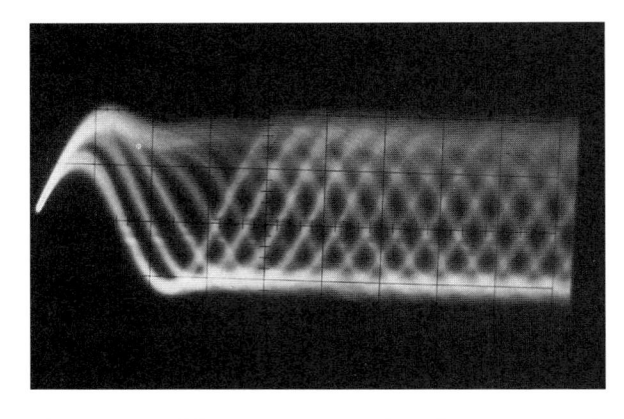

Figure 9.34 *RF eye pattern waveform from a poor quality disc*

Quality of the RF eye pattern waveform

The ideal waveform is shown in Fig. 6.13, but if the quality of the reflective surface is poor, and this can be in relation to the actual quality of the 'pits' or 'bumps', then the waveform will suffer accordingly. Fig. 9.34 shows the RF eye pattern waveform obtained from a poor quality disc that has been played without any major problems on many players, but which is illustrated to show how the quality of a disc may be determined by monitoring this waveform.

Excessive vertical bouncing of the waveform, compared to a known reliable test disc, can be due to excessive disc eccentricity, which again can result in 'jumping' and 'skipping'.

By compressing the timebase range on the oscilloscope to ~ 2 ms cm^{-1}, it is possible to actually monitor scratches on the surface of the disc whilst it is being played. This can give some idea as to how the player is being required to handle the 'glitches' that can be caused with scratched discs, as shown in Fig. 9.33.

APPENDIX 1: ANALOGUE AND DIGITAL PROCESSING

A wide range of publications are available, especially in the reference sections of the larger public libraries, that will cover in depth all the relevant intricacies concerning the principles of processing the analogue signal into digital, and back into its hopefully original analogue form once again.

Many factors have to be considered when endeavouring to analyse most of the concepts of digital processing, the majority of which is very mathematical. The following information is aimed at providing a basic outline of some of the processes involved, leaving the reader to investigate further depending upon the amount of personal and/or individual desire to proceed further.

Analogue to digital conversion – a few initial thoughts

When an analogue signal is converted, or quantized, into a digital form, that analogue signal is sampled at certain intervals, and each interval is stored for a brief period of time in a sample and hold circuit, whilst a digital processing circuit converts that level into a specific data word of so many 'bits'. Both circuits are driven by a clock circuit to provide a fixed reference source (Fig. A1.1).

As outlined in Chapter 1, the original analogue data are converted into 16 bit data words, which after further processing techniques are, in effect, implanted onto the disc, and then processed back into analogue form within the relevant circuits of the CD player. It would therefore seem logical to consider analogue to digital conversion techniques first, but logic does not necessarily prevail, as it can prove easier to get the 'feel' for this subject by considering digital to analogue first; there is also the fact that some types of A to D converters actually employ some D to A processing to enable the final digital data to be achieved.

Digital to analogue conversion

The 16 bit data words from the disc, which are developed within the decoder of the CD player from the complex information on the disc, are processed within a general block arrangement as illustrated in Fig. A1.2.

This relatively complex arrangement outlines the processing of the EFM signal from the disc; this is in fact digital information. With processing within the decoder, the original 16 bit data will be extracted from the disc information, but after application via the digital filter, the data words passing to the D to A converter will not usually be in 16 bit form. They will in fact comprise data words in excess of 16 bits due to oversampling techniques as described in the later sections of Chapter 1.

The fundamentally analogue output from the D to A converter is applied to a deglitching circuit to remove undesirable spikes that can develop from the D to A process.

Consider that the input to the D to A converter is a series of data words, with each data word comprising a certain number of bits (which for this example will be 16 bits), and that these bits may be identifed as ranging from the least significant bit (LSB), i.e. the lowest value area of the data word, to the most significant bit (MSB), i.e. the largest value area of the data word, as shown in Fig. A1.3.

In order that analogue information can be obtained from the digital data it is necessary to consider the significance of each of the data bits when applied to the D to A converter, and this may be considered in the manner shown in Fig. A1.4.

Analogue and digital processing

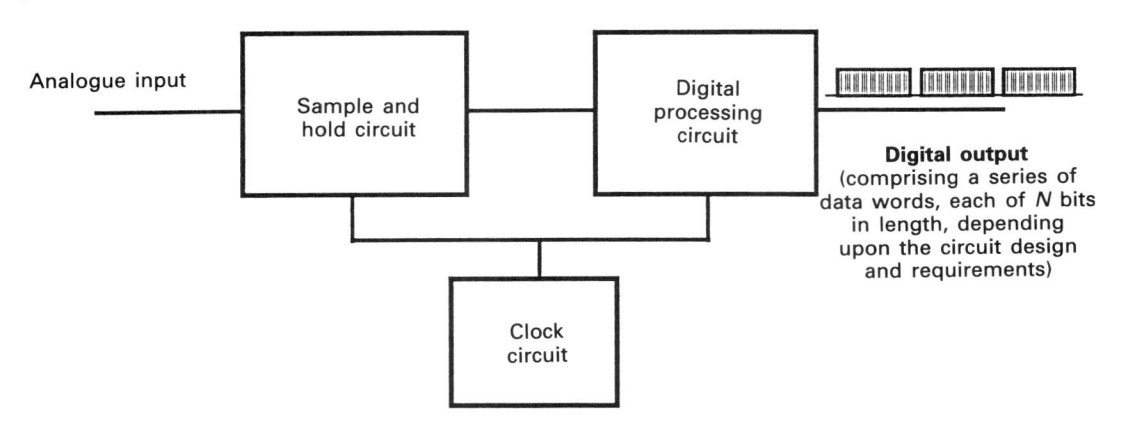

Figure A1.1 *Basic principle of analogue to digital conversion*

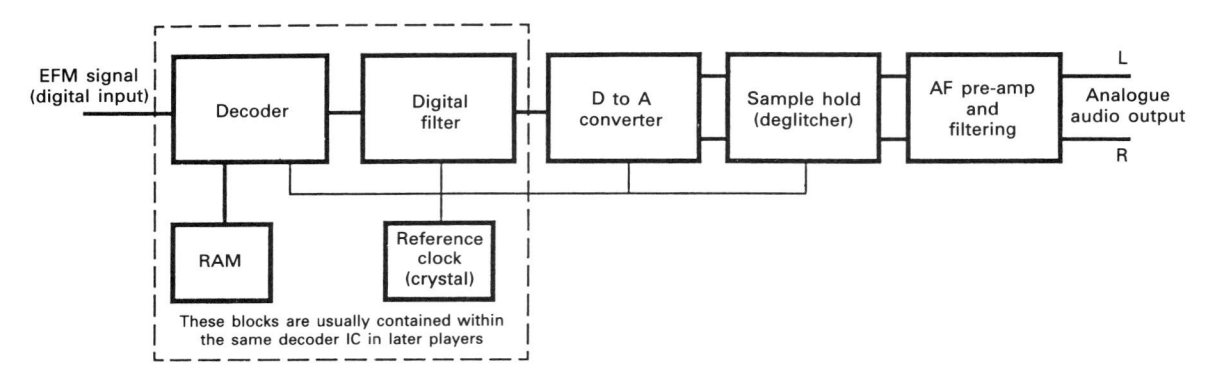

Figure A1.2 *Typical D to A processing arrangement of a CD player*

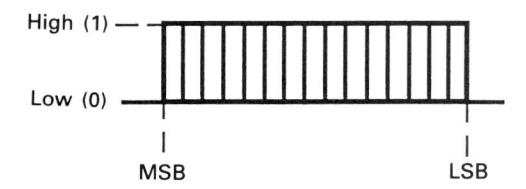

Figure A1.3 *Example of a 16 bit data word*

Fig. A1.4 provides an overall analysis of the binary system, where each bit of a 16 bit data word is represented as a specific power of '2', with the LSB being the lowest power or value, and the MSB being therefore the highest value.

With a 16 bit data word there are 65 536 possible combinations of the 16 bits from all the 0's through to all the 1's. The LSB will represent the

Power of '2'	2^{15}	2^{14}	2^{13}	2^{12}	2^{11}	2^{10}	2^9	2^8	2^7	2^6	2^5	2^4	2^3	2^2	2^1	2^0
Actual value	32768	16384	8192	4096	2048	1024	512	256	128	64	32	16	8	4	2	1
Fractional value	$\frac{1}{2}$	$\frac{1}{4}$	$\frac{1}{8}$	$\frac{1}{16}$	$\frac{1}{32}$	$\frac{1}{64}$	$\frac{1}{128}$	$\frac{1}{256}$	$\frac{1}{512}$	$\frac{1}{1024}$	$\frac{1}{2048}$	$\frac{1}{4096}$	$\frac{1}{8192}$	$\frac{1}{16384}$	$\frac{1}{32768}$	$\frac{1}{65536}$

MSB   LSB

Figure A1.4 *Binary number system*

Analogue and digital processing

		Fractional value	Fraction of 10 V (values rounded off)
MSB	1	1/2	5.0
	0	0	0
	0	0	0
	0	0	0
	1	1/32	0.31250
	1	1/64	0.15625
	0	0	0
	0	0	0
	1	1/512	0.01953
	1	1/1024	0.00976
	1	1/2028	0.00493
	1	1/4096	0.00244
	1	1/8192	0.00122
	0	0	0
	1	1/32768	0.00030
LSB	1	1/65536	0.00015
		Total value =	5.50843

Figure A1.5 *Numerical illustration of digital to analogue conversion*

smallest fraction or value in digital terms, whilst the MSB will represent half the total possible value. Each of the 1's that may be present in a 16 bit data word can be analysed to determine the actual proportion of the original maximum analogue value as follows. Consider the following 16 bit data word, with the MSB being the first digit, and the maximum analogue value for the purpose of this description being 10 V:

1000110011111011

When the power of 2 increases by one (i.e. 2^4 to 2^5), the numerical values these powers represent double (i.e. from 16 to 32), whilst if the power of 2 is reduced by one, the representative numbers will halve. It is these factors which contribute to the operation and accuracy of the digital to analogue conversion process.

Basic D to A converter

Fig. A1.6 illustrates the basic concept of a D to A converter, which comprises a series of current generators that are supplied from an accurate reference voltage source, a series of switches with each one linked to a current generator, and a current to voltage generator, which will provide an analogue voltage output as a result of the total current that has been created from the input data word which has closed relevant switches in relation to the 1's information of the data word.

The relation of each current generator is that the most significant bit (MSB) current generator must accurately provide half of the total required current from the reference source, and that the descending generators must again accurately provide half of the current provided by their preceding neighbour. This required accuracy is an essential factor of D to A conversion, and can prove extremely costly when related to domestic CD players, especially when considering the extremely small amounts of current that are required in relation to the least significant bits (LSBs).

Basic resistor ladder D/A converter

With a basic resistor ladder D to A converter (Fig. A1.7) the switches are closed in accordance with the 1's content of the data word to enable a total current to be produced, which will in turn develop an analogue voltage output that will ideally replicate the original analogue sample.

Unfortunately there are a couple of problems with this method, as well as the following R-2R resistor ladder circuit shown in Fig. A1.8.

1 It is difficult to produce resistors that have an absolute accuracy of each resistor being either exactly double or half the value of its neighbour, with virtually no variation in tolerance. For example, if R0 was 5K in value, then R1 must be 10K and R2 would equal 20K and so forth, through to R16 which would equal 327.68 MΩ.

Analogue and digital processing

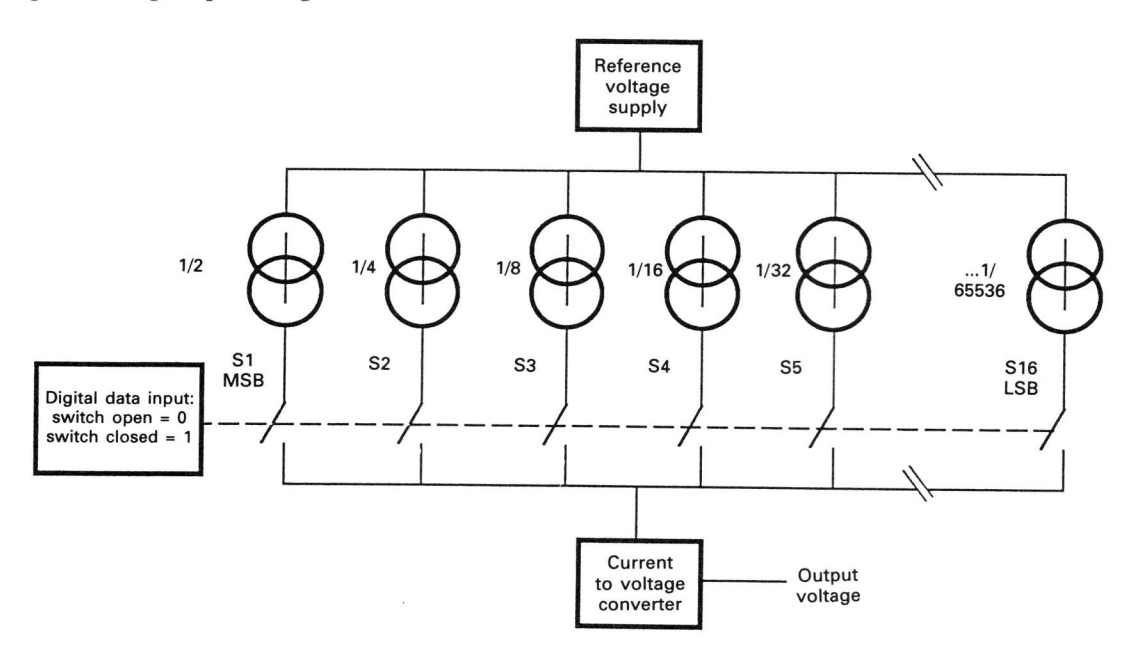

Figure A1.6 *Basic D to A conversion*

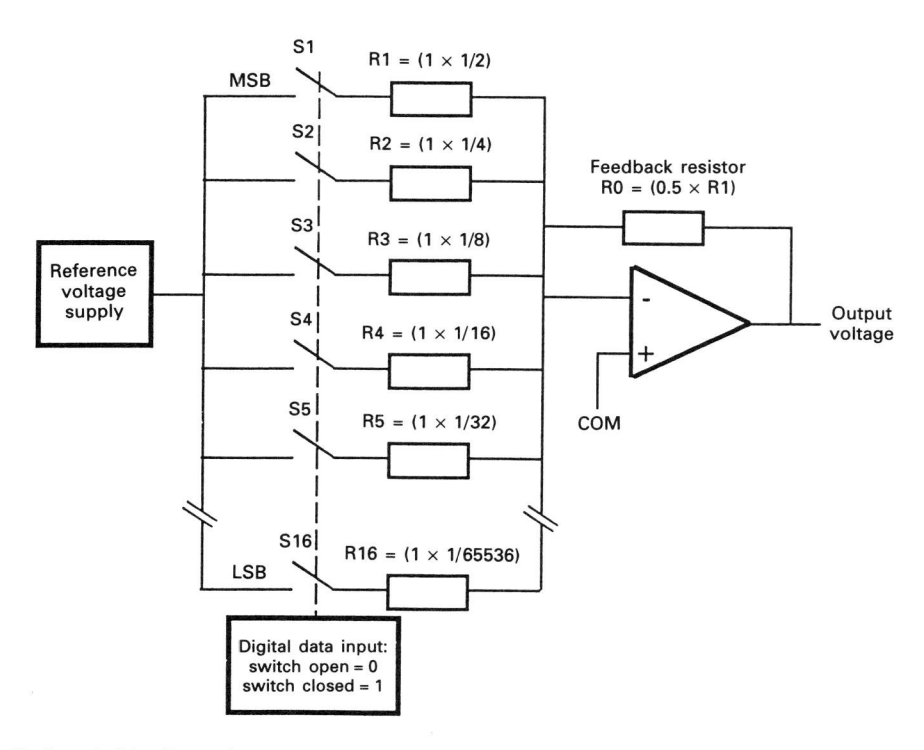

Figure A1.7 *Resistor ladder D to A converter*

Analogue and digital processing

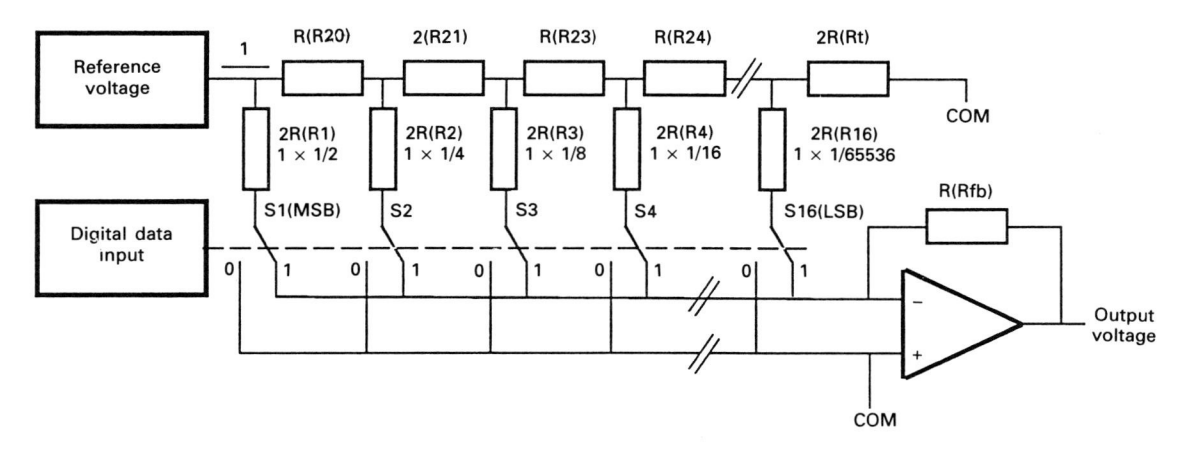

Figure A1.8 *R-2R resistor ladder D to A converter*

2 When a large number of bits are used in each data word, the range of current inputs from the MSB to the LSB can prove too much for the operational amplifier to handle, especially where the lower order bits are involved. For example, if the reference voltage is 10 V, then the current flow through R1 at 10K will be 1 mA, whilst the current through R16 at 327.68 MΩ would equal ~ 0.3 µA.

Basic R-2R resistor ladder D to A converter

The R-2R resistor ladder D to A converter (Fig. A1.8) comprises a set of series resistors, (R20), (R21), (R22), etc., and a set of shunt resistors, (R1), (R2), (R3), etc., and only two resistor values are required, e.g. R = 10K and 2R = 20K. If the current flowing through the chain of resistors R20 through to Rt is I, then through R1, $I = I \times 1/2$, and through R2, $I = I \times 1/4$ and so forth.

With the switches in the '0' position the lower end of the resistors R1 through to R16 are connected to common or ground, and when set to the '1' position the 'virtual earth' input of the operational amplifier will still maintain the same effect, but of course will comprise the total current which will relate to the 1's information of the data word input.

This method of D to A conversion is in fact easier to manufacture, as only two specific values of resistor are required, i.e. R and 2R.

Typical D to A converter

Fig. A1.9 illustrates the basic D to A converter arrangement used in many players, and is based on a circuit which utilises the Sony D to A converter IC. This form of D to A converter is often referred to as an integrating dual slope D to A converter. It comprises two constant current generators, one providing I_0 which is related to the eight most (or higher) significant bits, and the other providing i_0 which is related to the eight least (or lower) significant bits.

The current switches are similar to those described in previous basic D to A circuits, with the resultant current being passed via the integrating circuit, to the relevant left or right sample and hold circuit. Generally the four main inputs to this circuit will comprise: DATA, bit clock, word clock (WDCK) and left/right clock (LRCK).

The DATA will be the 16 bit data word, which will operate the relevant current switches to provide a total current that relates to the analogue value of each data word. The bit clock is a constant clock frequency at the relevant data rate to enable the data to be processed correctly. The word clock (WDCK) enables the higher and lower order 8 bit sections of each data word to be processed as required, whilst the left/right clock (LRCK) ensures that the left and right data words are routed through to their respective circuits to provide the final left and right analogue outputs (Fig. A1.10).

Analogue and digital processing

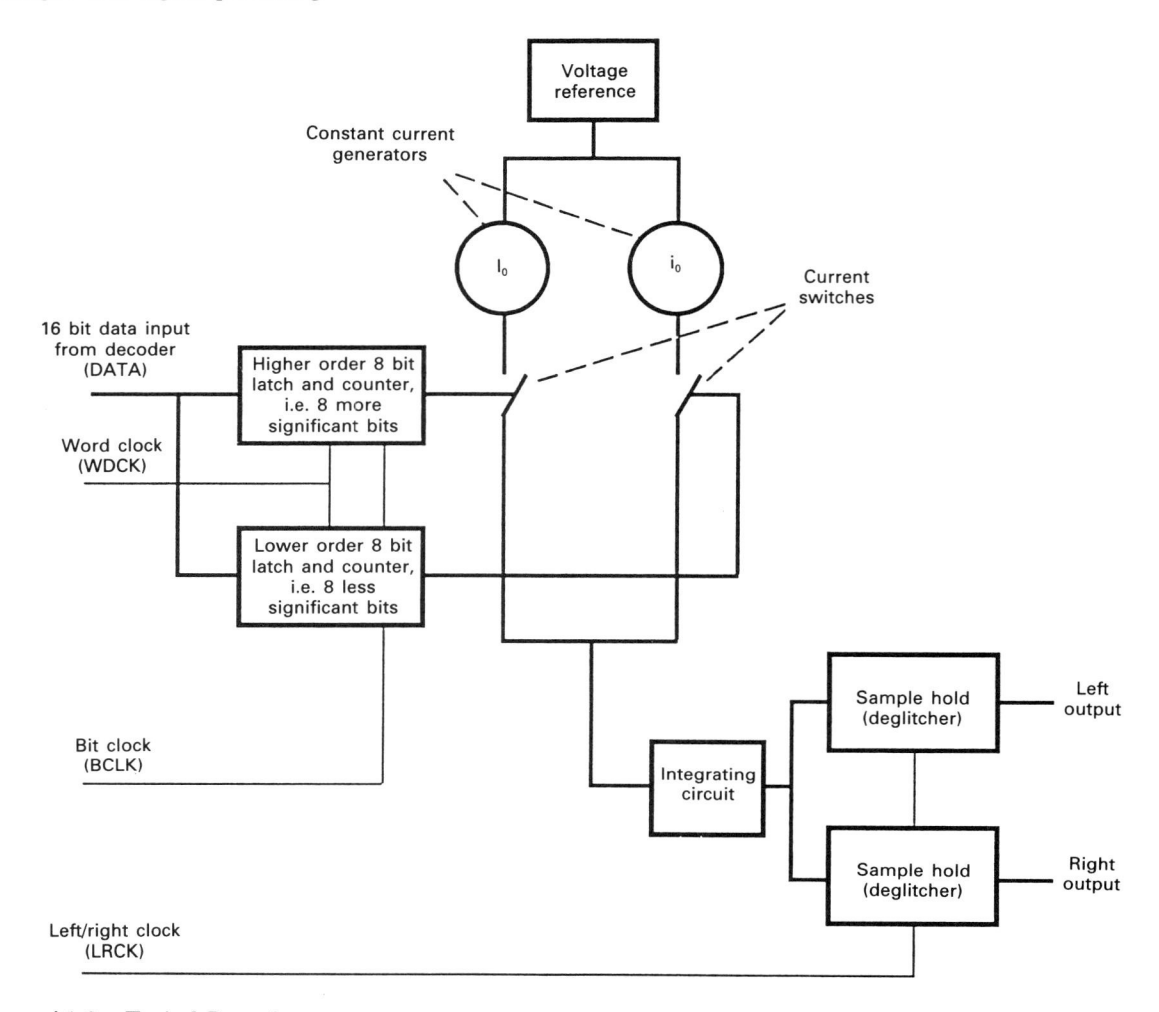

Figure A1.9 *Typical D to A converter*

Sample and hold circuits

Whilst both the D to A and the A to D conversion circuits operate in a digital manner, it is essential that some form of sample and hold circuits are utilised to help these circuits operate efficiently.

In the D to A process the sample and hold circuit is used to remove undesirable interference, frequently referred to as 'glitches', which can occur at the output of the D to A circuit after the integration of each data word.

With respect to the A to D process, the sample and hold circuit provides a 'sample' of the input analogue signal, and then 'holds' it whilst that value is converted into a digital data word.

Basically the sample and hold circuit may be represented as shown in Fig. A1.11, and comprises a couple of switches and a capacitor, with S1 switching the analogue sample to the capacitor, whilst S2 will discharge the capacitor when the digital conversion has been completed, and before the next sample is selected.

Sampling an analogue signal for A to D conversion

The analogue signal is sampled at regular intervals, and each sample is stored in the capacitor C, whenever S1 is closed, whilst S2 discharges the capacitor in readiness for the next sample (Fig. A1.12).

Analogue and digital processing

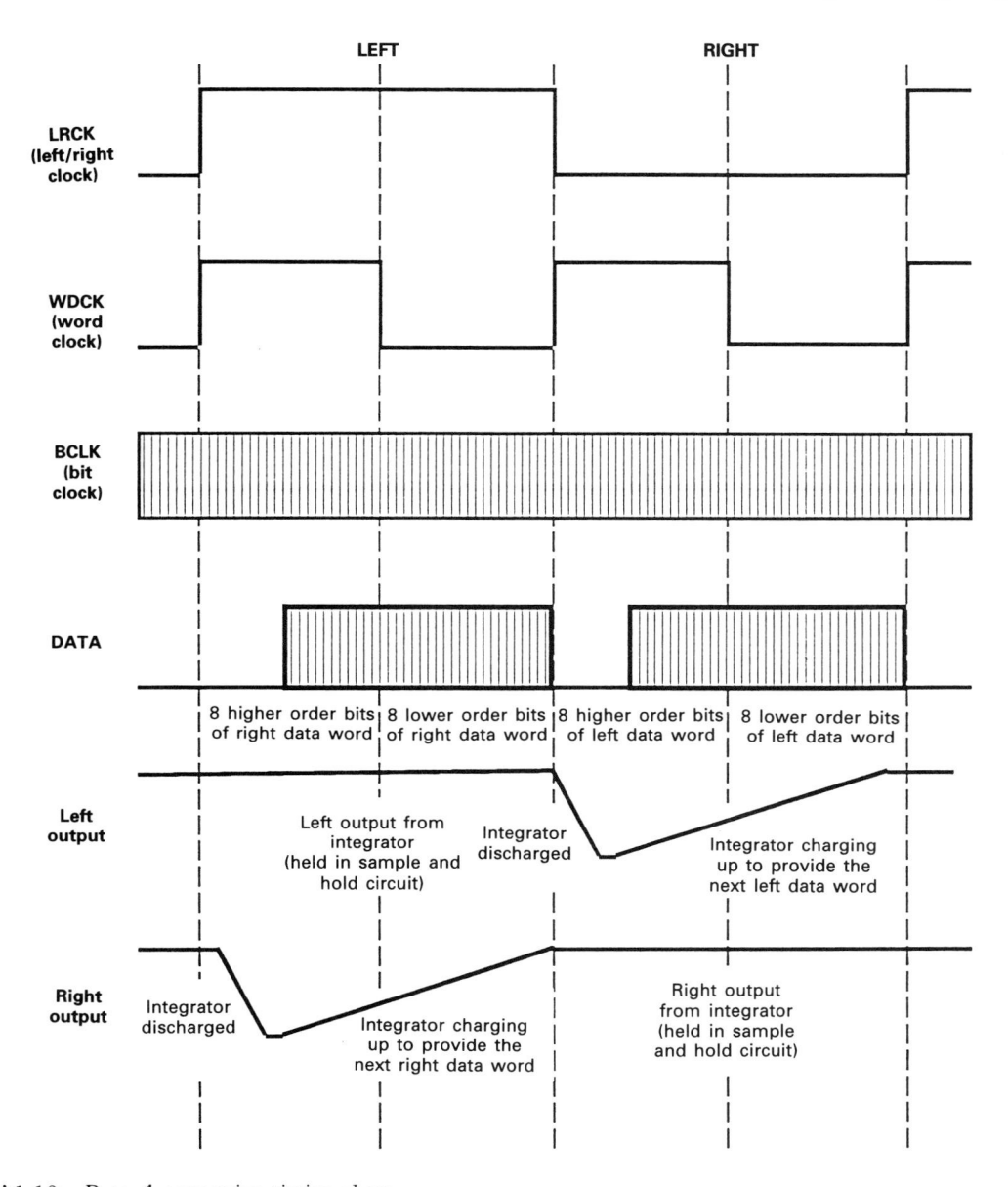

Figure A1.10 *D to A conversion timing chart*

Whilst each sample is held in the capacitor C, the specific value of each sample will be turned into a digital data word. In practice the switching pulses for both switches will be high in frequency terms, whilst the change of each pulse into a data word of so many bits will be even higher, with the final data frequency being related to the number of bits being used for the data words.

The switches S1 and S2 are usually CMOS type semi-conductor switches which are practical for high frequency operation.

Sampling an analogue signal after D to A conversion

When the digital data word is transformed or quantised back into its ideally analogue value, it is possible to develop interference pulses as the D

185

Analogue and digital processing

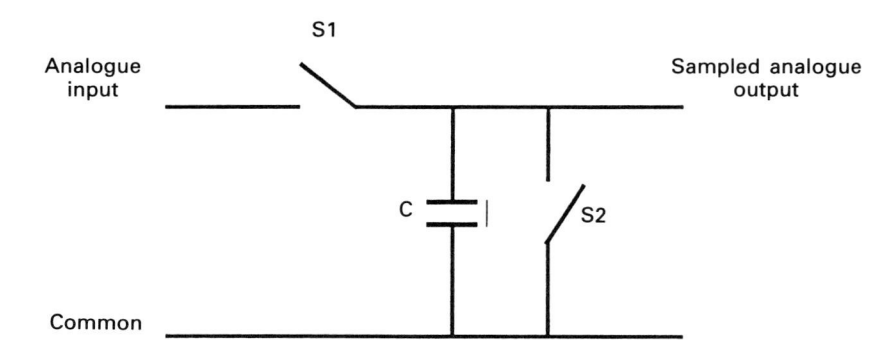

Figure A1.11 *Sample and hold circuit*

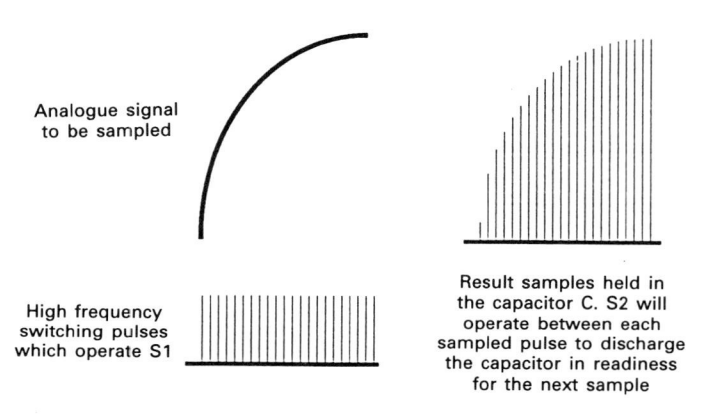

Figure A1.12 *Basic sampling of the analogue signal*

to A converter settles or stabilises from one data word to the next. This interference can frequently be referred to as erroneous data, quantising noise, or even glitches, and can cause unnecessary noise and possibly distortion of the required analogue signal output. The effect of this can be minimised by using a sample and hold circuit which is frequently referred to as a 'de-glitcher' (Fig. A1.13).

Analogue to digital conversion

Having considered the basic concepts of D to A conversion, and also some of the principles of sample and hold techniques, it is now convenient to consider some of the basic concepts of A to D conversion.

Basically the requirement is to sample an analogue value at regular intervals, and turn those samples into data words, each of which comprises a certain number of bits. The actual number of bits is related to the accuracy that is required of a particular system or method of A to D conversion, and also the frequency range that the system may be required to cover; in the case of CD technology, the required audio range is in the region of 0–20 kHz, with 16 bits in each data word, with each sample of the analogue signal being carried out at a sampling frequency of 44.1 kHz. In order that an appreciation can be achieved of how an analogue signal is turned into a digital form, consider the following method of A to D conversion.

Successive approximation analogue to digital converter

Assume the reference voltage = 10 V, and the analogue input = 6.6 V (Figs A1.14 and A1.15).

Basically the control logic section causes each clock pulse to operate the shift register which applies a series of data bus lines to the D to A converter via the output latches. The data bus lines will

Analogue and digital processing

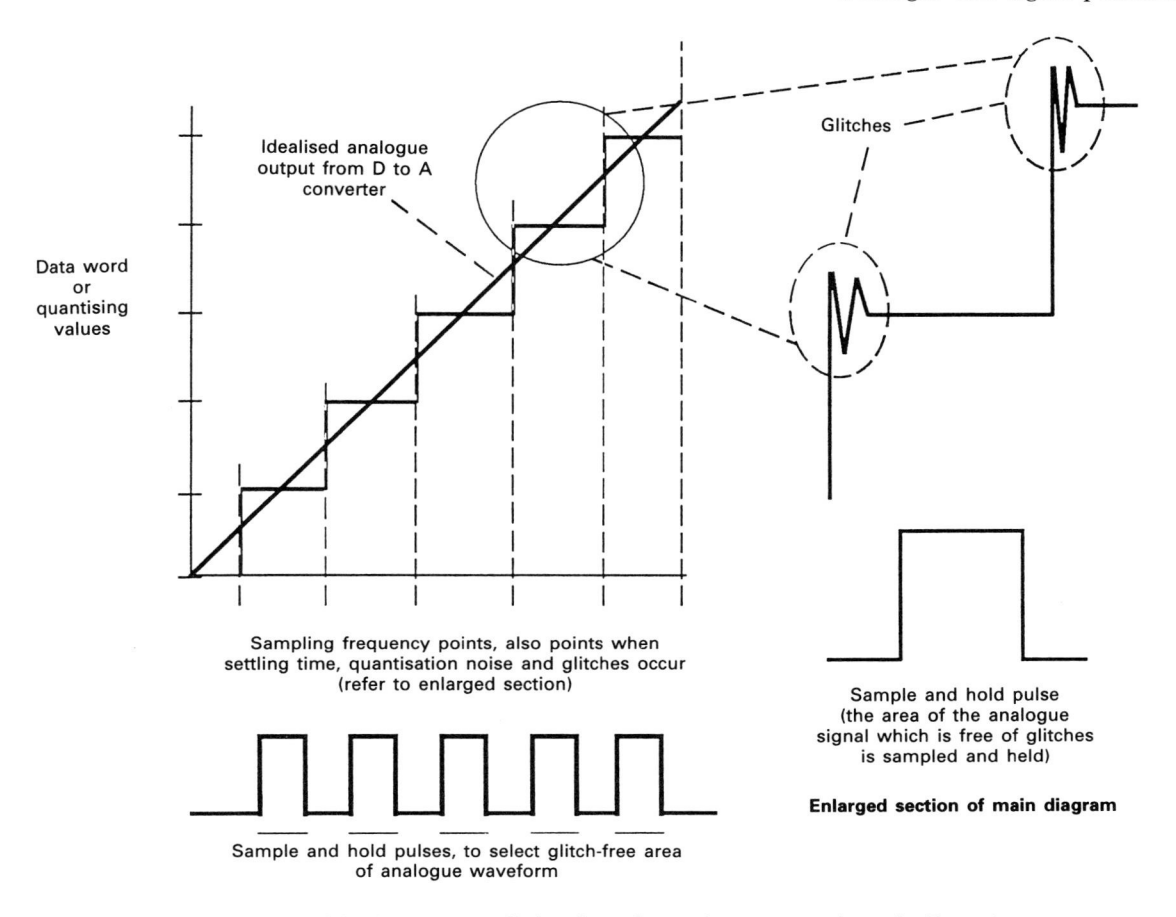

Figure A1.13 *Using sample and hold to remove glitches from the analogue output from the D to A converter*

provide the relevant digital data word on completion of the analysis of the sampled analogue input.

The D to A converter provides successively reducing levels of the input reference voltage of 10 V, which is applied to the converter stage, the output of which will change either high or low depending upon whether the D to A output is greater than, or less than, the input analogue input of 6.6 V.

The sequence of operations is as follows:

— *Stage 1.* The start signal is sent, and the first clock pulse sets the data output B1–B8 to 10 000 000, which is equal to 5 V from the D to A converter, to the input (V in) of the comparator. The comparator output remains high as the analogue input of 6.6 V is greater than 5 V. B1 is latched to '1' and held.

— *Stage 2.* The second clock pulse is sent and B2 is set to '1'. Data output B1–B8 is now 11 000 000 = 7.5 V. V in is now greater than 6.6 V, and therefore the comparator output is now low. B2 is now latched to '0' and held.

— *Stage 3.* The third clock pulse sets B3 to '1'. B1–B8 is now 10 100 000 = 6.25 V. V in is now less than 6.6 V, and therefore the comparator output is high. B3 remains latched to '1' and held.

— *Stage 4.* The fourth clock pulse sets B4 to '1'. B1–B8 is now 10 110 000 = 6.825 V. V in is now greater than 6.6 V, and therefore the comparator output is low. B4 is now latched to '0' and held.

— *Stage 5.* The fifth clock pulse sets B5 to '1'. B1–B8 is now 10 101 000 = 6.562 V. V in is now less than 6.6 V, and therefore the comparator output is high. B5 remains latched to '1' and is held.

— *Stage 6.* The sixth clock pulse sets B6 to '1'. B1–B8 is now 10 101 100 = 6.718 V. V in is now

Analogue and digital processing

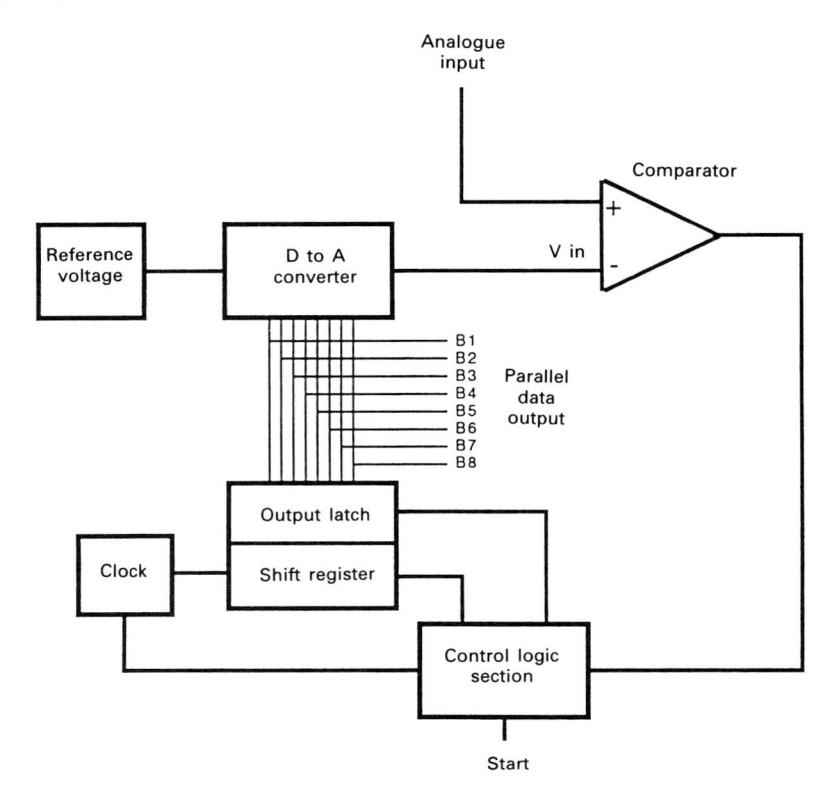

Figure A1.14 *Successive approximation A to D converter*

Power of '2'	2^7	2^6	2^5	2^4	2^3	2^2	2^1	2^0
Actual value	128	64	32	16	8	4	2	1
Fractional value	$\frac{1}{2}$	$\frac{1}{4}$	$\frac{1}{8}$	$\frac{1}{16}$	$\frac{1}{32}$	$\frac{1}{64}$	$\frac{1}{128}$	$\frac{1}{256}$
Fractional value of V. Ref = 10V (lower values rounded up)	5V	2.5V	1.25V	0.625V	0.312V	0.156V	0.078V	0.039V

Figure A1.15 *8 bit binary table*

greater than 6.6 V, and therefore the comparator output is low. B6 is now latched to '0' and is held.
— *Stage 7.* The seventh clock pulse sets B7 to '1'. B1–B8 is now 10 101 010 = 6.64 V. V in is now greater than 6.6 V, and therefore the comparator output is high. B7 is now latched to '0' and held.
— *Stage 8.* The eighth clock pulse sets B8 to '1'. B1–B8 is now 10 101 001 = 6.601 V. V in is now greater than 6.6 V, and therefore the comparator is high. B8 is now latched to '0' and held.
— *Stage 9.* The next clock pulse completes the operation and the final data word of 10 101 000 is the data word output. The system awaits the next start signal to commence the operation on the next analogue input. Engineers studying this section are invited to calculate data outputs for alternative analogue inputs, especially for inputs that are less than half the reference voltage potential.

Binary counter A to D converter

Compared to the previous description, the binary counter A to D converter (Fig. A1.16), though fundamentally similar in design, is much simpler in operation.

Again a reference voltage is applied to the D to A converter, but the control logic causes the clock to commence operating a binary counter which commences counting from zero (i.e. B1–B8 = 00 000 000) upwards until an output from the D to A converter produces a voltage input (V in) to the comparator which is greater than the analogue input, at which time the comparator will

[Figure A1.16 diagram: Binary counter A to D converter — shows Reference voltage → D to A converter → V in to Comparator (+ input: Analogue input); Comparator output → Control logic section; Clock → Binary counter → Parallel data output B1–B8 (feeding D to A converter); Control logic section ← Start.]

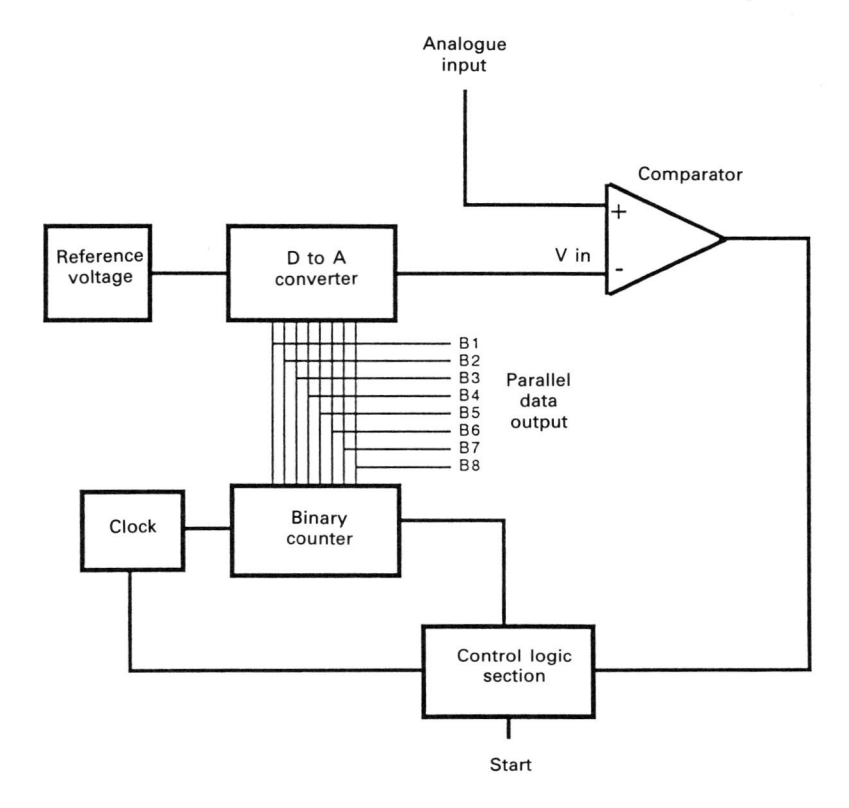

Figure A1.16 *Binary counter A to D converter*

change the output level from high to low, which in turn causes the binary counter to stop the counting sequence, and the data output, B1–B8, at that point must approximate to the value of the analogue input.

Dual slope A to D converter

The circuit shown in Fig. A1.17 essentially comprises an integrator which will enable the capacitor C to charge up to the analogue value from the sample and hold circuit, a comparator which controls a clock driving a binary counter, and a logic control stage to ensure the complete circuit operates in the correct logical sequence.

Before a start signal is applied, S2 will be closed to ensure that the capacitor C is discharged. When the start signal is applied, S2 will open, whilst S1 will switch over to enable the analogue input from the sample and hold circuit to charge up the capacitor towards the analogue level, during the time ($t0$–$t1$). As the charging commences (Fig. A1.18), the clock pulses will drive the binary counter for a set period of time which is also ($t0$–$t1$).

At time $t1$, S1 will change over to the reference voltage input, the binary counter will be set to zero, and as the capacitor C changes its charge towards the reference voltage, the binary counter will now commence counting, until the point at which C is discharged, at time ($t2$), when the comparator output will change over, stopping the clock operation.

The binary counter is held at that moment and will be an indication of the digital value of the anlogue signal input.

Accuracy of analogue and digital processing circuits

Whether a designer is considering analogue to digital conversion or vice versa, the accuracy of the conversion circuits is extremely important. Whilst the equipment for A to D conversion in

Analogue and digital processing

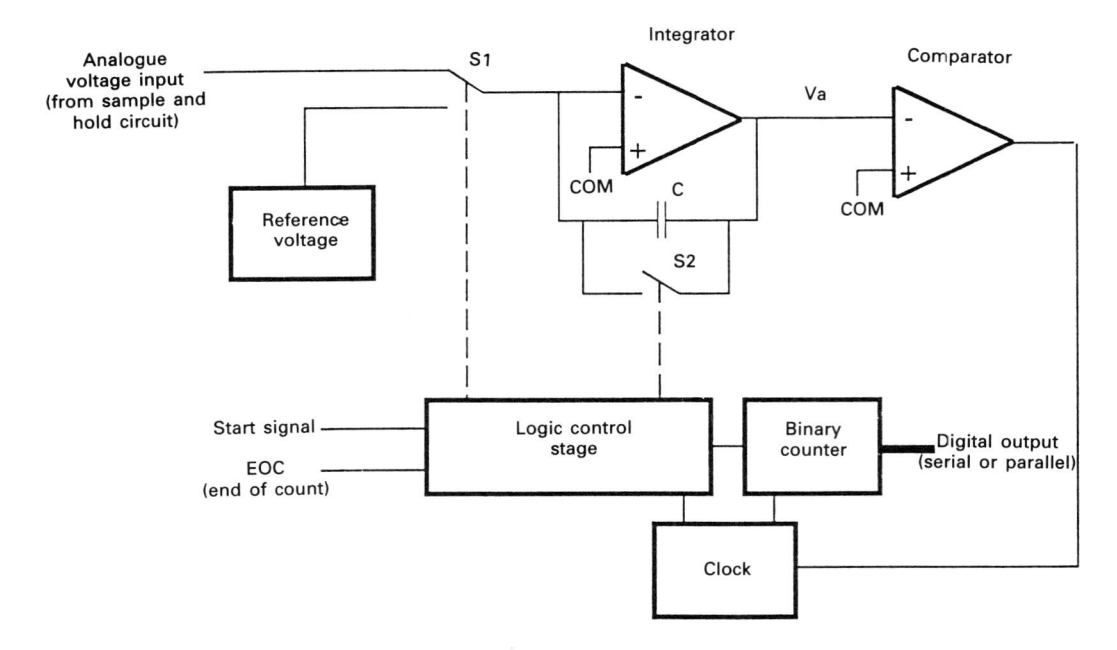

Figure A1.17 *Dual slope A to D converter*

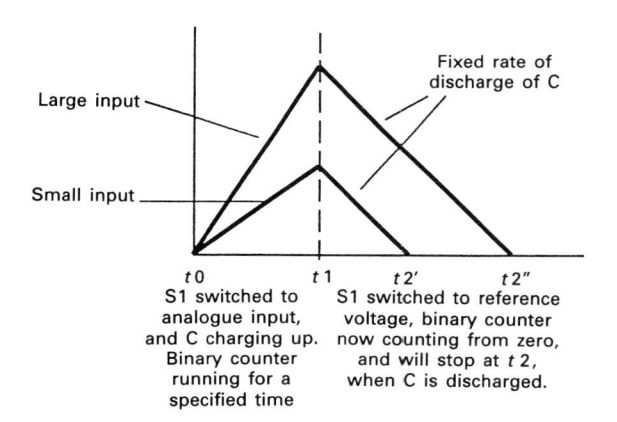

Figure A1.18 *Timing sequence of the dual slope A to D converter*

the recording studios may achieve an excellent quality of accuracy due mainly to a high standard of equipment which will no doubt result in accurate conversion of the analogue signal into its digital counterpart, it is in the CD player that absolute accuracy may not be achieved in view of the fact that the retail price of a player will most certainly not reflect the true cost of the technology needed to achieve the accuracy that may be required.

Because D to A converters for CD players are difficult to design whilst maintaining cost restraints, the latest trends towards the use of one bit, pulseflow, and bit stream techniques have resulted in a very effective method of D to A conversion which generally maintains a high degree of accuracy at a much reduced cost involvement, despite the technology involved, generally due to modern mass production methods.

Analogue to digital accuracy problems

The ideal response of the A to D converter is shown in Fig. A1.19, which illustrates the ideal condition of specific voltage levels, enabling a different data word to be provided, i.e. 5 V = 100 and 8.75 V = 110, etc. If a different level of voltage were to provide an incorrect data word, due to non-linearity, then the incorrect analogue value would be obtained during the D to A conversion process.

This problem can be reduced by using an increased number of steps between the minimum and maximum analogue levels, which in turn would require data words of more bits.

In CD technology where 16 bit data words are used, there are 65 536 variations of the 16 bit

Analogue and digital processing

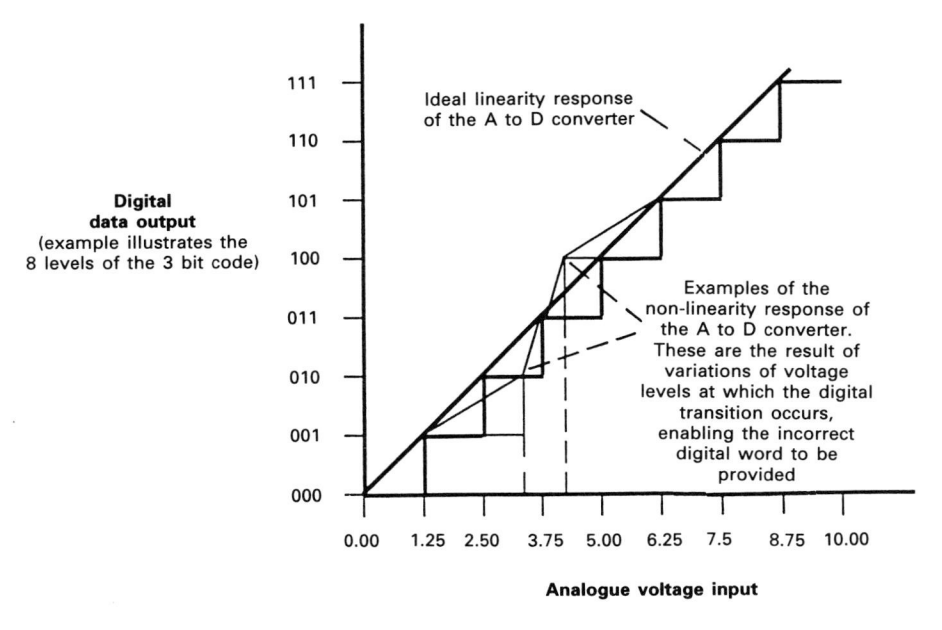

Figure A1.19 *Linearity requirements of the A to D converter*

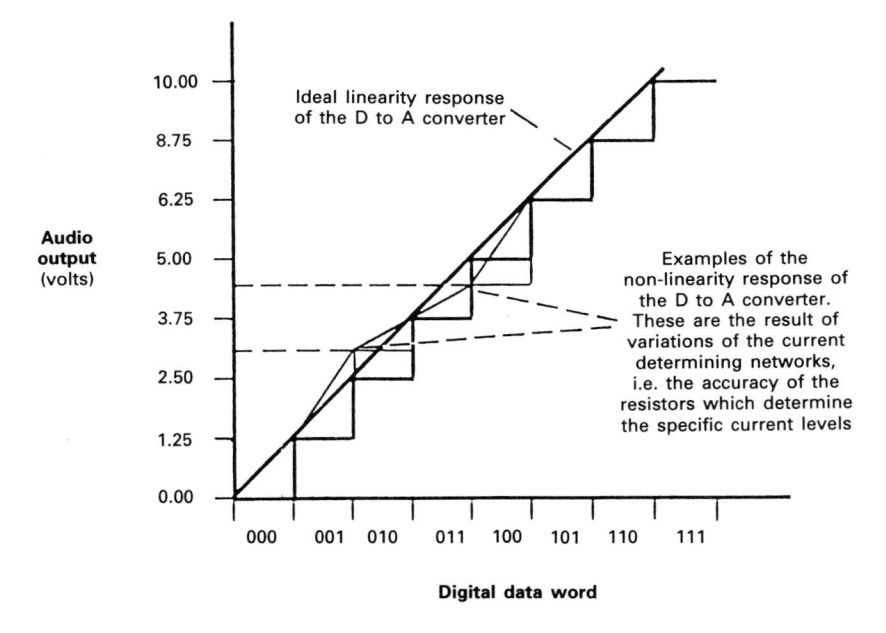

Figure A1.20 *Linearity requirements of the D to A converter*

191

Analogue and digital processing

Analogue voltages
These levels are related to the values of the resistors R in the chain. Each resistor can be considered to be related to a specific data word in the D to A conversion stage. Therefore if any resistor is not accurate in providing a specific potential, then in this basic illustration all other potentials can be affected, and can indicate problems that can occur in the accuracy requirements of the D to A converter

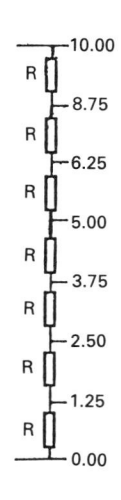

```
─┬─ 10.00
R │
 ─┤─ 8.75
R │
 ─┤─ 6.25
R │
 ─┤─ 5.00
R │
 ─┤─ 3.75
R │
 ─┤─ 2.50
R │
 ─┤─ 1.25
R │
─┴─ 0.00
```

Figure A1.21 *Fundamental illustration of the accuracy requirements of the D to A converter*

codes from all the 0's through to all the 1's, which implies there are the same number of intermediate steps from the minimum to maximum values of the analogue signal which is probably much smaller in value than the 10 V range illustrated in Fig. A1.19.

Digital to analogue accuracy problems

The ideal response of the D to A converter is shown in Fig. A1.20, which illustrates the ideal condition of providing specific voltage levels which are the result of a range of current determining circuits, e.g. resistors, with each specific current being related to a particular data word. As illustrated in Fig. A1.21, if one or more of the resistors in the chain is not absolutely accurate in order to produce a defined voltage, then the voltages throughout the chain will not be at the required level.

In similarity to the A to D accuracy problem this problem can be reduced by using an increased number of steps between the minimum and maximum analogue levels, which in turn requires data words of more bits. But unfortunately the cost of producing such accurate resistors can prove to be extremely high, especially for the domestic product range of CD players.

APPENDIX 2: HEALTH AND SAFETY

The various aspects of health and safety have made apparent the many hazards within the working environment where service engineers are involved. There are areas which must be considered to ensure that the health, safety and even livelihood of employee, employer and also the customer must be considered at all times.

Service manuals cover many factors within the servicing of most domestic products, and a wide range of reference books are available on this subject. A brief selection is listed at the end of this appendix; these together with many more are usually available from public libraries.

Ideally all workshop facilities should include essential items such as isolating transformers, and also some form of residual current circuit breaker (RCCB) system. Unfortunately neither of these items will be available to the engineer when confronted with servicing products in the customer home. Also, some of these items have to be serviced in the most inappropriate positions (field service engineers have many an interesting tale to tell in this respect).

This subject area can only be briefly covered in this appendix and will relate generally to servicing CD players as it may be assumed that most workshop environments may already include acceptable facilities for servicing other types of domestic product.

The main area of concern when servicing CD players is the area concerning laser radiation, and players usually contain labels that relate to relevant cautions and the class of the laser product (see Fig. 6.6).

As previously mentioned in the chapters concerning Adjustments and Fault Diagnosis (Chapters 6 and 9), care should be taken concerning exposure to the laser, especially when the cabinet top cover has been removed, and the optical assembly is operated in a manner that can enable the laser beam to be viewed.

There is in existence a publication that specifically refers to laser products titled:

Radiation safety of laser products, equipment classification, requirements and user's guide.
Publication 825 First edition 1984.
International Eletrotechnical Commission – IEC Standard.

This is a Swiss publication, which is printed in English and French. It is useful to be aware of pages 41 and 43 in particular; relevant extracts are as follows:

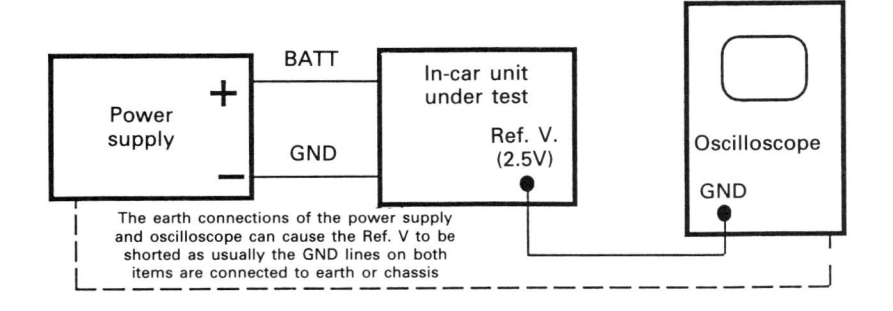

Figure A2.1

Health and safety

9.2 Definition of laser classes.
Laser products are grouped into four general classes for each of which accessible emission limits (AELs) are specified. Class 1 lasers are those that are inherently safe (so that the maximum permissible exposure level cannot be exceeded under any condition) or are safe by virtue of the engineering design.
Page 41.

Author's note: Class 1 lasers are used within CD and laserdisc players. They are in effect the lowest power level and are frequently adjusted to levels in the region of 0.12 to 0.23 mW. However, information is available from a publication referred to as BS 4803 pt 3 with certain amendments made in 1984 which provides calculations which refer to the Maximum Permissible Laser Emission (MPE) of Class 1 lasers as being 0.625 mW, which in fact is the maximum acceptable from these lasers under fault conditions, as well as the maximum adjustable condition.

Class 3B lasers may emit visible and/or invisible radiation . . . CW lasers may not exceed 0.5 W.
Note: Direct intrabeam viewing near these devices is always hazardous.
Viewing unfocused pulsed laser radiation by diffuse reflection is not hazardous and under certain conditions CW laser beams may be safely viewed via a diffuse reflector.
These conditions are:
a. a minimum viewing distance of 13 cm.
b. a maximum viewing time of 10 secs.
Page 43.

Author's note: In view of the conditions mentioned above with respect to a CW laser of 0.5 W, where CW means that it is permanently on such as in CD players when playing the disc, the recommended viewing distance mentioned in Chapter 6 is 50 cm, extreme safety being applied if it is considered necessary to view a Class 1 laser. Regarding the maximum viewing time of 10s, how long does a person have to look at any type of lamp to determine whether it is on or not? It is unlikely that 10s is required. As previously mentioned, the author never recommends anyone to view the laser beam of a CD player, but that is a decision for the person involved.

It is hoped that the foregoing regarding lasers in CD players may create confidence in their non-lethality, and like many things in life, excess is not good for the individual and therefore sensible precautions must apply.

Whenever a new type of optical assembly is introduced in a CD player, the laser incorporated in that new type is always tested to extremely strict conditions by BEAB to ensure that the required safety standards are maintained.

Equipment safety

The major reference regarding most safety aspects related to mains-operated electronic and related equipment for household and similar general use is the BSI publication BS415, and the Health and Safety Executive have issued an information sheet that relates to portable electrical equipment and refers to the Electricity at Work Regulations of 1989, where one of the regulations, 4(2), states:

As may be necessary to prevent danger, all systems shall be maintained so as to prevent, so far as is reasonably practicable, such danger.

The word 'system' as defined in the Regulations includes portable electrical equipment, and the effect of this is to require duty holders to maintain the portable electrical equipment in a safe condition. Typical of portable electrical equipment in the workshop is the oscilloscope and regular checks should be carried out to maintain such items in a safe condition. The reason for mentioning the oscilloscope in particular is that it is the item of equipment that is frequently operated with the earth connection removed, particularly for servicing television receivers.

However, on some models of in-car CD players it frequently occurs that a reference potential, to which the common connection of the oscilloscope may be connected during adjustment procedures, can be shorted to ground or chassis via the chassis and earth connection circuit that can be developed between the oscilloscope and the power supply, as illustrated in Fig. A2.1.

Leakage current checks

Some service manuals also give details of leakage tests on mains-powered electrical equipment, and there generally three recommended methods.

1 Usually for USA models, i.e. 120 V supply. Measure the leakage current to a known earth by connecting a leakage current tester between the earth ground and all exposed metal parts of the appliance (input/output terminals, screwheads, metal overlays, control shafts, etc.). Plug the appliance directly into a 120 V a.c. 60 Hz outlet, and turn the a.c. power switch on. Any current

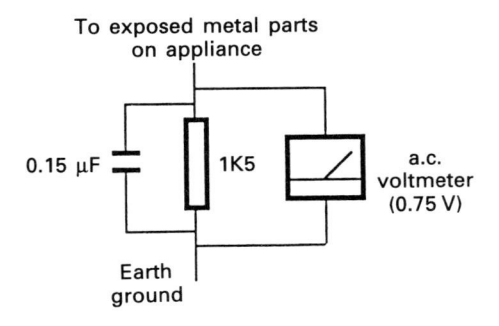

Figure A2.2

measure must not exceed 0.5 mA. *Any measurements not within these limits are indicative of a potential shock hazard and must be corrected before returning the appliance to the customer.*

2 (This information is available in Sony service manuals.) The a.c. leakage from any exposed metal part to earth ground and from all exposed metal parts to an exposed metal part having a return to chassis must not exceed 0.5 mA. The leakage current can be measured by any one of three methods:

(i)   A commercial leakage tester, such as the Simpson 229 or RCA WT-50A. Follow the manufacturers' instructions to use these instruments.

(ii) A battery-operated a.c. milliammeter. The Data Precision 245 digital multimeter is suitable.

(iii) Measuring the voltage drop across a resistor by means of VOM or battery-operated a.c. voltmeter. The 'limit' indication is 0.75 V, so analogue meters must have an accurate low voltage scale. The Simpson 250 and Sanwa SH-63 Trd are examples of passive VOM that is suitable. Nearly all battery-operated digital multimeters that have a 2 V a.c. range are suitable (see Fig. A2.2).

3 A high voltage insulation tester is another method to be recommended, and is a specialised unit which supplies an output in the region of 1.5–2.0 kV which should be connected between either of the mains connections and any exposed metal parts, and the test carried out for 1 min. During this period of testing no current in excess of 2 mA or 700 µA a.c. should flow.

Product safety notices

Most service manuals emphasise the importance of replacing defective components with the manufacturer's recommended replacement, especially where those components are specifically identified in the service manual as having specialised safety characteristics.

Improperly performed repairs can adversely affect not only the safety of a product, but also its reliability, and may void the warranty or service contract if undertaken by an unauthorised servicing agent.

Furthermore, whilst productivity figures are a major requirement in many service operations, safety must be the main criterion when servicing any form of electrical appliance, and if an accident occurs which results in death, fire, injury or other form of physical damage as a result of an inferior repair, especially the replacement of a safety component with a non-recommended type or non-safety type, the relevant engineer who serviced the product can be held responsible, and will therefore be liable to prosecution as well as responsible for the costs of the resulting loss.

Manufacturers ensure at the design stage that their appliances comply with the relevant recommendations of the specific safety standard BS415, which enables BEAB approval to be obtained. Therefore it is irresponsible for an appliance to be serviced in a manner that infringes any of the safety codes which manufacturers have been encouraged over the years to provide, and which ensure that products are manufactured to the highest safety requirements.

Solder and related toxicity

It is interesting to note that some service manuals occasionally contain a reference to the lead content in solder and the fumes that are produced by the process of soldering. Generally the service manuals that usually contain such information relate to products intended for the American market, but the information may well be of general interest to engineers and an example is as follows:

> WARNING: Lead in solder used in this product is listed by the California Health and Welfare agency as a known reproductive toxicant which may cause birth defects or other reproductive

harm (California Health and Safety Code, Section 25249.5). When servicing or handling circuit boards and other components which contain lead in solder, avoid unprotected skin contact with the solder. Also, when soldering do not inhale any smoke or fumes produced.

Some recommended publications available regarding health and safety

BS415 – British Standards Institute.
Usually available in the reference section of public libraries and also from BSI Sales, Linford Wood, Milton Keynes MK14 6LE.

Maintenance of Portable Electical Equipment – HSE Information Sheet.
HSE Information Centre, Broad Lane, Sheffield S3 7HQ. Tel. (0742) 892345.

Broadhurst, A. (1990) *Health and Safety*. Pitman, London.
A NatWest Business Handbook, which provides a range of extremely useful information on general health and safety recommendations.

INDEX

Accufocus system, 77, 78
A/D conversion, 179, 184–92
  accuracy, 190–2
  aliasing frequencies, 3, 4
  binary counter, 188, 189
  dual slope, 189, 190
  linearity requirements, 30, 31, 190–2
  low pass filter, 3, 33, 34
  Nyquist criterion, 3
  quantisation errors, 3
  sample and hold, 179, 184–6
  successive approximation, 186–8
Adjustment procedures, 109–34, 146, 147
Aliasing frequencies, 3, 4
Antistatic precautions, 116
APC, 41, 76
Asymmetry, 20, 82
ASY, 20, 82
Audio pre-amplifier

Beam splitter prism, 37, 42–7, 57
Binary counter, 188
Binary number system, 180, 188, 189
Bitstream, 33–5
Blurred eye pattern, 120, 130, 167
Bumps *see* Pits

Carriage servo, 22, 23, 37, 68, 69,
  Carriage motor, 69 *see also* Tracking servo
CD player block diagram, 19
Collimation lens, 37, 39, 53
Compact disc
  basic technology, 2–17
  care of, 5, 6, 175, 176
  comparison to vinyl, 1
  construction, 10, 11
  recording process, 10, 11
  some basic facts, 1
Concave lens, 39, 42
Constant angular velocity, 1
Constant linear velocity, 1

Control word, 8–10
Coupling bits, 14
CPU interface, 25, 136, 137
CRCC, 25 *see* Cyclic redundancy check code
Critical angle prism, 42, 46, 58
Crystal oscillator, 24, 33, 34, 72
Current generator, 32, 181, 182
Current to voltage generator, 32, 181, 182
CX 20108, 73
CX 20109, 73
CX 23035, 96, 97
CXA 1081, 76, 77
CXA 1082, 73, 84
CXA 1372, 73
CXA 1471, 73, 103
CXD 1125, 73, 96
CXD 1135, 73, 96
CXD 2500, 73, 98, 136
CXD 2515, 73, 100, 102
Cyclic redundancy check code, 25
Cylindrical lens, 40, 42, 57

D.C.-D.C. convertor, 145, 146
D/A conversion, 26–35
  accuracy, 190–2
  basic arrangement, 32, 181–3
  binary number system, 180, 188
  bitstream, 33–5
  current generators, 32, 181, 182
  de-glitcher, 180, 184
  digital filtering, 27, 30
  glitches, 31
  multi stage noise shaping, 32
  noise shaping, 32
  non linearity errors, 30–4, 190–2
  one bit, 33, 34
  oversampling, 26–35
  pulse density modulation, 34
  pulse width modulation, 33
  resistor ladder, 181, 182
  sample and hold, 179, 184–6
  R-2R resistor ladder, 183

Index

timing chart, 185
zero cross distortion, 31
De-glitcher, 180, 184
De-interleaving, 9
Decoder, 25, 96
Decoder I.C.'s
  CX23035, 96, 97
  CXD 1125, 73, 96
  CXD 1135, 73, 96
  CXD 2500, 73, 98, 136
  CXD 2515, 73, 100, 102
  SAA 7010, 73
  SAA 7020, 73
  SAA 7210, 73
  SAA 7310, 73
  TC 9221, 73, 107
  UPD 6375, 108
Defect circuit, 82–4
Delay line, 77, 78
Diffraction grating, 38, 40, 112, 132–4
Diffraction grating adjustment, 132–4
Digital filter, 26–35
  1Fs, 26, 27
  2Fs, 27, 28
  4Fs, 27–9
  8Fs, 32
  oversampling, 26–9
Digital filtering, 27
Digital VCO, 100, 101
Disc does not rotate, 152
Disc motor *see* Spindle motor
Disc motor servo *see* Spindle motor servo
Disc rotation
  does not rotate, 151
  rotates at an excessive speed, 154
  constant angular velocity, 1
Distorted output, 154, 173–6
Dither, 34, 35
Dual slope, 184, 190

EFM, 10–14
EFM comparator, 20, 21, 79, 82
EFM waveforms, 83, 168
Eight to fourteen modulation, 10–14 *see also* EFM
Electrical adjustments, 112
Equipment safety, 193, 194
Error correction, 5
  linear interpolation, 8
  muting, 6
  previous word hold, 7
Eye pattern comparisons, 120, 122
Eye pattern waveform, 20, 83, 109, 120, 122

Fault diagnosis, 147, 150–78
  basic operating sequence, 151
  basic operational checks, 153, 160
  carriage checks, 165, 171, 172
  cleanliness of the lens, 156
  disc does not rotate, 152
  disc rotates at an excessive speed, 154
  disc rotates at an acceptable speed, 154
  distorted output, 155, 173–6
  eye pattern quality, 155, 172, 176–8
  focus servo checks, 161–3, 165
  initial checks, 151
  jumping, 155, 176–8
  laser diode checks, 163–8
  lateral adjustment, 172
  mechanical checks, 160, 161
  no audio output, 155, 173–6
  noisy output, 155, 173–6
  player appears dead, 152
  power supplies, 156
  radial checks, 165, 171
  skipping, 155, 176–8
  spindle servo checks, 172
  symptom check list, 152
  system control, 157
  tangential adjustment, 172
  test modes, 163
  tracking servo checks, 168–71
Five element photo diode array, 53, 54, 58, 68
Focus balance, 112, 123
Focus bias, 112, 123
Focus coil, 56, 86, 178
Focus correction circuit, 87, 88
Focus error amplifier, 19, 114
Focus error signal, 48, 49
Focus gain, 112, 114, 123
Focus offset, 56, 112, 114, 115
Focus servo, 22, 55, 56
  accufocus system, 77, 78
  block diagram, 22, 56–9, 114
  focus coil, 22, 56–9, 114
  focus correction circuit, 87, 88
  focus error, 48, 49, 125
  focus error amplifier, 19, 57, 58, 76, 79, 114
  focus error signal, 48, 49
  focus gain, 56, 112, 114, 123
  focus offset, 56
  focus zero cross, 85
  FOK, 55, 77, 137
  FOK amplifier, 20, 21, 77, 79
Focus servo system, 22, 84
  block diagram, 22, 56–9, 114
Focus zero cross, 85

Index

Four element PD array, 37, 39, 42–52, 57–60, 66, 67, 68
Frame information, 17

Glitches, 31, 186

Health and safety, 193–6
Hi-speed dubbing, 99, 100
Hologram prism, 53, 58, 59

In-car CD players, 141–8
Initial checks, 151
Integrated circuits, 73, 74
    CX 20108, 73
    CX 20109, 73
    CX 23035, 73
    CXA 1081, 73
    CXA 1082, 73
    CXA 1372, 73
    CXA 1471, 73, 103
    CXD 1125, 73
    CXD 1135, 73
    CXD 2500, 73
    CXD 2515, 73
    M51593, 73
    M52103, 73
    SAA 7010, 73
    SAA 7020, 73
    SAA 7210, 73
    SAA 7310, 73
    TA8137, 103
    TC 9220, 73
    TC 9221, 73, 107
    TDA 5708, 73
    TDA 8808, 73, 82
    TDA 8809, 73, 82
    UPC 1347, 105
    UPD 6375, 108
Interleaving, 8
    Parallel/serial convertor, 9
    Serial/parallel convertor, 9
    Interpolation, 27–9

Jumping, 155, 176–8

Laser beam precautions, 116, 117
Laser diode, 28, 41, 76
Laser diode current, 118–20

Laser power, 109, 112, 116–20
Laser power control, 41, 74, 76, 116–20
Laser power meter, 110, 117, 118
Laser warnings, 117
Latch enable clock, 173–6
Lateral adjustment, 112, 127
LDON, 40, 41, 74, 76
Leakage current checks, 194, 195
Least significant bit, 31, 32, 174–84
LEC *see* Latch enable clock
Left right clock, 173–6
Linear drive, 70, 71
Linear interpolation, 8
Linearity requirements, 30, 31, 190–2
Low pass filter, 3, 33, 34
LRCK *see* Left right clock
LSB *see* Least significant bit

M51593, 73
M52103, 73
Magazines, 141–3
Magic three, 158 *see* System control
MASH *see* Multi stage noise shaping
Master system controller, 136
Mechanical adjustments, 112, 126
MIRR *see* Mirror
Mirror, 81
Mirror amplifier, 21, 79, 81
Monitor diode, 41, 76
Most significant bit, 31, 32, 179–84
MSB *see* Most significant bit
Multi disc mechanism control, 141–4
Multi stage noise shaping, 28, 32–4
Multimeter, 110
Muting, 6

No audio output, 155, 173–6
Noise shaping, 28–35 *see also* Multi stage noise shaping
Noisy output, 155, 173–6
Non linearity errors, 30–4, 190–2
Nyquist criteria, 3

Objective lens, 37, 39–41, 53
One bit D/A, 33, 34
Operating sequence, 150
Optical assemblies
    PEA 1030, 114, 118, 129–31
    PEA 1179, 114, 118
    Philips, 37

Index

Pioneer '92, 48, 53, 54, 114, 118
PWY 1003, 114, 118, 129–31
PWY 1011, 114, 118, 129–31
Optical assembly, 18
  beam splitting prism, 42, 44–7, 57
  cleanliness of the lens, 156
  collimation lens, 37, 39, 40, 53
  concave lens, 39, 42
  critical angle prism, 42, 46
  cylindrical lens, 40, 42, 57
  diffraction grating, 38, 40
  hologram prism, 53
  laser beam precautions, 116, 117
  laser diode, 28, 41, 76
  laser power, 109, 112, 116–20
  laser warnings, 117
  monitor diode, 41, 76
  objective lens, 37, 39–41, 44, 53
  photo diode array, 37, 39, 42–60, 66, 67, 78, 114
    five element, 53, 54, 58, 68
    four element, 37, 39, 43–52, 57–60, 66, 67, 78
    six element, 40, 43, 52, 68, 114
  polarised filter, 40
  polarised prism, 40
  prism, 37, 41
  quarter wave plate, 40, 41
  semi transparent mirror, 39, 41
  single beam type, 37, 39
  three beam type, 40, 41
Oscilloscope, 110
Oscilloscope checks, 112, 147
Output expander, 136 *see also* System control
Oversampling, 26–35
  dither, 34, 35
  multi stage noise shaping, 32
  noise shaping, 32
  1Fs, 26, 27
  2Fs, 27, 28
  4Fs, 27–9
  8Fs, 32

Parallel/serial convertor, 9
PD array, 19, 20, 37, 39, 42–60, 66, 67, 78, 114
PEA 1030, 114, 118, 129–31
PEA 1179, 48, 53, 54, 114, 118
Phase comparator, 25, 72, 95
Phase difference method, 46, 51, 67
Phase locked loop, 95, 126
Philips optical assembly, 37
Photo diode array, 19, 20, 37, 39, 42–60, 66, 67, 78, 114

Photo diode processor I.C.'s, 73–5
  TDA 5708, 73
  TDA 8808, 73–5
Photo-diode balance, 112, 123
Pit length, 11–17
Pit quality, 20, 21
Pits, 11–17, 20, 21
Player appears dead, 151
PLL *see* Phase locked loop
Polarised filter, 40
Polarised prism, 40
Power supplies, 145, 146
Previous word hold, 7
Prism, 37, 41
Product safety notices, 194
Pulse density modulator, 34
Pulse width modulator, 33
Push pull method, 46, 50, 66
PWY 1003, 114, 118, 129–31
PWY 1011, 114, 118, 129–31

Quantisation errors, 3
Quantisation noise, 27–9
Quarter wave plate, 40, 41

R.F. amplifier, 18–20, 76, 77
R.F. amplifier I.C.'s
  CX 20109, 73
  CXA 1081, 76, 77
  CXA 1471, 104
  M51593, 73
  M52103, 73
  UPC 1347, 105
R.F. eye pattern, 117, 120, 122
R.F. eye pattern waveform, 20, 120, 122, 130
  quality of, 177
R.F. level, 112, 123
R.F. offset, 112, 116, 117
Radial adjustment, 127
Radial servo, 22
Radial error processor, 82–5
Radial error processor I.C.'s
  TDA 5708, 73
  TDA 5709, 73
  TDA 8808, 73, 82
  TDA 8809, 73, 82
Radial tracking system, 60–5
RAM, 25, 135, 140, 179
Recording process, 10–11
Reset, 158
ROM, 136, 140

Index

SAA 7010, 73
SAA 7020, 73
SAA 7210, 73
SAA 7310, 73
Sample and hold, 180, 184–6
Sampling frequency, 4
Sampling points, 3, 27–9
Semi transparent mirror, 39, 41
Serial/parallel convertor, 9
Service modes, 111
Servo auto sequencer, 137
Servo circuits, 22–5 see also Carriage servo
    Focus servo
    Radial servo
    Spindle servo
    Tracking servo
Servo control
Servo I.C.'s
  CX 20108, 73
  CXA 1082, 84, 86,
  CXA 1372, 106
  TC 9220, 105
Servo systems, 55–72
Single beam device, 36
Single beam optical device, 39
Six element PD array, 40, 43, 52, 68, 114
Skipping, 154, 176–8
Slave system controller, 136
Sled servo see Carriage servo
Solder toxicity, 194
Some basic facts, 1
Speed control, 69, 71, 72
Spindle motor, 72, 93
Spindle servo, 22, 24, 25, 69–72, 87–95
Spindle servo block diagram, 24
Spurious data, 174
Sub code see Control word
Sub code decoder, 25
Successive approximation, 186–8
Summation amplifier, 19, 20, 77
Sync. word, 16, 17, 98
Synchronous detector, 64
System control, 25, 135

Table of contents, 137
Tangential adjustment, 112, 127, 128
TA 8137, 103
TC 9220, 73
TC 9221, 73, 107
TDA 5708, 73
TDA 5709, 73
TDA 8808, 73

TDA 8809, 73
Test discs, 109
Test equipment, 110
Test jigs, 110
Test modes, 109, 110, 113, 147, 148
Three beam device, 38, 48, 52, 68
Timing generator, 25
TOC see Table of contents
Tools, 109
Track detail, 17
Tracking balance, 60, 68, 112, 121–3
Tracking coil, 60
Tracking error, 46, 48, 52, 80
Tracking error amplifier, 19, 80
Tracking error signal, 46, 48, 52
Tracking error waveform, 125, 132, 169
Tracking gain, 60, 66, 68, 112, 123
Tracking offset, 60, 68, 112, 113
Tracking servo, 22, 23, 59, 87
  block diagram, 23, 60, 65, 66, 68, 116
  phase difference method, 46, 51, 67
  push pull method, 46, 50, 66
  radial tracking system, 60–5
  tracking balance, 60, 68, 112, 121–3
  tracking coil, 60, 66–9
  tracking error, 46, 48, 52, 69, 80
  tracking error amplifier, 50, 66, 68, 77, 80
  tracking error waveform, 125, 132, 169
  tracking gain, 60, 66, 68, 112, 123
  tracking offset, 60, 68, 112, 115
Turntable height, 112, 127
Turntable height adjustment, 127, 128
Turntable motor see Spindle motor
Turntable servo see Spindle servo

UPC 1347, 105
UPD 6375, 108

Variable pitch, 99–101
Variable speed, 99
VCO, 24, 25, 72, 95, 96, 125, 126
VCO block diagram, 95, 126
VCO frequency, 112, 125
VCO timing generator, 24, 95
Voltage controlled oscillator see VCO

Waveforms
  bit clock, 174, 185
  blurred eye pattern, 120, 130, 167

Index

    data, 174–6, 185
    decoder outputs, 174
    efm waveform, 83, 168
    eye pattern, 20, 83, 117, 120, 122, 130, 178
    focus error, 125
    illustrating glitches, 187
    latch enable clock, 175, 176
    left right clock, 174–6, 185
    poor quality eye pattern, 178
    reset, 159
    scratch on disc, 177
    spindle motor, 92, 94
    spurious data, 174
    tracking balance, 123, 125, 132, 170
    tracking error, 123, 125, 132, 170
    tracking servo noise, 125
    vco waveform, 168
    word clock, 185
WDCK *see* Word clock
Wobble frequency, 63–5
Word clock, 173–6

Zero cross distortion, 30, 31